ELECTROLYTIC DISSOCIATION

PHYSICAL CHEMISTRY

A Series of Monographs

Edited by

ERIC HUTCHINSON and P. VAN RYSSELBERGHE

Department of Chemistry, Stanford University, Stanford, California

Volume I.	W. JOST: Diffusion in Solids, Liquids, Gases. Third Printing, with Addendum, 1960
Volume II.	S. MIZUSHIMA: Structure of Molecules and Internal Rotation. 1954
Volume III.	H. H. G. JELLINEK: Degradation of Vinyl Polymers. 1955
Volume IV.	M. E. L. McBAIN and E. HUTCHINSON: Solubilization and Related Phenomena. 1955
Volume V.	C. H. BAMFORD, A. ELLIOTT, and W. E. HANBY: Synthetic Polypeptides. 1956
Volume VI.	GEORGE J. JANZ: Estimation of Thermodynamic Properties of Organic Compounds. 1958
Volume VII.	G. K. T. CONN and D. G. AVERY: Infrared Methods. 1960
Volume VIII.	C. B. MONK: Electrolytic Dissociation. 1961

In preparation

P. J. HOLMES (*ed.*): The Electrochemistry of Semiconductors

ELECTROLYTIC DISSOCIATION

By

C. B. MONK
University College of Wales
Aberystwyth, Wales

1961
ACADEMIC PRESS, LONDON AND NEW YORK

ACADEMIC PRESS INC. (LONDON) LTD.
17 OLD QUEEN STREET
LONDON, S.W.1

U.S. Edition published by
ACADEMIC PRESS INC.
111 FIFTH AVENUE
NEW YORK 3, NEW YORK

Library of Congress Catalog Card Number 60-16911

Printed in Great Britain at
THE ABERDEEN UNIVERSITY PRESS

PREFACE

MONOGRAPHS on electrochemistry usually emphasize particular aspects of the behaviour and properties of electrolytes in solution and those who are unfamiliar with the associated mathematical discussions may find that these demand considerable time and effort to appreciate. At the other end of the scale are the elementary treatments given in textbooks of general physical chemistry. It is the aim of the present book to follow a course between these two extremes and at the same time pay most attention to the properties of electrolytes which are mostly concerned in the study of their dissociation in solution. By adopting these policies it is hoped that useful material has been furnished for those who are preparing for Honours or equivalent examinations in chemistry and for those who are commencing research work involving ionic equilibria.

A general classification of electrolytes is to regard them as being "strong," meaning that they are completely or extensively dissociated in aqueous solution, and as "weak," meaning that the extent of dissociation at working concentrations is very small. Until recently, it has been widely held that many of the strong electrolytes (which include most salts, inorganic acids and soluble inorganic hydroxides) dissociate completely at all concentrations. This was a natural conclusion to form after a satisfactory interpretation of the effects of interionic forces of attraction and repulsion had been achieved by Debye and Hückel. They obtained an expression for activity coefficients that was successful in accounting for the effects of low concentrations of certain salts upon the solubilities of certain sparingly soluble salts, could reproduce experimental activity coefficients derived from potentials of some dilute electrolyte solutions; the changes in rate constants of certain ionic reactions with concentration could likewise be explained. In addition, their treatment provided a foundation for Onsager to derive an expression which accounts for the decrease in conductance with concentration of dilute solutions of simple electrolytes.

On the other hand, electrolytes of higher valency have conductances which are lower than can be explained by Onsager's equation, their activity coefficients (in dilute solutions) are lower than expected and specific, usually enhanced solubility results are obtained in the presence of polyvalent ions. These features mean that either the interionic attraction treatments are inadequate or that complete dissociation does not take place when polyvalent ions are present in solution. It is certainly true that the theoretical treatments are not perfect; in the early versions, "limiting" expressions were derived which included the major approximation that ions were regarded as point charges, or in other words their finite size was ignored. This was soon at least partly remedied with respect to activity coefficient expressions, and in recent years the relevant modifications have been introduced into treatments of conductance.

However, since many ions, particularly cations, are often solvated in solution, and there is as yet no exact way of determining their exact effective sizes when this occurs, there is a limit to the extent to which formulae can be modified to deal with precise assessments of dissociation equilibria. Nevertheless there is a wealth of experimental information which can be successfully interpreted in terms of incomplete dissociation in solutions of many "strong" electrolytes and in spite of the limitations of the theoretical treatments a fairly good quantitative estimate of ionic equilibria is possible. These topics form the subject-matter of the second half of the book, and the general conclusion is that the number of electrolytes which dissociate completely in water is comparatively small; even among the 1 : 1-valent electrolytes there are a number for which there is evidence of ion-pairing.

In a broad sense, there are two main schools concerned with studies of electrolytic dissociation. In one group the derivation of "true" or "thermodynamic" dissociation constants is one of the major goals. It means that measurements must be confined to very dilute concentrations where the interionic attraction theories are valid or where any errors arising from them are kept to a minimum. The restrictions imposed often mean very accurate measurements must be obtained by sensitive equipment and very pure materials used since the calculations depend on very small differences in the measurements or on the concentrations of ion-pairs or complexes derived from them. In most of this work only the first product of ion-association, namely the interaction of cations and anions in a 1 : 1 ratio is considered. Another limitation is that the values of large dissociation constants can depend markedly upon the chosen or assessed values of the ionic radii.

The second group of workers, with which the Scandinavian countries are particularly associated, prefer to study wide ranges of cation : anion ratios, keeping the total salt concentration, or rather the total ionic strength, constant. In this way not only are significant concentrations of the major associating species formed, but further stages of association are also revealed. Of necessity, the total ionic strength is kept high so that the evaluation of thermodynamic constants is precluded. In addition it is assumed that ion-association between the "inert" salt that is used to maintain a constant ionic strength and the ions of the electrolyte under investigation is negligible.

Both types of investigations are described and discussed although only "mononuclear" species are considered; hydrolytic and polynuclear species have not been dealt with and neither have fused electrolytes, irreversible electrode reactions or oxidation-reduction systems. In addition extensive tables of data on dissociation constants, activity coefficients, etc., have not been included since such information is readily available in a number of other books and reviews which are mentioned in the text at suitable places.

It is a pleasure for the author to express his appreciation of the sustained interest shown, and the advice and guidance given by Professor C. W. Davies over a number of years. In addition, Dr. Mansel M. Davies and Dr. W. J. Orville-Thomas are thanked most warmly for their valuable help with a number of topics. The author would also like to acknowledge the help he has gained by consulting the works of others, particularly those of Professors Harned and Owen, Professors Robinson and Stokes, and Professor Guggenheim.

C. B. MONK

The Edward Davies Chemical Laboratories,
Aberystwyth, Wales, 1961

CONTENTS

DEFINITIONS OF PRINCIPAL SYMBOLS

A, A', A''	Debye-Hückel constants
B, B', B''	Debye-Hückel constants
C	molarity
$\overline{C}_p$	partial molal heat capacity
D	absorption (spectrophotometry) ; diffusion coefficient
E	electromotive force
F	faraday charge
G	free energy
G^*	"cell constant"
H	molal heat content; magnetic field
I	ionic strength $= \frac{1}{2}\sum m_i z_i^2$; light intensity
J	relative partial molal heat capacity
K, K_{diss}	thermodynamic dissociation constant
K'	concentration dissociation constant
K_A, $K_{ass.}$	association constant
L	relative partial molal heat content; ligand (general)
L^*	specific conductance
M	molecular weight; cation (general)
N	Avogadro's number; mole fraction
P	pressure
Q	empirical parameter
R	gas constant per g mole; resistance
S	entropy; solubility
T	temperature in ° K; transference number
V	volume; volts
X	electric field
a	activity; ion-size parameter ($\mathring{a}$ if in Å)
b	parameter of Bjerrum's theory
c	concentration in g equivalents per l.
c^*	velocity of light
d	density; path-length (spectrophotometry)
e	charge on electron; ion-charge (e_i, e_j)
f	activity coefficient
h	hydration number
k	Boltzmann constant
k_n	concentration association or formation constant; rate constant
k'	2.3026 RT/F
l	litre; length
m	molality (g moles per 1,000 g solvent)
n	number of Faradays; number of ions per c.c
$\bar{n}$	ligand number
p	partial pressure
r	radius
s	area of cross-section; total no. of species.

t	time, ° C.
u	velocity of sound; ion mobility
y	molar activity coefficient
z	valence
α, α^*	coefficient in theory of conductance
α_C	degree of dissociation
β, β^*	coefficient in theory of conductance; coefficient of compression
β_n	complexity or stability constant
γ	molal activity coefficient
ϵ	dielectric constant; molar extinction coefficient (spectrophotometry)
η	viscosity
θ	freezing-point depression
κ	function in Debye-Hückel theory
λ	molal freezing-point depression
λ_i, λ_j	equivalent conductance of ions
μ	chemical potential; 1,000Å
ν	number of ions per molecule
ν_i, ν_j	number of ions of type i or j per molecule
ρ	density
ϕ	practical osmotic coefficient
ψ	electrical potential
ω	ion mobilities
Γ	ion concentration as defined by $\sum c_i z_i^2$
Λ	equivalent conductance
$\sum$	summation

CHAPTER 1

GENERAL INTRODUCTION

Many of our current ideas about the properties and behaviour of electrolytes in solution rest on theories and experimental studies that have been developed and pursued during the present century. One of the most important conclusions from this work is that the constituent parts or ions of an electrolyte may exist anywhere between two extreme states; they may be entirely independent of one another (completely dissociated) or they can be linked together to form molecules, ion-pairs or complexes (undissociated). The extent of dissociation depends on many factors such as the nature, charges and sizes of the ions, the characteristics of the solvent and the temperature. There are many ways of studying and interpreting dissociation, especially in solutions at equilibrium, and it is the chief aim of the present book to describe the more important of these for those who seek introductory material and for those who intend to undertake fresh investigations of this particular phenomenon.

Before concentrating on current material and theoretical treatments, some of the chief contributions of the earlier workers must be mentioned for many of the older ideas have not been completely demolished by more recent ones, and many of the experimental methods in use today have been practised for many years, although they have naturally been modified with the passage of time, particularly through the development of electronics.

A convenient starting-point is provided by Faraday's laws of electrolysis, which were given in 1833. Both he and Daniell assumed that electrolysis consists of the transport of electricity by charged particles, i.e., "ions," which are discharged at the electrodes. The first suggestion that ions are produced by the simple act of dissolving an electrolyte in water can be ascribed to Clausius (1857). He also envisaged a form of electrolytic dissociation, namely a dynamic equilibrium between free ions produced by bond disruption, and unbroken molecules of solute. The much more comprehensive views on dissociation envisaged by Arrhenius in 1883 contain concepts on which many of present-day treatments are founded. From studies of the conducting powers of solutions he came to the conclusion that an electrolyte can vary in the extent to which it supplies ions. Many inorganic salts, acids and bases, when dissolved, form such good conducting solutions that they must be largely present as free ions, while substances such as many organic acids and bases, being such poor conductors, provide very few ions. Arrhenius made the important point that in all cases of incomplete dissociation the degree of dissociation, or fraction dissociated, α_c, depends on the concentration; it increases towards unity as the concentration tends towards zero, or in other words, as the dilution tends to become infinite.

Shortly after Arrhenius had made these propositions, van't Hoff reported the result of applying the gas laws to dilute solutions. With non-conducting solutes (non-electrolytes), osmotic pressures are expressed by $PV = RT$, but with electrolytes, the observed osmotic pressure P is greater than expected. The freezing-point depressions and boiling-point elevations of such solutions are also greater than those derived from the formulae which apply to non-electrolytes. Van't Hoff modified the expressions to make them apply to ionic solutions by introducing an empirical parameter, e.g., $PV = iRT$. The value of i was derived from the ratio (actual measurement/measurement calculated from "ideal" formula). At first it was thought that i was constant for a specific valence type, but in fact it varies slightly with concentration. By using van't Hoff's findings together with his own conductance data, Arrhenius calculated what he thought were degrees of dissociation. His method of deriving α_c values from conductance depends on the conducting power of 1 g equivalent of electrolyte. The resistance of a uniform conductor, R_t, is expressed by

$$R_t = l/L^*s \tag{1.1}$$

where l is the length and s the area of cross-section, while L^*, the specific conductance, is the reciprocal of R_t when l and s are 1 cm and 1 cm^2 respectively. For electrolytes, the conductance of 1 g equiv. is defined by

$$\Lambda = 1{,}000L^*/c \tag{1.2}$$

where Λ is the equivalent conductance and c is the stoichiometric concentration in g equivs. per l. In a specific solvent at a fixed temperature Λ usually increases as c decreases, and since the conducting power depends on the number of ions present, Arrhenius regarded this variation as being due to increasing dissociation. In an infinitely dilute solution, Λ would reach its maximum, Λ_∞, (nowadays Λ_0 is used, referring to zero concentration), and at real concentrations,

$$\alpha_c = \Lambda/\Lambda_0 \tag{1.3}$$

If the law of Mass Action is applied to ionic solutions, then for the equilibrium $ML \rightleftharpoons M^+ + L^-$, the concentration dissociation constant K' is given by

$$K' = [M^+][L^-]/[ML] = \alpha_c^2 c/(1 - \alpha_c) \tag{1.4}$$

where [ ] expresses concentrations in g ions or g molecules per l. Combination of (1.3) and (1.4) gives what is termed Ostwald's dilution law,

$$K' = \Lambda^2 c/\Lambda_0(\Lambda_0 - \Lambda) \tag{1.5}$$

Under fixed conditions, Λ_0 is constant so (1.5) can be written

$$\Lambda c = U_1/\Lambda - U_2 \qquad (U_1 = K'\Lambda_0^2;\ \ U_2 = K'\Lambda_0) \tag{1.6}$$

It follows that a plot of Λc against $1/\Lambda$ should be a straight line, the slope and intercept of which give Λ_0 and K'. An alternative way of deriving Λ_0 so that individual values of α_c and hence of K' could be derived is to use the observation that with very dilute solutions of highly dissociated electrolytes, a straight line is obtained on plotting Λ against $\sqrt{c}$. Other empirical methods have been devised[1,2] but this is the only one supported by modern theory. It does not apply to weak electrolytes, such plots being curved and very steep. For these substances, Kohlrausch's "law of the independent migration of ions" can often be applied, namely

$$\Lambda_0 = \lambda_i^0 + \lambda_j^0; \quad \Lambda = \lambda_i + \lambda_j \tag{1.7}$$

where λ_i^0, λ_i and λ_j^0, λ_j refer to cation and anion conductances. Other accurate methods are described in Chapters 3 and 8.

To obtain values of $\lambda_i^{(0)}$ or $\lambda_j^{(0)}$, transference (or transport) number measurements are needed. The transference number $T_i^{(0)}$, $T_j^{(0)}$ of ions i and j is the fraction of the total current carried by these ions and it is related to their relative speeds and valencies. Hittorf, from about 1850 onwards spent many years collecting such data; a more precise method than his is now available (see Chapter 2).

The velocity of an ion depends on the applied electric force and on the viscosity of the solution. A potential gradient (V) of 1 volt per cm is taken as the unit of applied force and under such conditions let v_i and v_j be the velocities in cm/sec of the cation and anion respectively. The specific conductance L^* is the current in amps (coulombs per sec) which would flow across two opposite faces of a 1-cm cube of solution; this follows by putting $l = s = 1$ in (1.1) and from Ohm's law ($R_t = V/\text{amps}$) so that $R_t = 1/L^* = 1/\text{amps}$. Now 1 g equiv. of an ion carries 1 Faraday (F) of coulombs, so $Fc\alpha_c v_i/1{,}000$ amp are carried by the cations across the unit cube while the anions carry $Fc\alpha_c v_j/1{,}000$ amp. Also, from (1.2) and (1.7),

$$\Lambda = \lambda_i + \lambda_j = 1{,}000L^*/c = 1{,}000(L_i^* + L_j^*)/c \tag{1.8}$$

where L_i^* and L_j^* are the specific conductances of the cations and the anions. Thus

$$\lambda_i = 1{,}000L_i^*/c = 1{,}000Fc\alpha_c v_i/1{,}000c = F\alpha_c v_i \tag{1.9a}$$

and

$$\lambda_j = F\alpha_c v_j \tag{1.9b}$$

The transference numbers T_i and T_j are expressed by

$$T_i = v_i/(v_i + v_j); \quad T_j = v_j/(v_i + v_j); \quad T_i + T_j = 1 \tag{1.10}$$

whence, from (1.7) and (1.9),

$$T_i = \lambda_i/\Lambda; \quad T_j = \lambda_j/\Lambda; \quad T_i^0 = \lambda_i^0/\Lambda_0; \quad T_j^0 = \lambda_j^0/\Lambda_0 \tag{1.11}$$

The determination of T_i^0 and T_j^0, the values at $c = 0$, is dealt with in Chapter 3.

Some values of K' for acetic acid, obtained by the method of Arrhenius, are given in Table 1.1. These are derived from some of the most accurate conductances available,[3] and are based on $\lambda^0_{\mathrm{H}} = 349.8$, $\lambda^0_{\mathrm{Ac}} = 40.9$.

TABLE 1.1

K' Values for Acetic Acid in Water. Method of Arrhenius

10^4c	1.114	2.184	10.283	24.14	59.12	200	300
Λ	127.75	96.493	48.146	32.217	20.962	11.566	7.358
$10^5K'$	1.769	1.768	1.783	1.788	1.797	1.805	1.807

The constancy of K' is quite good. Similar information is obtained with other weak electrolytes so that the hypothesis of incomplete dissociation in the form suggested by Arrhenius appears to be fully substantiated. Since Λ also decreases with increasing c with strong electrolytes, it should also be possible to derive K' values for these, but as Table 1.2 shows, K' for[4] KCl is far from constant. Davies[1] has pointed out that K' is not truly constant even with weak electrolytes; this is apparent on a critical examination of Table 1.1; K' slowly increases as c increases.

TABLE 1.2

K' Values for KCl in Water. Method of Arrhenius ($\Lambda_0 = 149.98$)

10^4c	5.506	15.636	37.404	70.663	92.057	130.10
Λ	147.74	146.26	144.35	142.49	141.57	140.23
$10K'$	0.14	0.62	0.94	1.3	1.5	1.8

Inter-Ionic Forces

The failure to obtain consistent K' values with electrolytes such as KCl was referred to at the time as the "anomaly" of the strong electrolytes, and around 1890 to 1910 a number of suggestions as to the cause were made. Some believed that the formation of complex ions, e.g. KCl_2^-, K_2Cl^+ was the explanation, and others thought that ion-solvation was the answer. A more attractive proposition that was given is that α_c is not given by $\alpha_c = \Lambda/\Lambda_0$, but that the values of strong electrolytes decrease as c increases because of the forces of attraction and repulsion between ions. Jahn (in 1900) contended that these forces cause mobilities to be concentration-dependent whereas the contrary is implied in the theory of Arrhenius. Noyes *et al.*[5] showed that transference numbers also indicate that mobilities depend on concentrations. Thus the view that strong electrolytes could be completely dissociated at all concentrations while weak electrolytes attain this state only at infinite dilution soon gained favour, and a number of theoretical treatments which took interionic forces into account were devised by Hertz, Ghosh, Milner and others[1,2,6] between 1910 and 1920. These, for various reasons, had their limitations, and the first successful attempt is that of Debye and Hückel, which was published in 1923. A detailed account of this is left to Chapter 3. Their application of this to conductance[8] was not too convincing, but after modification and extensive development by Onsager,[9] together with Fuoss,[12] expressions were devised that can account remarkably well for the decrease in with c for very dilute solutions of a number of low valence type strong electrolytes, so that there is now strong evidence that they are completely ionized in solvents such as water. Again, details and specific examples are left to Chapters 3 and 8.

Activities

Besides discussing conductances, Jahn also pointed out that the "active masses" of ions, which are used on applying the law of Mass Action to ionic equilibria, are not necessarily equal to their concentrations. This is because interionic forces prevent ions from having free and random movement. This effect influences not only equilibria involving ions and their undissociated species but also numerous physical properties of ionic solutions such as freezing- and boiling-points, vapour pressures, viscosities, heat contents and volumes, diffusion, solubilities, potential differences, and others. In all cases the active masses are expected in terms of "activities", which, roughly speaking, are fractions of concentrations.

Precise definitions of activities and methods of determining them from experimental data were evolved by Lewis and Randall[10] between about 1900 and 1920 using thermodynamic principles and thermodynamic theorems of Gibbs. Starting with a simple version of the Lewis treatment, if ΔG is the free energy change on transferring one g mol. of solute from a solution where its molar concentration is n_A to one where it is n_B, then for "ideal" solutions,

$$-\Delta G = RT \ln (n_A/n_B) \tag{1.12}$$

Since an ionic solution is not ideal, Lewis replaced concentrations by activities, a, the ratio of the activities in the two solutions being defined by

$$-\Delta G = RT \ln (a_A/a_B) \tag{1.13}$$

As this ratio is identical with the ratio of the concentrations in an ideal solution, the law of Mass Action can be applied to the ionic solutions if activities are used, but in order to assign numerical values to the activities, a standard of reference is required. Before defining this, exact definitions must be introduced, since the values of ΔG and other thermodynamic functions depend, among other factors, on the proportions of solute and solvent. This is met by defining partial molal quantities. The partial molal free energies (or "chemical potentials"), $\bar{G}$, of a solvent (1) and solute (2) are defined by

$$\bar{G}_1 = (\partial G/\partial n_1)_{n_2,T,P}; \quad \bar{G}_2 = (\partial G/\partial n_2)_{n_1,T,P} \tag{1.14}$$

where G is the total free energy of the system, and the terms outside the brackets are kept constant.

If $\bar{G}_1^0$ and $\bar{G}_2^0$ are defined as the free energies in arbitrary solutions which are in some chosen standard state, then for the solvent, the activity, a_1, is defined by

$$\bar{G}_1 - \bar{G}_1^0 = RT \ln a_1 \tag{1.15}$$

The conventional standard state for the solvent is the pure solvent at the same T and P as the solution. It is necessary to define activities for solvents since they may not be ideal although when considering dilute solutions, the difference between activity and concentration is negligible. At higher concentrations this is not so; the actual activity can be deduced as follows. If p_1^0 is the vapour pressure of the pure solvent and p^0 is its partial vapour pressure in a solution composed of n_1 moles of solvent and n_2 of solute, then Raoult's law for an ideal solution is

$$(p_1^0 - p_1)/p_1^0 = n_2/(n_1 + n_2), \text{ or } p_1 = p_1^0 n_1/(n_1 + n_2) = p_1^0 N_1 \tag{1.16}$$

where N_1 is the mole fraction of solvent. A similar argument applies to the solute. In order to apply (1.16) to actual solutions which only obey the law at infinite dilution, Lewis introduced the concept of "fugacity", f^*. This is an ideal or corrected vapour pressure, and

$$f_1^*/f_1^{0*} = N_1; \quad f_2^*/f_2^{0*} = N_2 \tag{1.17}$$

It has been pointed out[11] that the vapour pressures of most electrolyte solutions are small and little different to those of the solvent, so vapour pressures can be used for f^*. Accordingly, it is sufficient to write

$$a_1 = p_1/p_1^0 \tag{1.18}$$

Turning to the electrolyte, the expression comparable to (1.15) is

$$\bar{G}_2 - \bar{G}_2^0 = RT \ln a_2 \tag{1.19}$$

but the activity a_2 incorporates the activities of both cations and anions, and these may differ. To deal with this, (1.19), in terms of 1 g ion may be written, for a cation, i,

$$\bar{G}_i - \bar{G}_i^0 = RT \ln a_i \tag{1.20}$$

and likewise for an anion, j. Then if 1 g mol. of electrolyte gives a total of ν ions, ν_i of these being cations and ν_j being anions of valencies z_i and z_j respectively,

$$\bar{G}_2 - \bar{G}_2^0 = \nu_i(\bar{G}_i - \bar{G}_i^0) + \nu_j(\bar{G}_j - \bar{G}_j^0) \tag{1.21}$$

so that

$$RT \ln a_2 = RT(\nu_i \ln a_i + \nu_j \ln a_j); \quad a_2 = a_j^{\nu_j} a_i^{\nu_i}. \tag{1.22}$$

Activities may be equated to concentrations by introducing "activity coefficients." Since three different concentration scales are in use, the various terms and symbols denoting these concentrations and coefficients will be listed.

(a) Molality $= m =$ moles/kg of solvent. The ionic molality, m_i, is $\gamma_i m$, and if a_i' is the ion activity,

$$a_i' = \gamma_i m_i. \text{ Similarly } a_j' = \gamma_j m_j \tag{1.23}$$

on the molal scale, where γ_i and γ_j are molal activity coefficients.

(b) Molarity $= C =$ moles/l. of solution. The ionic molarity C_i is $y_i C$, and if a_i'' is the ion activity,

$$a_i'' = y_i C_i. \text{ Similarly } a_j'' = y_j C_j \tag{1.24}$$

on the molar scale, where y_i and y_j are molar activity coefficients.

(c) Mole fraction $= N =$ moles of solute/(moles of solute + moles of solvent). The ionic mole fraction N_i is $f_i N$, and if a_i''' is the ion activity,

$$a_i''' = f_i N_i. \text{ Similarly } a_j''' = f_j N_j \tag{1.25}$$

From (1.22) and (1.23),

$$a_2' = (\gamma_i m_i)^{\nu_i} (\gamma_j m_j)^{\nu_j} = (\gamma_i \nu_i)^{\nu_i} (\gamma_j \nu_j)^{\nu_j} m^{\nu} \tag{1.26}$$

For example, with a 1 : 1 electrolyte, $a_2' = \gamma_i \gamma_j m^2$, with a 1 : 2 electrolyte, $a_2' = \gamma_i (2\gamma_j)^2 m^3$, and so on. Similar types of expressions can be written by taking (1.24) and (1.25).

The physical impossibility of having solutions containing only cations or anions means that the individual activities and activity coefficients of ionic species cannot be studied separately. For this reason, formulae such as (1.26) need modifying. This is done by defining mean values, e.g.,

$$a_2 = a_{\pm}^{\nu}, \quad m^{\nu} \nu_i^{\nu_i} \nu_j^{\nu_j} = m_{\pm}^{\nu}, \quad \gamma_i^{\nu_i} \gamma_j^{\nu_j} = \gamma_{\pm}^{\nu} \tag{1.27}$$

For example, the mean molal activity coefficient of a 1 : 2 electrolyte is related to the ionic values by $\gamma_{\pm}^3 = \gamma_i \gamma_j^2$.

On diluting an electrolyte solution, interionic forces are weakened as the ions are more widely separated. Hence ion activities and concentrations tend to become equal as the concentration approaches zero. This gives the clue to the chosen standard state for activity coefficients, namely that mean ionic activity coefficients are regarded as tending towards unity as the concentration tends towards zero. This is regardless of the concentration scale.

Ionic Strengths

Since the activities of ions depend on concentrations, when a solution contains several electrolytes, they exert a mutual influence. Because of this, Lewis and Randall[10] introduced a "rule" in 1921 which states that in dilute solutions, the activity coefficient of an ion depends on the "total ionic strength" of the solution, I, where

$$I = 0.5\sum m_i z_i^2 \tag{1.28}$$

where m_i represents ion molality in general and z_i represents ion valency, and the summation is for all the ions present. For example, in a mixture of 0.2 M NaCl and 0.3 M La $(ClO_4)_3$, $I = 0.5(0.2 + 0.2 + 2.7 + 0.9) = 2.0$ (assuming complete dissociation). Although this rule arose from a comparison of the activity coefficients calculated from measurements with individual and mixed electrolytes, it is an integral part of the Debye–Hückel interionic attraction theory.

The Evaluation of Experimental Activity Coefficients

A number of methods are available for the calculation of activity coefficients from experimental data; many of these were devised by Lewis.[10] They include the potentials of e.m.f. cells, freezing-points, vapour pressures, solubilities, rates of ionic reactions, and distribution studies, and are dealt with in later chapters.

In general, it is assumed that the electrolytes under investigation are completely dissociated. If this is not so, due allowance should be made. Sometimes this is not done either because it is not known or realized that ion-association is occurring, and sometimes it is not possible to separate these two features. As a consequence there are instances where the data are questionable.

Calculated Activity Coefficients

Debye and Hückel[7,8] derived several formulae for the evaluation of activity coefficients. There is a limiting form that ignores ion sizes, namely

$$-\log f_{\pm} = A \mid z_i z_j \mid \sqrt{I} \tag{1.28}$$

where the factor A incorporates a number of physical constants. This form is applicable over a very limited range; even with 1 : 1 electrolytes in water it is about 3 % in error at $I = 0.001$. By introducing a parameter, a, to allow for the finite sizes of ions, if this is suitably chosen, the expression can reproduce experimental activity coefficients up to about $I = 0.1$, and by including a further term, $Q(I)$, the working range can be extended to quite high ionic strengths. A generalized form of the full expression may be written

$$-\log f_{\pm} = A\,|\,z_i z_j\,|\,\{\sqrt{I}/(1 + Ba\sqrt{I}) - Q(I)\} \qquad (1.29)$$

where B, like A, is the product of several physical constants while a, the average cation-anion radius, and Q, which is supposed to allow for such features as the effect of electrolytes on the dielectric constant of the medium, are empirical since they have to be selected to make (1.29) coincide with the experimental figures. The derivation of (1.28), (1.29) and other related treatments are to be found in Chapter 3.

True or Thermodynamic Dissociation Constants

For the equilibrium $ML \rightleftharpoons M^+ + L^-$, K_c, the true or thermodynamic dissociation constant is expressed by

$$K_c = a_M a_L / a_{ML} = [M^+][L^-]y_{\pm}^2/[ML]y_{ML} = \alpha_c^2 C y_{\pm}^2/(1 - \alpha_c)y_{ML} \qquad (1.30)$$

and a similar expression using K_m, m and γ can be written if molalities are used.

To establish K_c (or K_m), methods are required to derive α_c or the ratio $\alpha_c^2 C/(1 - \alpha_c)$, and the activity coefficients then calculated when I is known. So far, only the conductance method of Arrhenius has been mentioned as a way of finding α_c, and since the interionic attraction theory shows this to be inadequate, the conductance method requires adjustment. The proper treatments are in fact based on the equations of Onsager,[9] and Onsager and Fuoss;[12] these are described in Chapters 3 and 8. Other methods of studying dissociation include potential and pH measurements; visible, ultra-violet, Raman and nuclear magnetic resonance spectra; the solubilities of sparingly soluble salts; the distribution of ionic species between liquid phases or solution and ion-exchange resins; rates of ionic reactions, and freezing-point measurements. Accounts of these occupy a number of chapters. It will be noticed that many of these methods have been listed earlier on as ways of deriving values of experimental activity coefficients. The difference is that it is assumed that for this purpose dissociation is complete or that the extent of ion-association is known and is allowed for.

The evaluation of the necessary activity coefficients when calculating unknown dissociation constants is a source of considerable discussion. It is usually good enough to put $y_{ML} = 1$ when studying dilute solutions, but in order to derive $y_{\pm}$, the parameters a and Q have to be estimated from those which have been found suitable for fully dissociated electrolytes, or they are chosen so that K_c is constant over a range of ionic strengths. However, when K_c is large, two difficulties may arise. Firstly, in order to get accurate values of α_c, high concentrations may be required and there is the danger of going outside the working range of the activity coefficient expression. Secondly, it is possible to find a range of combinations of a and Q, each of which will lead to constant values of K_c over the range of concentrations investigated, but this varies as the parameter combination is varied. As a consequence, an uncertainty of $\pm 5\%$ in K_c (and even more), is quite common. One answer to this is to obtain measurements at ionic strengths which are as low as possible.

Another important aspect in the study of electrolytic dissociation is concerned with finding out the number of stages of association, i.e., the value of N in ML_N, and the relative values of the "complexity constants." This entails working with media of high constant ionic strengths so that the concentration range of the "ligand" L can be varied sufficiently. In such work it is customary to derive the complexity constants in terms of concentrations since the ionic strengths are too high for the calculation of activity coefficients. Many of the expressions that have been derived for sorting out the constants are described and illustrated in later chapters.

There are other interesting topics, which are dealt with in various sections. For instance, the cation-anion bonding in the associated form of a weak electrolyte is so strong that the two ions lose most of their ionic character, in contrast to the association products of the stronger electrolytes where ion-pairs result from electrostatic forces of attraction, and since such forces are long-range, the ions or even their solvation shells need not be in actual contact for the formation of an ion-pair.

REFERENCES

1. Davies, C. W. "The Conductivity of Solutions." Chapman and Hall Ltd., London, 2nd edn. (1933).
2. Harned, H. S. and Owen, B. B. "The Physical Chemistry of Electrolytic Solutions." Reinhold Publishing Corp., New York; Chapman and Hall Ltd., London, 3rd edn. (1958).
3. MacInnes, D. A. and Shedlovsky, T. *J. Amer. chem. Soc.* **54**, 1429 (1932).
4. Owen, B. B. and Zeldes, H. *J. chem. Phys.* **18**, 1083 (1950).
5. Noyes, A. A. and Sammett, G. V. *J. Amer. chem. Soc.* **24**, 944 (1902). Noyes, A. A. and Kats, Y. *ibid.* **30**, 318 (1908).
6. Falkenhagen, H. "Electrolytes." Translated by R. P. Bell, Oxford University Press, Oxford (1934).
7. Debye P. and Hückel, E. *Phys. Z.* **24**, 185 (1923).
8. Debye, P. and Hückel, E. *ibid.* **24**, 305 (1923).
9. Onsager, L. *ibid.* **27**, 388, (1926); **28**, 277 (1927); *Trans. Faraday Soc.* **23**, 341 (1927).
10. Lewis, G. N. and Randall, M. "Thermodynamics." McGraw-Hill Book Co. Inc., New York (1923).
11. Robinson, R. A. and Stokes, R. H. "Electrolytic Solutions." Butterworths Scientific Publications, London, 2nd edn. (1959).
12. Onsager, L. and Fuoss, R. M. *J. phys. Chem.* **36**, 2689 (1932). Fuoss, R. M. and Onsager, L. *ibid.* **61**, 668 (1957).

CHAPTER 2

THE MEASUREMENT OF CONDUCTANCES AND TRANSFERENCE NUMBERS

Conductances

1. Audio-Frequency A.C. Methods

The larger part of conductance data has been obtained with some form of Wheatstone bridge in which a cell containing electrolyte solution forms one arm. An elementary arrangement is shown in Fig. 2.1a. This bridge is fed by an oscillator, O, which supplies a signal in the audio-frequency range (around 550 to 5,000 cycles/sec), although it is common practice to use a note of about 1,000 cycles/sec only. R_3 and R_4 are known resistances, and R_1 is a variable resistance box. The cell resistance, R_2, is found by adjusting R_1 until the note heard in the phone P is at a minimum or zero. From simple theory, $R_2 = R_1R_4/R_3$, and if a D.C. signal were used, this would be the only requirement, but until recently this produced polarization errors. Modern D.C. methods are described later on, but with an A.C. signal, the potential difference and the current through each arm have to be considered. These are both time-dependent and may be out of phase with one another, so that to secure a bridge "balance", a second condition must be satisfied, namely that $\theta_1 + \theta_4 = \theta_2 + \theta_3$, where θ is the angle of phase by which the voltage leads the current (the numbers correspond to those of R). A full discussion on this and other features of conductance such as bridge design, the construction of resistance boxes and cells is given in a series of articles by Jones and his collaborators.[1] The bridge circuit which they devised is shown in Fig. 2.1b. R_1 is a specially designed resistance box; a single calibrated coil in series with this is used to get readings to the nearest 0.001 ohm when R_1 is below 100 ohms. R_3 and R_4 are each 1,000 ohms, and are connected by a resistance wire with a slide which is adjusted so that $R_3 = R_4$. C_1 is a pair of variable air condensers, one of 1,000 $\mu\mu$f, and the other a trimmer, that are adjusted to balance the capacitances of R_1 and of the cell, as shown by a maximum sharpness of the balance point. R_5 and R_6 are each 1,000 ohms with an earthed slide near their midpoint, g. This arrangement is termed a Wagner[2] bridge. The oscillator O is connected to the bridge by earthed screened cables which are long enough to place O at some distance from the bridge to reduce the pick-up of stray signals by the bridge network. It is essential that O be free of harmonics; this is true of good commercial instruments. Jones and Josephs[1] found that an interchange of the cables could alter the balance point; this is remedied by balancing the capacitances between each cable and earth by means of C_g.

A later version of the Jones bridge and resistance box has been described by Dike.[3] Ives and Pryor[4] have also described a modified Jones bridge; in this the output is fed to an amplifier.[5] This increases the bridge sensitivity. The amplifier has identical channels for two inputs, one from the bridge and the other from a few feet of aerial wire fixed above the bridge. The two signals are mixed, and this cancels out any interference from the bridge and amplifier.

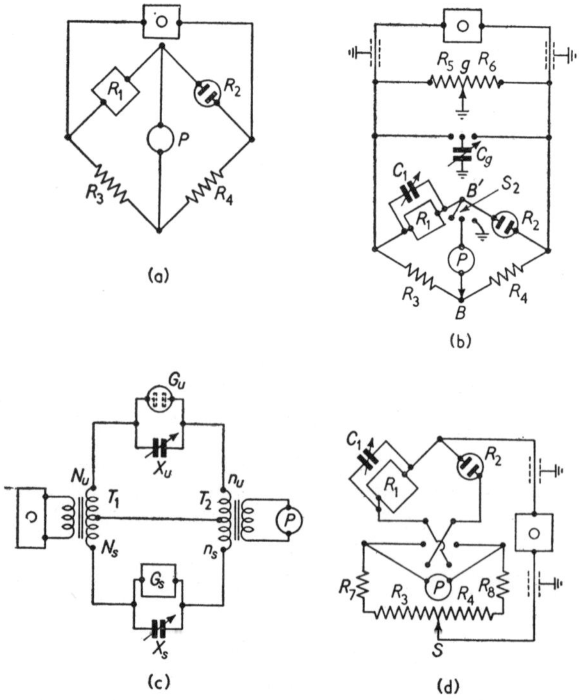

FIG. 2.1. A.C. Conductance bridge circuits.

To balance the Jones bridge, R_1 and C_1 are adjusted to an approximate balance, the phone P is disconnected by S_2 from B', and earthed. The slide, g, and C_g across R_5 or R_6 are adjusted till P is silent. This brings B to earth potential. S_2 is switched so that B' is rejoined, R_1 and C_1 are again adjusted to ensure that B' is at earth potential. This procedure avoids putting a charging current into any capacitance between P and the observer.

Instead of a phone, an oscilloscope may be used. Eisenberg and Fuoss[6] have given details about this, including coupling through an amplifier and phase shifter. By using polythene insulated cable they found it reduced the change in the balance point on altering the frequency from 0.5 to 10 kc/s.

A transformer ratio-arm bridge—and electrode-less cells (see later)—has been designed[7] for the study of concentrated solutions.[31] The basic circuit is shown in Fig. 2.1c, where u and s refer to unknown and standard sides of the bridge. T_1 is a voltage transformer; O is connected to its primary, and the secondary is tapped to give variable numbers of N_s and N_u turns. The primary of T_2, the current transformer, is tapped at (variable) n_s and n_u turns; the detector is joined to its primary. G_u is the unknown resistance, G_s is a standard, and X_u and X_s are variable air condensers. At balance, $G_u = G_s N_s n_s / N_u n_u$. The practical details and refinements that have been introduced are too many to deal with here.

Fig. 2.1d shows a bridge devised by Davies.[8] R_3 and R_4 comprise a metre bridge wire (about 3 ohms), R_7 and R_8 are fixed resistances (about 15 ohms), S is a slide made of a fine wire fixed across a transparent block. The other circuit details are much the same as those featured in Fig. 2.1b. Davies has tested the earthing arrangement used by Jones and Josephs[1] and has found that its omission causes an error of less than 0.01 % provided $R_1 >$ about 300 ohms. R_1 and R_2 can be interchanged by a wax block-mercury pool commutator, so two balance points of S are obtained. The bridge wire is calibrated over this range by putting another box at R_2, and plotting R_1/R_2 against the balance point differences. To measure the solvent, R_7 or R_8 is shorted, the commutator left in one position and S balanced for different settings of a box R_2; R_1/R_2 is plotted against cms.

Since the temperature coefficient of conductances is over 2 % per °C, a bath with a control of $\pm$ 0.005°C is required for an accuracy of 0.01 %. This is easily achieved by suitable commercial equipment.[9] A room with thermostat control is also needed. The bath fluid in which the cell is placed should be kerosene or a mobile transformer oil. Jones and Josephs[1] found that water is unsuitable; the bridge reading varies considerably with the signal frequency.

Many forms of cell have been designed. For very dilute solutions (0.0001 to 0.005 N) Davies[8] has used a modified Hartley-Barrett type[10] (Fig. 2.2a). The body is of fused quartz, with a capacity of about 300 ml. The cap is of pyrex and so are the leads carrying the electrodes and mercury contacts. The cap also has an arm so that CO_2-free, solvent-saturated, gas (air, O_2, N_2) can be passed over the surface of the solution, and a tube with a loose cap, wide enough for inserting a weight pipette for adding small amounts of stock solution. The cap and body have location marks so that they can be joined reproducibly. The electrodes are of rigid platinum sheets (about 2 $\times$ 2.5 cm), spaced about 3 mm apart, and having holes to ensure circulation of the solution, which is always sufficient to submerge the electrodes by at least

4 cm. Shedlovsky[11] has criticized this "dipping electrode" cell on the grounds that errors arise through the proximity of the electrode leads. Davies[8] has examined this point; he found that if $R_1 > 300$ ohms, the mean error in thirty measurements was less than 0.015 %.

Shedlovsky has also devised a cell for "dilution runs" at low concentrations. This incorporates recommendations of Jones and Bollinger.[1] It is sketched in Fig. 2.2b. The body is a litre fused quartz flask, and the electrodes, which are located in a subsidiary arm, are truncated cones of platinum foil. A similar cell based on an Erlenmeyer flask has been constructed and used by the Kraus school;[12] they have described this in detail and have given many

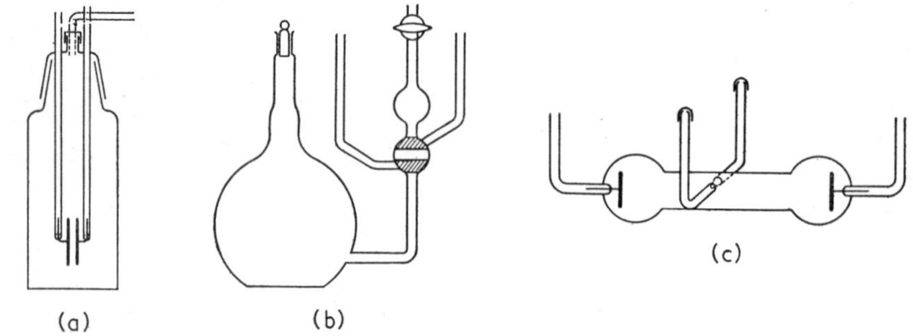

FIG. 2.2. A.C. conductance cells.

useful points about conductance techniques. Ives and Pryor,[4] and Feates *et al.*[13] have described an ingenious double cell in which the two cells have "cell constants" (see 2.2) in the ratio 4 : 1. Since measurements are based on the resistance difference between these two cells, polarization errors are largely removed. The true resistance R_1 is given by

$$R_1 = (f^A \Delta R_2^A - f^B \Delta R_2^B)/(f^A - f^B) \tag{2.1}$$

where f^A and f^B are two frequencies of the signal (3,880 and 1,150 c/s), while ΔR_2^A and ΔR_2^B are the cell resistance differences at these frequencies. This procedure gave a cell constant to $\pm$ 0.005 % with 4 to 106×10^{-4} N KCl.

The usual cell to bridge connections consist of wires dipping into mercury pools. Ives[13] regards the "four leads" method[14] used in platinum resistance thermometry as superior. Two leads are connected to each electrode and taken to a reversing switch (mercury cups). Let (a, b), (c, d) be the two, X the cell resistance, R_2' and R_2'' the box resistance for the two switch positions. If (i) a and d connect to the bridge and box, c connects to the phone and the box is at R_2', and then (ii) a and d are reversed when b connects to the phone and the box is at R_2'', $X = (R_2' + R_2'')/2$.

"Pipette" cells, such as that in Fig. 2.2c are convenient for concentrated solutions. The two arms by which the cell is filled and emptied are bent at 90° in their middles so that the arms and electrode leads are kept far apart to reduce capacitance effects.[15]

The electrodes of cells are usually coated with fine Pt or "platinum black." To form this, the electrodes are immersed in a solution of 0.3 % $PtCl_4$, 0.025 % $PbAc_2$ and 0.025 N HCl, and a current of about 10 mA/cm^2 passed for about 10 min, reversing its direction every 10 sec. This coating reduces polarization and ensures that a sharp balance can be achieved. Parker[17] considered that such coatings can cause adsorption errors, since he found that he obtained readings which were very frequency dependent when using them. Jones and Bollinger[1] showed the real cause to be the close spacing of the electrode leads which form an extra capacitance (this, however, should be compared with the observations of Davies[8] mentioned earlier). Absorption errors can arise under certain conditions, for Kraus and his co-workers[12,18] have found these by using large electrodes with Pt black coatings in the study of every low ionic concentrations in organic solvents. It is then better to use uncoated electrodes.

Internal electrodes may be dispensed with, although high frequency signals are needed[41] if the bridge is the conventional type. The cell is placed inside a coil or condenser plates, and the electrolyte causes frequency changes depending on the ions and their concentrations. This arrangement has found application in analytical chemistry. The transformer bridge[7] mentioned previously, which works with audio-frequencies may also be used with electrode-less cells. N_s and n_u of Fig. 2.1c are reduced to a single turn and both transformers are wound on toroidal cores moulded into an inert insulating material. The cores are linked by the cell, which is tube-shaped.

Pure solvents are required in conductance studies especially when the electrolyte concentrations are low, since even when the solvent is pure its conductance is a significant fraction of the whole. Until recently, "conductivity" water could only be prepared by means of special stills,[19] which have a tin condenser in which the distillate is condensed and collected out of contact with CO_2 and NH_3. Alkaline $KMnO_4$, or $NaHSO_4$ is added to the boiler to either retain CO_2 or to expel it in the first fractions. Nowadays a simpler process is to pass ordinary distilled water through a column of mixed ion-exchange resins packed in a hard glass column over a plug of cotton wool. Some suitable resins are [20] one part "Amberlite" IR-120 and two parts IRA-400 or[21] the mixed resin "Amberlite" MB-2*. These can give water with $L^* = 0.2$ to 0.05×10^{-6} ohm^{-1} (provided CO_2 is excluded), although traces of colloidal polyelectrolytes may be present.[22] Before filling, cells such as that in Fig. 2.2a are swept out with CO_2-free gas, and after sufficient solvent has been added, the cell is shaken gently until its resistance becomes constant before adding solution. The cell should also be shaken before taking readings of solutions, not only to ensure the attainment of equilibrium but also in case temporary adsorption has occurred.[23]

From (1.1),

$$L^* = (l/s)/R_t = G^*/R_t \qquad (2.2)$$

where G^* is termed the "cell constant." The value of R_t is usually derived from

$$1/R_t = 1/R_s - 1/R_w \qquad (2.3)$$

where R_s is the resistance of the solution and R_w is that of the solvent, but as Davies has shown,[24] if R_w is due to H^+ and HCO_3^- ions, it may be modified; neutral salts cause increased ionization of H_2CO_3 while strong acids suppress it.

Cell constants are found by means of KCl solutions of known specific conductance. Of the several sets of available data,[25] those of Jones and Bradshaw[26] are taken as the present-day standards. These were derived by firstly finding the D.C. resistance of mercury at 0°C in a cell "Z" which had widely spaced electrodes. The resistance of an international ohm is defined

TABLE 2.1.

Specific Conductances of KCl Solutions [26] in ohm^{-1} cm.$^{-1}$.

0° C	18° C	25° C
1.0D = 71.1352 g KCl, 1 kg solution (vacuum wts.)		
0.065176	0.097838	0.111342
0.1D = 7.41913 g KCl, 1 kg solution (vacuum wts.)		
0.0071379	0.0111667	0.0128560
0.01D = 0.745263 g KCl, 1 kg solution (vacuum wts.)		
0.00077364	0.00122052	0.00140877

as[27] that of a column of mercury 106.300 cm long, and of 14.4521 g mass at 0°C. Since the density is 13.5951 g/c.c, if $l = 106.30$, s of (2.2) is 14.4521/(13.5951 × 106.30), whence $L^* = 10629.63$/ohm. cm, and G^* of "Z" is obtained. Next, the mercury was replaced by H_2SO_4 of a suitable resistance (about 6 N) and its L^* value obtained. This was then used to find G^* of a smaller cell, "Y", which in turn was used to find L^* of a "demal" KCl solution (= 1.0 D of Table 2.1), and from this G^* of a still smaller cell, "N" was found, and "N" used to find L^* of 0.1 D and 0.01 D KCl solutions.

In 1937, Jones and Prendergast[28] redetermined the L^* values of 1.0, 0.1 and 0.01 N KCl. Gunning and Gordon[29] have calculated that these give G^* values which are 0.05 % and 0.02 % higher when using 0.1 N and 0.01 N KCl respectively than those given by the standards of Jones and Bradshaw.

When G^* is very small (0.05 or less), a 0.01 D solution is too concentrated since the cell resistance is too small for accuracy. A number of semi-empirical conductance equations for KCl have been devised for calibrating such cells. The best of these at 25°C is[30]

$$\Lambda = 149.93 - 94.65\sqrt{c} + 58.74c \log c + 198.4c \qquad (2.4)$$

which is based on a weighted average of 9 sets of figures by various workers[30] which are all based on the Jones and Bradshaw standards. The factors of the last three terms are derived from modern conductance theory (Chapter 3). If G^* is determined at 25°C and measurements then made at other temperatures, allowance for the expansion of the glass and the platinum may be needed.[31]

2. Direct Current Methods

It has been shown by Gordon[29] that D.C. conductance measurements can attain the precision of A.C. methods. The principles can be explained by means

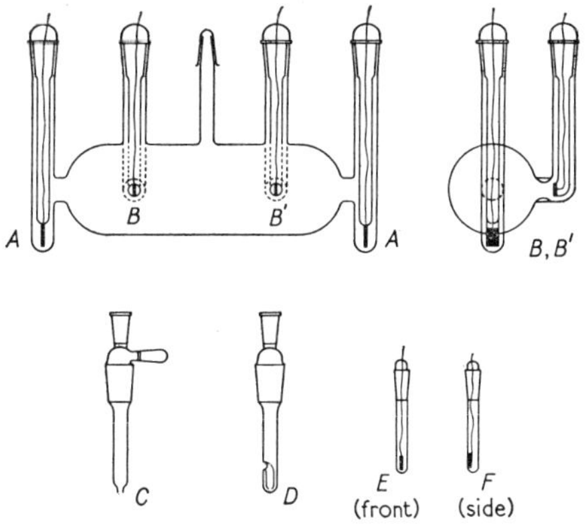

FIG. 2.3. Gordon's D.C. conductance cell and electrodes, and (C to F), the probe compartments and electrodes of Elias [33] for B, B'.

of Fig. 2.3. A constant current device[32] giving about 1.3 mA feeds D.C. through a standard 500 ohm resistance and through the cell via the electrodes A, A'. These are "reversible" electrodes (see Chapter 4), of thick Pt foil, silver plated and coated with fused AgCl or AgBr. They are located in wells connected by narrow tubes to the main cell body. The "potential probes" B, B' are about 10 cm apart. Each is a Pt disc mounted vertically and covered with fused glass except for a vertical slit one mm wide. The strip is silver plated and part of this is converted to AgCl or AgBr by electrolysis in dilute HCl or HBr. In theory, these strips should be points, since if their size is appreciable there is a potential drop across their surfaces. All four electrodes are fitted into ground-glass sockets with locating marks to ensure reproducible placing in the cell. When measuring the solvent, bright Pt electrodes are used.

The limitations imposed by the use of reversible electrodes, i.e., the electrolyte anion is common to that of the electrodes, have been overcome by various modifications such as that described by Ives[33] who has designed a cell so that it has regions of current reversal where there is no potential gradient. The probes are located in these regions. Elias[33] has tackled the problem by replacing B, B' of Fig. 2.3 by tubes containing Pt/Ag/AgCl electrodes and KCl solution (or AgBr, KBr). The tubes have outlets places so that the KCl can only enter the main solution by diffusion. The use of such liquid junctions permits the use of irreversible Pt/Ag current electrodes at A, A', and none of the electrodes has to be reproducibly situated.

The P.D. across the 500 ohm resistance, E_s, is measured by potentiometer (Chapter 4), and so also is E_c, the P.D. across B, B'. A second set of readings is taken with the current reversed; the average eliminates any static bias potential between the probes. Then, for a given cell,

$$E_s/E_c = G^*/L^* \tag{2.5}$$

G^* is evaluated by means of the standard KCl solutions mentioned a short way back.

3. High Voltage (or Field) Measurements

The applied field in ordinary conductance is about 1 V/cm, and consequently ion velocities are only about a few cm/h (from (1.9), $v_i = \lambda_i/F$). When fields of the order of kV/cm are used, the velocities increase several hundred-fold, and tend to reach a limiting value. These features were discovered and investigated by Wien.[34] In general, the consequent conductance increases depend on the concentration and ion valencies.

The application of high fields tends to produce temperature rises and to cause electrolysis; to avoid these effects, the field must be applied in short pulses of about microsecs duration. Wien and others developed successful methods for this purpose, which have been described in detail by Eckstrom and Schmelzer[35] but since then (1939) methods of pulsed power generation by electronic methods have been improved considerably. Patterson[36] has applied these in the construction of a suitable bridge for following the Wien effect up to 200 kV/cm, with an accuracy of about 0.1 %; the apparatus is too complex to describe here.

4. High Frequency Measurements

Frequencies in the megacycle range (Mcs) also produce conductance increases. Wien,[37] who demonstrated this effect, Sack and others,[38] devised ways of investigating this. Their work was made in the 40 to 1 metre range. Since wavelength in metres = $c^*/100\,\bar{\omega}$, where c^* is the velocity of light in cm/sec, $\bar{\omega}$ is the frequency = 7.5 to 300 Mcs. More recent electronic developments are available for studying this aspect of conductance; examples are to be found in articles by e.g., Lane and Saxton,[39] Hasted and Roderick.[40]

Transference Numbers

The first accurate quantitative information on the relative speeds of ions was obtained by Hittorf. His well-known apparatus[42] requires the determination of the concentration changes which take place around the electrodes through passing a known quantity of direct current. Washburn,[43] and MacInnes and Dole[44] have been responsible for a number of improvements to the apparatus and their results are probably the most accurate that can be obtained by this method. It has proved difficult to get accurate answers for dilute solutions since the required chemical determinations are subject to marked relative errors in these regions. This limitation has been largely overcome by Steel and Stokes[45] with a form of the apparatus which contains a mixing bulb large enough to contain all the solution subjected to electrolysis and having a calibrated arm so that the volume can be read accurately. There are also side-arms with electrodes so that both before, and after mixing the product, conductance measurements can be made. Then by reference to known data, the concentrations can be estimated. Concentration changes of 10 % can be determined to ± 0.1 %; the main uncertainty is the solvent conductance which is sensitive to traces of impurities.

1. Moving Boundary Methods

For a long while it was known[46] that the speeds of ions subjected to a D.C. field could be ascertained by observing the movement of the junction of two electrolyte solutions. MacInnes and Longsworth[47,48] are associated with the efforts which have produced from this the most accurate way of determining transference numbers; their 1932 review[49] gives an excellent account of the practical and theoretical development up to this time. The main points can be introduced as follows:

Consider a tube (Fig. 2.4a) containing electrolyte solutions AR and BR where R is a common anion, with the solutions meeting at a common boundary *ab*. If D.C. is passed between a suitable anode at the bottom of the tube, and a suitable cathode at the top, then the cations A and B will move up the tube while R will move downwards. Suppose that when F coulombs (one faraday) have passed that the boundary has moved to *cd*, i.e., the volume

V in *abcd* is now occupied by BR. Considering now a reference plane MN beyond *cd*, the solution containing A which passed MN is Vc_A litres, where c_A is the concentration of AR in g equivs. Thus for AR, the solution under investigation,

$$T_A = Vc_A \tag{2.6}$$

where T_A is the transference number of A. AR is termed the "leading" solution and BR is termed the "following" or indicator solution.

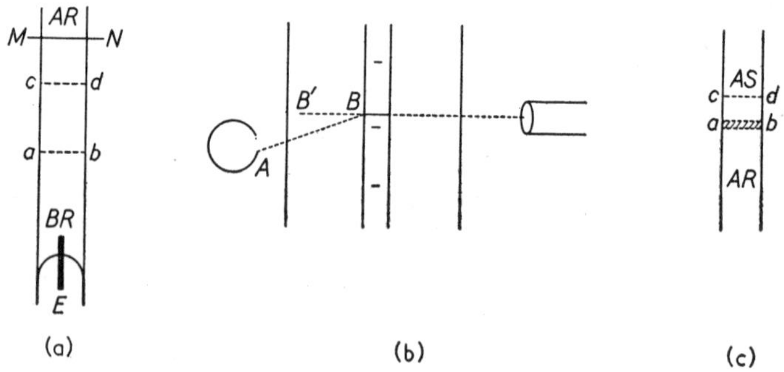

FIG. 2.4. Moving boundary.

If the boundary moves through a volume v for the passage of f coulombs, $f = it$, where i is the current in amps, and t is the time in secs, so since $v/V = f/F$,

$$T_A = Vc_A(v/V)/(it/F) = vc_AF/it \tag{2.7}$$

This assumes that there have been no disturbances such as diffusion or mixing at the boundary. To achieve this, solution AR must be less dense than that of BR, and the mobility of B must be less than that of A (assuming that the boundary rises). The motion of the boundary is followed by cathetometer using differently coloured solutions where possible or else making use of the refractive index change at the boundary; Fig. 2.4b shows the arrangement for this, the angle ABB′ being set at the critical angle for total reflection.

Several ways of forming sharp boundaries between leading and indicator solutions have been invented. MacInnes and Brighton[47] devised a "sheared" boundary technique which can be explained by means of Fig. 2.5a. A is a tube connected to an electrode vessel and fits into a glass disc A′. The graduated tube in which the boundary moves, B, fits into the disc B′. Each disc contains a recess opposite the end of the tube in the other disc. A and B are filled with AR and BR respectively, an extra drop hanging at the ends of the tubes. On rotating the disc, excess solution is sheared off and a sharp boundary forms when the tubes coincide. A second method[50] is to maintain an air bubble between the two solutions (Fig. 2.5b) until they are in position, and to then withdraw the bubble. A third, relatively simple way[51] of boundary formation can be explained by Figs. 2.5c, 2.5d. A large stopcock connects F and G, F being connected to the tube in which the boundary moves. This tube has a flask at its upper end for the leading solution. E is joined to the indicator solution container. G and H are tubes through which the solutions are forced into their respective compartments. By clockwise rotation of the tap through 90°, the boundary tube is connected to the indicator solution, the boundary forming where E enters the stopcock. The electrodes are placed in the vessels joined to E and F.

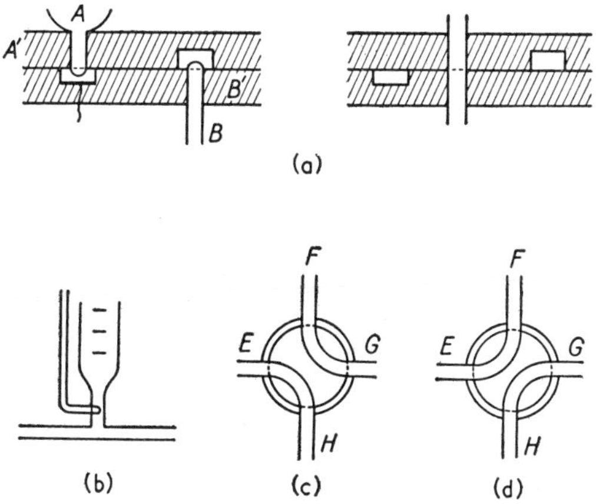

FIG. 2.5. Moving boundary.

Instead of forming a boundary between two solutions it is sometimes possible to form the indicator solution and boundary by generating a second electrolyte. For this "autogenic" boundary method,[52] a metal anode is dissolved on the passage of current, e.g., for the study of chlorides, Cd is converted to $CdCl_2$ solution underneath the leading solution and thus produces a rising boundary. For nitrates and perchlorates,[53] Ag is suitable. The anode may be placed in the recess of [54] disc B′ below (Fig. 2.5) with a wire sealed through B′ for the connection.

The graduated tube requires calibration. To do this,[55] mercury is drawn in from a weighing bottle, its weight noted and the length of the thread found by micrometer, using the graduation marks as references. This procedure is applied to the whole tube.

Although the quantity of electricity passed could be measured by coulometer, since a constant current is required to prevent mixing at the boundary, and since the quantity passed is small, it is more accurate to apply a steady known current over a known time. MacInnes *et al.* constructed an elaborate mechanical controller[56] but this has been superseded by electronic devices.[32,50,57,58]

Diffusion and convection tend to disturb the boundary but these tendencies are counterbalanced by potential gradients which restore the ions to their proper positions provided the ratio AR : BR is within certain limits. If c_B is the concentration of BR at the boundary (it is not necessarily the same as the bulk concentration of BR, c'_B) then, by theory,[59]

$$T_A/T_B = c_A/c_B \tag{2.8}$$

where T_B is the transference number of B. It so happens that if $c_B \sim c'_B$, the boundary remains sharp. For instance,[47] there is a possible tolerance of 10 % difference for the indicator LiCl used for the study of KCl. Gordon and Kay[60] have discussed and described a simple way of adjusting the indicator con centration to its correct region.

There is a small but real difference between transference numbers obtained by the Hittorf and the moving boundary methods. For the former, the value is for ions which cross a boundary fixed with reference to the water of the solution, while with the latter, the boundary is fixed with reference to the tube graduations. As a consequence, if any volume changes occur at the electrodes or in solution, corrections are required to put answers by the two methods on the same basis. If for instance, E in Fig. 2.4a is the anode, and a solution of ER forms as the results of electrode reactions, the volume correction ΔV is given by

$$\Delta V = \bar{V}_{ER} - V_E - T_A \bar{V}_{AR} \tag{2.9}$$

where the $\bar{V}$'s are partial molal volumes (see Chapter 6). In this, $\bar{V}_{ER} - V_E$ is the volume change due to the reaction of E and $T_A \bar{V}_{AR}$ is the volume displaced by solution ER. The complete derivation of the ΔV correction was first obtained by Lewis;[61] full details have been given by MacInnes and Longsworth[49] and examples are to be found in articles by Longsworth[62] and by Gordon.[51] Some of these are shown in Table 2.2 (the cathode is AgCl in all cases). Generally, the correction is small, but it depends on the concentrations. The corrected expression is, from (2.7, 2.9),

$$T'_A = (V - \Delta V)c_A = vc_A F/it - \Delta V c_A \tag{2.10}$$

As with conductances, it is essential to work with pure solvents and to exclude CO_2 (at least, when the solutions are relatively dilute), and to make a solvent correction since a fraction of the current is carried by the solvent ions. Longsworth[54] has shown that through this, the fully corrected expression is

$$T''_A = T'_A(1 + L^*_w/L^*_s) \quad (w = \text{solvent},\ s = \text{solution}) \tag{2.11}$$

Example 2.1. For[54] 0.1987 KNO_3, T_+ observed = 0.5117, $\bar{V}_{AgNO_3} = 28.8 + 2.5\sqrt{c}$ (in ml.), $\bar{V}_{KNO_3} = 38.0 + 3.5\sqrt{c}$, $V_{Ag} = 10.3$, whence

$\Delta V_+ = -0.64$ ml. Thus $V_+c/1{,}000 = -0.0001_3$. The solvent correction was negligible, so $T''_+ = 0.5118$. The corresponding determination for the anion was $T''_{NO_3} = 0.4878$, so the average of $T''_K = 0.5120 \pm 0.0002$.

The possibilities of what is termed the "analytical" boundary method were investigated by Brady.[63] The apparatus comprises a central tube partitioned by a sintered glass disc (*ab* of Fig. 2.4c). The solution under investigation, AR, is placed below *ab*, and a solution AS is placed above. The

TABLE 2.2

Leading Soln.	Indicator Soln.	Type	Boundary	ΔV Type	Anode
KBr	$CdBr_2$	Autogenic	Cation	(a)	Cd
KI	Tetraiodo-fluorescein	Sheared	Anion	(b)	Ag
KNO_3	$Ba(NO_3)_2$	Sheared	Cation	(c)	Ag
$CaCl_2$	$CdCl_2$	Autogenic	Cation	(d)	Cd
NH_4Cl	NH_4IO_3	Sheared	Anion	(e)	Ag

Type (a): $\Delta V_+ = 0.5\,\overline{V}_{CdBr_2} - 0.5\,V_{Cd} - T_+\overline{V}_{KBr}$
Type (b): $\Delta V_- = \overline{V}_{KCl} - V_{AgCl} + V_{Ag} - T_-\overline{V}_{KI}$
Type (c): $\Delta V_+ = \overline{V}_{AgNO_3} - V_{Ag} - T_+\overline{V}_{KNO_3}$
Type (d): $\Delta V_+ = 0.5\,\overline{V}_{CdCl_2} - 0.5\,V_{Cd} - 0.5\,T_+\overline{V}_{CaCl_2}$
Type (e): $\Delta V_- = V_{Ag} - V_{AgCl} + T_+\overline{V}_{NH_4Cl}$

roles of leading and indicator solution of the moving boundary method are reversed here. Through the passage of current, the boundary moves from its initial position at *ab* to some unspecified position *cd*. The arrangement under discussion is designed to determine T_R in AR and for this purpose the quantity of $R(= M_R)$ that passes above the disc (assuming an upward boundary movement) is determined for the passage of a known amount of constant current. From (2.7) and (2.8),

$$vc_{AS} = T_s it/F \tag{2.12}$$

$$T'_R/c'_{AR} = T_s/c_{AS} \tag{2.13}$$

so

$$c'_{AR}v = itT'_R/F \tag{2.14}$$

In general, $c'_{AR} \neq c_{AR}$ and a slow-moving secondary boundary will remain in the neighbourhood of *ab*. This will sweep out a volume

$$(c_{AR} - c'_{AR})v' = it(T_R - T'_R)/F \tag{2.15}$$

If the secondary boundary is stationary or moves downwards,

$$M_R = c'_{AR}v = itT'_{AR}/F \tag{2.16}$$

but if it moves upwards, from (2.14) and (2.15),

$$M_R = c'_{AR}v + (c_{AR} - c'_{AR})v' = itT_R/F \qquad (2.17)$$

Where there is ambiguity about the direction of this movement, the general shape of the T_R, c curve must be ascertained. By using a leading solution of known properties, the value of c'_{AR} can be approximated from (2.13), so that (2.15) can decide which concentration the observed transference number refers to. The ambiguity may also be resolved by addition of radioactive tracers to AS.

Spiro and Parton[64] have improved Brady's apparatus and have determined T_{Ag} in 0.10 and 0.01 N $AgNO_3$, making the ΔV and solvent corrections of (2.10) and (2.11). The estimated uncertainties are 0.1 % and 0.2 % respectively. The chief weakness is an analytical one (cf. the Hittorf method), being concerned with estimating low concentrations of indicator solution in a large excess of leading solution.

REFERENCES

1. Jones, G. and Josephs, R. C. *J. Amer. chem. Soc.* **50,** 1049 (1928). Jones, G. and Bollinger, G. M. *ibid.* **51,** 2407 (1929); **53,** 411, 1207 (1931).
2. Wagner, K. W. *Elektrotech. Ztg.* **32,** 1001 (1911). See Luder, W. F. *J. Amer. chem. Soc.* **62,** 89 (1940), and Wolff, H. N. *Rev. sci. Instrum.* **30,** 1116 (1959).
3. Dike, P. H. *Rev. sci. Instrum.* **2,** 379 (1931).
4. Ives, D. J. G. and Pryor, R. H. *J. chem. Soc.* 2104 (1955).
5. Ives, D. J. G. and Pitman, R. W. *Trans. Faraday Soc.* **44,** 644 (1948).
6. Eisenberg, H. and Fuoss, R. M. *J. Amer. chem. Soc.* **75,** 2914 (1953).
7. Calvert, R., Cornelius, J. A., Griffiths, V. S. and Stock, D. I. *J. phys. Chem.* **62,** 47 (1958).
8. Grindley, J. and Davies, C. W. *Trans. Faraday Soc.* **25,** 133 (1929). Davies, C. W. *J. chem. Soc.* 432 (1937).
9. *J. chem. Educ.* **36,** A131, A199 (1959).
10. Hartley, H. and Barrett, W. H. *J. chem. Soc.* **103,** 786 (1913).
11. Shedlovsky, T. *J. Amer. chem. Soc.* **54,** 1411 (1932).
12. Daggett, H. M. Jr., Blair, E. J. and Kraus, C. A. *ibid.* **73,** 799 (1951).
13. Feates, F. S., Ives, D. J. G. and Pryor, J. H. *Trans. electrochem. Soc.* **103,** 580 (1956). See Ref. 11, Ch. 1.
14. Smith, F. E. *Phil. Mag.* **24,** 541 (1912). Mueller, E. F. *J. Res. nat. Bur. Stand.* **13,** 547 (1916/17).
15. Jones, G. and Bickford, C. F. *J. Amer. chem. Soc.* **56,** 604 (1934).
16. Jones, G. and Bollinger, D. M. *ibid.* **57,** 280 (1935).
17. Parker, H. C. *ibid.* **45,** 1366, 2017 (1923).
18. Cox, N. L., Kraus, C. A. and Fuoss, R. M. *Trans. Faraday Soc.* **31,** 749 (1935).
19. Bourdillon, R. *J. chem. Soc.* 791, (1913). Stuart, J. M. and Wormwell, F. *ibid.* 85 (1930). Kraus, C. A. and Dexter, W. B. *J. Amer. chem. Soc.* **44,** 2468 (1922).
20. Davies, C. W. and Nancollas, G. *Chem. & Ind.* (*Rev.*) 129 (1950).
21. Shedlovsky, T. and Kay, R. L. *J. phys. Chem.* **60,** 151 (1956).
22. Schenkel, J. H. and Kitchener, J. A. *Nature, Lond.* **182,** 131 (1958).
23. James, J. C. and Monk, C. B. *Trans. Faraday Soc.* **46,** 141 (1950).

24. Davies, C. W. *ibid.* **25,** 129 (1929).
25. See Ref. 2, Ch. 1.
26. Jones, G. and Bradshaw, B. C. *J. Amer. chem. Soc.* **55,** 1780 (1933).
27. *J. Res. nat. Bur. Stand.* **60,** 29 (1916).
28. Jones, G. and Prendergast, M. J. *J. Amer. chem. Soc.* **59,** 731 (1937).
29. Gunning, H. E. and Gordon, A. R. *J. chem. Phys.* **10,** 126 (1942).
30. Lind, J. E., Zwolenk, J. J. and Fuoss, R. M. *J. Amer. chem. Soc.* **81,** 1557, (1959).
31. See Ref. 11, Ch. 1.
32. Le Roy, D. J. and Gordon, A. R. *J. chem. Phys.* **6,** 398 (1938).
33. Ives, D. J. G. and Swaropa, S. *Trans. Faraday Soc.* **49,** 788 (1953). Elias, L. and Schiff, H. I. *J. phys. Chem.* **60,** 595 (1956). Graham, G. D. and Maass, O. *Canad. J. Chem.* **36,** 315 (1958).
34. Wien, M. *Ann Phys. Lpz.* **83,** 327 (1927); **85,** 795 (1928); *Phys. Z.*, **28,** 834 (1927); **29,** 751 (1928).
35. Eckstrom, H. C. and Schmlezer, C. *Chem. Rev.* **24,** 367 (1939).
36. Gledhill, J. A. and Patterson, A. *Rev. sci. Instrum.* **20,** 960 (1949); *J. phys. Chem.* **56,** 999 (1952).
37. Wien, M. *Ann. Phys. Lpz.* **83,** 840 (1927).
38. Falkenhagen, H. "Electrolytes," translated by R. P. Bell, Oxford University Press, Oxford (1939).
39. Lane, J. A. and Saxton, J. A. *Proc. roy. Soc.* **213A,** 400, 473 (1952); **214A,** 531 (1952).
40. Hasted, J. B. and Roderick, G. W. *J. chem. Phys.* **29,** 17 (1958).
41. Blake, C. G. "Conductimetric Analysis at Radio-Frequency." Chapman and Hall, Ltd., London (1950); Reilly, C. N. "New Instrumental Methods in Electrochemistry." Interscience Publishers, Inc., New York (1954).
42. Taylor, H. S. "A Treatise on Physical Chemistry." Van Nostrand, New York (1931).
43. Washburn, E. L. *J. Amer. chem. Soc.* **31,** 330 (1909).
44. MacInnes, D. A. and Dole, M. *ibid.* **53,** 1357 (1931).
45. Steel, B. J. and Stokes, R. H. *J. phys. Chem.* **62,** 450 (1958). Ref. 11, Ch. 1.
46. Ref. (2), Ch. (1).
47. MacInnes, D. A. and Brighton, T. B. *J. Amer. chem. Soc.* **47,** 994 (1925).
48. Cady, H. P. and Longsworth, L. G. *ibid.* **51,** 1656 (1929). Longsworth. L. G. *ibid.* **52,** 1897 (1930).
49. MacInnes, D. A. and Longsworth, L. G. *Chem. Rev.* **11,** 172 (1932).
50. Hartley, G. S. and Donaldson, G. W. *Trans. Faraday Soc.* **33,** 457 (1937); Ref. 11, Ch. 1.
51. Allgood, R. W. Le Roy, D. J. and Gordon, A. R. *J. chem. Phys.* **8,** 418 (1940).
52. Franklin, E. C. and Cady, H. P. *J. Amer. chem. Soc.* **26,** 499 (1904).
53. Covington, A. K. and Prue, J. E. *J. Chem. Soc.* 1567 (1957).
54. Longsworth, L. G. *J. Amer. chem. Soc.* **54,** 2741 (1932).
55. MacInnes, D. A., Cowperthwaite, I. A. and Huang, T. C. *ibid.* **49,** 1710 (1927).
56. MacInnes, D. A., Cowperthwaite, I. A. and Blanchard, K. C. *ibid.* **48,** 1909 (1926).
57. Spedding, F. K., Porter, P. E. and Wright, J. M. *ibid.* **74,** 2778 (1952).
58. Hopkins and Covington, A. K. *J. sci. Instrum.* **34,** 20 (1957).
59. Kohlrausch, F. *Ann. phys. Chem.* **62,** 209 (1897). Ref. 11, Ch. 1.
60. Gordon, A. R. and Kay, R. L. *J. chem. Phys.* **21,** 131 (1953).
61. Lewis, G. N. *J. Amer. chem. Soc.* **32,** 862 (1910).
62. Longsworth, L. G. *ibid.* **57,** 1185 (1935).
63. Brady, A. P. *ibid.* **70,** 911, 914 (1948).
64. Spiro, M. and Parton, H. N. *Trans. Faraday Soc.* **48,** 263 (1952).

CHAPTER 3

THE INTERIONIC ATTRACTION THEORY AND ITS APPLICATION TO ACTIVITY COEFFICIENTS, CONDUCTANCES AND TRANSFERENCE NUMBERS

SOME of the observations that led to the recognition of the importance of the forces of attraction and repulsion between ions in solution have been mentioned in Chapter 1. Unlike non-ionic liquids in which the forces between particles are weak and short-range, and where thermal motions produce a purely random distribution of the constituents, in ionic solutions the interionic forces are long-range and tend to oppose the establishment of a random distribution. Because of the ionic forces, an ion is more likely to be surrounded by ions of opposite charge, forming an "ionic atmosphere." In fact, if it were not for thermal motions, an electrolyte solution would have the ordered structure of ionic crystals. The actual distribution is between these two extremes of randomness and order.

A successful quantitative theory must be able to allow the actual ionic distribution to be calculated. Now these are dependent on ionic forces which may be both internal and external (e.g., an applied electric field), and in turn, the internal forces due to the ions themselves are largely dependent on the ionic distribution. This mutual dependence poses a difficult problem, and as related in the first chapter, the first successful solution was found by Debye and Hückel in 1923. Among other things, their treatment gave an expression whereby activity coefficients of electrolytes can be calculated. Improvements and extensions to their theory in connection with this and other aspects of electrolytes have since followed. A number of these will be dealt with in connection with particular topics.

Activity Coefficients

1. Debye-Hückel Activity Coefficient Expressions

Assuming that an ion of charge e_j is in a solvent of dielectric constant ϵ, and that the ion has zero radius, then at a distance r from the ion, the potential is $e_j/\epsilon r$. In an electrolyte solution this simple condition does not hold because an ion becomes surrounded by ions of opposite charge, forming an ionic atmosphere. The potential due to the ion and its atmosphere is related to the charge density by Poisson's equation:

$$\Delta\psi = \nabla \cdot \nabla\psi = -4\pi\rho/\epsilon \tag{3.1}$$

where ψ is the potential, ρ is the charge density, and

$$\nabla\psi(= \text{grad}\ \psi) = (\partial\psi/\partial x) + (\partial\psi/\partial y) + (\partial\psi/\partial z) \tag{3.2}$$

$$\Delta\psi(= \text{div grad}\ \psi) = (\partial^2\psi/\partial x^2) + (\partial^2\psi/\partial y^2) + (\partial^2\psi/\partial z^2) \tag{3.3}$$

Equation (3.1) means that at any point located by the co-ordinates, x, y, z, the total outward flux at the point is proportional to the charge density at the point. For electrolyte solutions, the charge density, or charge per c.c, at a distance r from a j ion is

$$\rho_j = \textstyle\sum_1^s n_{ji}^0 e_i \tag{3.4}$$

where n_{ji}^0 is the time-average concentration of i ions in a volume element in the vicinity of a j ion, and where s represents the various species of ions. If ψ_j^0 is the potential at a distance r from a j ion, then

$$\nabla \cdot \nabla\psi_j^0 = -(4\pi/\epsilon)\textstyle\sum_1^s n_{ji}^0 e_i \tag{3.5}$$

It is assumed that n_{ji}^0 is determined by the Maxwell–Boltzmann distribution law, i.e.,

$$n_{ji}^0 = n_i \exp{(-E_{ji}/kT)} \tag{3.6}$$

where n_i is the average or the bulk concentration of i ions, and E_{ji} is the potential energy of an ion i in the vicinity of a j ion. Likewise, for a j ion in the vicinity of an i ion,

$$n_{ij}^0 = n_j \exp{(-E_{ij}/kT)} \tag{3.7}$$

The potential energy term E_{ji} can be replaced by that for the electrical potential energy, $\psi_j^0 e_i$. Then from (3.5) and (3.6),

$$\nabla \cdot \nabla\psi_j^0 = -(4\pi/\epsilon)\textstyle\sum_1^s n_i e_i \exp{(-\psi_j^0 e_i/kT)} \tag{3.8}$$

It has been assumed that the potential around an ion is proportional to its charge whereas the right-hand side of (3.8) contradicts this. To overcome this discrepancy, use is made of the exponential series

$$\exp x = 1 + x + (x^2/2!) + (x^3/3!) + \ldots \tag{3.9}$$

so that for small values of $x = e_i\psi_j^0$,

$$\nabla \cdot \nabla\psi_j^0 = -(4\pi/\epsilon)\textstyle\sum_1^s n_i e_i\{1 - (e_i\psi_j^0/kT)\} \tag{3.10}$$

Since, as a whole, an electrolyte solution is electrically neutral, $\sum e_i n_i = 0$, and since $e_i = z_i e$, where z_i represents ion valency and e is the electron charge, (3.10) becomes

$$\nabla \cdot \nabla\psi_j^0 = \kappa^2\psi_j^0 \tag{3.11}$$

where the following definition is incorporated:

$$\kappa^2 = (4\pi e^2/\epsilon kT)\textstyle\sum_1^s n_i z_i^2 \tag{3.12}$$

It must be emphasized that (3.11) is meant to apply accurately only to dilute solutions, where $z_i = 1$, and in media such as water where ϵ is large.

In the absence of an applied field, the electrolyte is said to be unperturbed, the ionic atmosphere possessing spherical symmetry, so spherical co-ordinates can be used with $r^2 = x^2 + y^2 + z^2$. The result is that

$$\nabla . \nabla\psi_j^0 = r^{-2}(\partial/\partial r)(r^2\partial\psi_j^0/\partial r) = \kappa^2\psi_j^0 \tag{3.13}$$

Substituting $u = \psi_j^0 r$ gives

$$\partial^2 u/\partial r^2 = \kappa^2 u \tag{3.14}$$

the general solution of which is

$$u = A^* \exp(-\kappa r) + B^* \exp \kappa r$$

or

$$\psi_j^0 = (A^* \exp - \kappa r)/r + (B^* \exp \kappa r)/r \tag{3.15}$$

When $r = \infty$, ψ_j^0 must be zero, whence $B = 0$ since $(\exp \kappa r)/r$ does not tend to zero as $r \to \infty$. Also, when $\kappa = 0$, the concentration is zero so the potential is simply that of a single ion, namely $z_j e/\epsilon r$ (see beginning of section). Thus from (3.15),

$$A^* = z_j e/\epsilon \tag{3.16}$$

Consequently,

$$\psi_j^0 = z_j e(\exp - \kappa r)/\epsilon r = z_j e/\epsilon r - z_j e(1 - \exp - \kappa r)/\epsilon r \tag{3.17}$$

The last term is therefore the potential of the ionic atmosphere at a distance r. Calling this $(\psi_j^{0*})_r$, from (3.15), since $B^* = 0$.

$$(\psi_j^{0*})_r = (A \exp - \kappa r)/r - z_j e/\epsilon r \tag{3.18}$$

Now an ion is not a point charge, and if a is its actual radius, for the field to be continuous when $r = a$, a necessary condition is

$$\partial\psi_j^0/\partial r = - z_j e/\epsilon r^2 \tag{3.19}$$

i.e., the field of the ion and its atmosphere must equal the field of the ion. From (3.18), on differentiating and introducing the stated conditions,

$$- (A^* \exp - \kappa a)/a^2 - (A^*\kappa \exp - \kappa a)/a + z_j e/\epsilon a^2 = 0$$

whence

$$A^* = (z_j e \exp \kappa a)/\epsilon(1 + \kappa a) \tag{3.20}$$

It is seen that this differs from (3.16) through allowing for ion size, and substituting in (3.15), again with $B^* = 0$, gives

$$\psi_j^0 = \{z_j e(\exp \kappa a)(\exp - \kappa r)\}/\{\epsilon r(1 + \kappa a)\} \tag{3.21}$$

and by analogy with (3.18),

$$(\psi_j^{0*})_r = \{z_j e(\exp \kappa a)(\exp - \kappa r)\}/\{\epsilon r(1 + \kappa a)\} - z_j e/\epsilon r \tag{3.22}$$

When $r = a$, this becomes

$$(\psi_j^{0*})_a = - z_j e\kappa/\epsilon(1 + \kappa a) \tag{3.23}$$

For a single ion, the extra free energy, or chemical potential, $\Delta\bar{G}_j$, of the solution, compared with that of an uncharged particle in solution is[2] equal to the work required to charge the ion in a field $(\psi_j^{0*})_a$, i.e.,

$$\Delta\bar{G}_j = \int_0^{z_j e} \psi_j^{0*} e_j de_j = -(z_j^2 e^2 \kappa)/2\epsilon(1+\kappa a) \tag{3.24}$$

By assuming that an ionic solution differs from an ideal one solely because of the extra free energy due to interionic forces, if $N\Delta\bar{G}_j$ (electrical) refers to this extra free energy of one mole of j ions, for the solution as a whole,

$$\bar{G}_j = \bar{G}_j(\text{ideal}) + N\Delta\bar{G}_j(\text{electrical}) \tag{3.25}$$

Substitution of (1.19) and (1.25) converts this into

$$\bar{G}_j^0 + RT\ln f_j + RT\ln N_j = \bar{G}_j^0 + RT\ln N_j + N\Delta\bar{G}_j(\text{electrical}) \tag{3.26}$$

whence, from (3.24),

$$\ln f_j = N\Delta\bar{G}_j(\text{electrical})/RT = -z_j^2 e^2 N\kappa/2\epsilon RT(1+\kappa a) \tag{3.27}$$

Individual ionic activity coefficients cannot be measured so (3.27) must be converted into a suitable form both for testing and for practical purposes by using a definition such as (1.27). The mean rational activity coefficient of an electrolyte, $f_\pm$, when one "molecule" dissociates into ν ions, ν_i being cations of valence z_i and ν_j being anions of valence z_j is obtained via the electroneutrality relationship $\nu_i z_i = -\nu_j z_j$. These factors lead to

$$\ln f_\pm = -|z_i z_j| e^2 N\kappa/2\epsilon RT(1+\kappa a) \tag{3.28}$$

Substituting for κ by (3.12), for n_i by

$$n_i = NC_i/1000 \tag{3.29}$$

and using the definition

$$\sum C_i z_i^2 = \Gamma \tag{3.30}$$

where Γ is termed the "ional" concentration, (3.28) is converted into

$$\log f_\pm = -A|z_i z_j|\sqrt{\Gamma}/(1+Ba\sqrt{\Gamma}) \tag{3.31}$$

where

$$A = (N\pi e^6)^{\frac{1}{2}}/2.3026(1{,}000k\epsilon T)^{\frac{3}{2}} \tag{3.32}$$

$$B = (4\pi Ne^2)^{\frac{1}{2}}/(1{,}000k\epsilon T)^{\frac{1}{2}} \tag{3.33}$$

in which $k = R/N$ and a is now the mean radius of cation + anion. In dilute solutions, $C_i = m_i d_0$, where m_i is the ionic molality and d_0 is the solvent density, whence, from (1.28),

$$2d_0 I = \Gamma \tag{3.34}$$

and from (1.23) and (1.27),

$$\log\gamma_\pm = -A'|z_i z_j|\sqrt{d_0 I}/(1+B'a\sqrt{d_0 I}) = -A''|z_i z_j|\sqrt{I}/(1+B''a\sqrt{I}) \tag{3.35}$$

For dilute aqueous solutions, $d_0 \simeq 1$, but beyond $I = 0.2$, it is more accurate to calculate $f_\pm$ and to then convert this to $\gamma_\pm$, if required, via (3.48). Substituting the numerical values of the constants, (Appendix),

$$A = 1.2903 \times 10^6/(\epsilon T)^{\frac{3}{2}}, \quad A' = A\sqrt{2} = 1.8247 \times 10^6/(\epsilon T)^{\frac{3}{2}} \tag{3.36}$$

$$B = 3.556 \times 10^9/(\epsilon T)^{\frac{1}{2}}, \quad B' = B\sqrt{2} = 5.029 \times 10^9/(\epsilon T)^{\frac{1}{2}} \tag{3.37}$$

At 25°C in water,

$$A = 0.3600, \qquad A' = 0.5091, \qquad A'' = 0.5083 \tag{3.38}$$

$$10^8 B = 0.2324, \quad 10^8 B' = 0.3287, \quad 10^8 B'' = 0.3282$$

The mean radius or distance of closest approach of cation to anion, a, as explained in Chapter 1, cannot be rigorously assessed owing to ion-solvation, but there are various ways of getting reasonable estimates, especially for many of the 1 : 1, 1 : 2 or 2 : 1 valent electrolytes; these methods are mentioned in later sections. In many instances a can be taken as 3 to 4Å, so in water at 25°C, $Ba \simeq 0.7$, $B'a \simeq 1.0$. So for very dilute solutions (up to $I \simeq 0.005$) the denominators of (3.31) and (3.35) are not far from unity, i.e., a limiting form of (3.31) is

$$\log f_\pm = -A \mid z_i z_j \mid \sqrt{\Gamma} \text{ or } \log \gamma_\pm = -A'' \mid z_i z_j \mid \sqrt{I} \tag{3.39}$$

Example 3.1. Considering (a) 0.002 M NaCl, (b) 0.002 M $CaCl_2$, (c) a mixture of 0.002 M NaCl and 0.001 M $CaCl_2$, at 25°C in water, from (3.30), $\Gamma =$ (a) 0.004, (b) 0.012, (c) 0.010. From (3.36) and (3.39), $f_\pm =$ (a) 0.949, (b) 0.834, (c) 0.847—for $CaCl_2$.

Even (3.31) and (3.35) fail to agree with activity coefficients calculated from experimental data beyond $I \simeq 0.05$, but by inclusion of an empirical parameter, Q, the range of conformity can often be extended to quite high concentrations (cf. Table 3.1). The equation then has the forms

$$\log f_\pm = -A \mid z_i z_j \mid \{\sqrt{\Gamma}/(1 + Ba\sqrt{\Gamma}) - Q'\Gamma\} \tag{3.40}$$

$$\log \gamma_\pm = -A'' \mid z_i z_j \mid \{\sqrt{I}/(1 + B''a\sqrt{I}) - Q''I\} \tag{3.41}$$

When using these in connection with incompletely dissociated electrolytes, the correct values of a, Q' and Q'' must be selected. If the dissociation constant K is small, I can be kept small, and the parameters of similar completely dissociated electrolytes can be used without causing much error. If K is large, I cannot be kept small without sacrifice of accuracy in the determination of α_c, and a wide combination of a and Q' or Q'' can be chosen; each particular combination gives a constant value of K over a range of ionic strengths, but the actual value of K depends on the chosen parameters. This feature is exemplified and discussed in later chapters.

Assuming that there is little error by taking those parameters which in (3.40) and (3.41) predict the experimental $f_\pm$ and $\gamma_\pm$ data of fully ionized materials, then general formulae can be written. Guggenheim[3] has found that the following form applies to 1 : 1 electrolytes up to $I = 0.1$ (Q'' is characteristic):

$$\log f_{\pm} = - A \mid z_i z_j \mid \{\sqrt{I}/(1 + \sqrt{I}) - Q''I\} \tag{3.42}$$

while Davies[4] has shown that

$$\log f_{\pm} = - 0.50 \mid z_i z_j \mid \{\sqrt{I}/(1 + \sqrt{I}) - 0.2I\} = - \phi(I) \tag{3.43}$$

is satisfactory for 1 : 1 and 2 : 1 electrolytes in water at 25°C up to $I = 0.1$, within about 2 %.

2. Other Formulae for the Calculation of Activity Coefficients

One explanation of the need for the extension term $Q'\Gamma$ or $Q''I$ is that it accounts for short-range ion-solvent interactions and changes in the dielectric constant of the medium as the concentration increases. On the other hand, since only the first two terms of the expansion series (3.9) are used in the derivation of (3.10), it is unwise to attribute too much physical significance to the Q parameter. In (3.10), terms of the 3rd, 5th, and higher orders are neglected for symmetrical electrolytes ($\mid z_i \mid = \mid z_j \mid$), while with unsymmetrical types the even order terms are also omitted. La Mer and Gronwall[5] have endeavoured to remove these approximations, but the resulting expressions are elaborate in spite of neglecting terms beyond the 5th order; the calculations are eased by suitable Tables.[5,6]

A graphical method for including the higher terms of (3.9) was devised by Müller.[7] An outline of this has been given by Fowler and Guggenheim,[8] and Guggenheim[9] has applied it to discuss 1 : 1 and 2 : 2 valent electrolytes. The procedure of Müller deserves more recognition than it has received.

Modified forms of the Maxwell–Boltzmann distribution and other distribution formulae[10] have been examined by Robinson and Stokes,[11] who consider that these additional treatments do not offer any significant improvement to that of Debye and Hückel. Furthermore, if $a > 3.6$ Å and I is small, the neglect of the higher terms in the D—H version has little effect. Kirkwood[12] has also confirmed this; he used statistical mechanics to investigate the extent of the approximations. His expression for the potential, corresponding to (3.23), involves complex integrals which Falkenhagen and Kelbg,[13] Bell and Levine[14] have managed to simplify. Mayer[15] has also used statistical mechanical analysis in connection with electrolyte theory. He derived an activity coefficient expression by way of his "cluster" theory of imperfect gases. Poirier[16] has carried out the lengthy calculations required in four instances; comparison with experimental information shows the valid regions to be: $NaCl = 0.2$ M, $CaCl_2 = 0.05$ M, $LaCl_3 = 0.01$ M, $ZnSO_4 = 0.01$ M.

Even these limited ranges depend on the choice of the parameter, a. Accordingly the Meyer treatment cannot be deemed superior to that of Debye and Hückel.

The activity coefficients of all 2 : 2, 2 : 3 and higher valent electrolytes, as derived from experimental investigations, are all lower than those calculated. Assuming that the approximations made in the D—H treatment are not of major significance for polyvalent electrolytes (at least for dilute solutions), the discrepancies can be attributed to incomplete dissociation.

3. Activity Coefficient Inter-Relationships

There is sometimes a need to convert activity coefficients from one concentration scale to another. The derivation of the appropriate formulae follows from the following considerations. From (1.20 to 1.27), the free energy of an electrolyte in solution can be written

$$\bar{G} = \bar{G}_N^0 + \nu RT \ln Nf_{\pm} = \bar{G}_C^0 + \nu RT \ln Cy_{\pm} = \bar{G}_m^0 + \nu RT \ln m\gamma_{\pm} \qquad (3.43a)$$

where

$$N = m/(\nu m + 1{,}000/M_1) = C/\{\nu C + (1{,}000d - CM_2)/M_1\} \qquad (3.43b)$$

in which N is the mole fraction of solute, M_1 and M_2 are the molecular weights of the solvent and solute respectively, d and d_0 are the densities of solution and solvent. Also,

$$C = md/(1 + mM_2/1{,}000); \quad m = C/(d - CM_2/1{,}000) \qquad (3.43c)$$

From (3.40), $\nu RT \ln f_{\pm} = \bar{G}_C^0 - \bar{G}_N^0 + \nu RT \ln y_{\pm} + RT \ln (C/N)$, and as $C \to 0$; from (3.41), $C/N \to 1{,}000d_0/M_1$, also $f_{\pm}$ and $y_{\pm} \to 1$, so

$$\bar{G}_C^0 - \bar{G}_N^0 = -\nu RT \ln (1{,}000d_0/M_1) \qquad (3.43d)$$

whence

$$\ln f_{\pm} = \ln y_{\pm} + \ln \{d/d_0 + C(\nu M_1 - M_2)/1{,}000d_0\} \qquad (3.44)$$

Similarly, as C and $m \to 0$, $m/C \to 1/d_0$, $y_{\pm}$ and $\gamma_{\pm} \to 1$, whence $\bar{G}_m^0 - \bar{G}_C^0 = \nu RT \ln d_0$, giving

$$\ln y_{\pm} = \ln \gamma_{\pm} + \ln (md_0/C) \qquad (3.45)$$

This, from (3.42) becomes

$$\ln y_{\pm} = \ln \gamma_{\pm} + \ln \{d_0/(d - CM_2/1{,}000)\} \qquad (3.46)$$

or

$$\ln \gamma_{\pm} = \ln y_{\pm} + \ln \{d(1 + mM_2/1{,}000)/d_0\} \qquad (3.47)$$

likewise

$$\ln f_{\pm} = \ln \gamma_{\pm} + \ln \{1 + (\nu M_1 m/1{,}000)\} \qquad (3.48)$$

Example 3.2. The experimental $\gamma_{\pm}$ values for aqueous 2.0 M $LaCl_3$ at 25°C is[17] 0.883, and $d = 1.392$. From (3.42), $C = 2.685$, and from (3.48), $f_{\pm} = 1.010$.

Example 3.3. The experimental $\gamma_{\pm}$ values for aqueous KCl at 25°C are[18] shown in the third column of Table 3.1, the fourth column lists the values of $f_{\pm}$ (exptl.) calculated via (3.48), the fifth column figures are based on the limiting form, (3.39), those of the sixth column are based on the generalized form, (3.43), and those of the last column are derived from (3.35, 3.36, 3.37) and $a = 3.5$ Å, together with the parameter 0.0165 obtained by comparison at 3 M, i.e., $-\log f_{\pm} = 0.3600\sqrt{\Gamma}/(1 + 0.2324 \times 3.50\sqrt{\Gamma}) - 0.0165\,\Gamma$. It is to be seen that the limiting equation is almost in error even at 0.005 M, the generalized form has less than 1 % error up to 0.1 M, while the form just given holds up to at least 3 M.

TABLE 3.1.

Activity Coefficients of KCl in Water at 25°C.

(1) m	(2)[19] C	(3) $\gamma_{\pm}$(exptl.)	(4) $f_{\pm}$(exptl.)	(5) $f_{\pm}$(calc.)	(6) $f_{\pm}$(calc.)	(7) $f_{\pm}$(calc.)
0.001	0.0009971	0.9650	0.9650	0.9635	0.965	0.965
0.002	0.001994	0.9516	0.9516	0.949	0.952	0.9515
0.005	0.004985	0.9270	0.9271	0.920_5	0.928	0.9265
0.01	0.009968	0.9015	0.9016	0.889_5	0.903	0.901
0.02	0.01993	0.8694	0.8696	0.847_5	0.873	0.869
0.05	0.04979	0.8164	0.8177	0.770	0.820	0.815
0.10	0.09943	0.769	0.772	0.691	0.776	0.768
0.20	0.1983	0.718	0.723	—	0.734	0.719
0.50	0.4914	0.649	0.661	—	—	0.657
1.0	0.9689	0.604	0.626	—	—	0.628
2.0	1.883	0.573	0.614	—	—	0.618
3.0	2.742	0.569	0.630	—	—	0.631
4.0	3.551	0.577	0.660	—	—	0.652

There are some interesting features about activity coefficients at high concentrations. These are mentioned in Chapter 5.

Conductance

1. The Onsager–Fuoss Limiting Conductance Treatment

Under equilibrium conditions, the time-average distribution of ions in solution produces symmetrical ionic atmospheres, but when an electric field is applied, disturbances upset this. The ion moves from the centre of its atmosphere and as a result, is subjected to a retarding force by the atmosphere. Thermal motions tend to restore the atmosphere to its symmetrical form and position, and the time that would be taken to achieve this is termed the time of relaxation. It is extremely small, e.g., in a 0.001 N solution it is about 10^{-7} sec. The average retarding force exerted on the central ion is termed the relaxation effect.

There is also another process that retards ion motion. It is termed the electrophoretic effect and arises from the tendency of an ion to drag along molecules of the solvent through a transference of motion. Furthermore, since ions of opposite charge move in opposite directions while the applied field operates, an ion does not move through a stationary medium, but against a stream of solvent molecules moving the other way.

Although both of these effects were considered in a treatment of conductance by Debye and Hückel, the first really successful and useful analysis was produced by Onsager,[20] and Onsager and Fuoss [21] between 1927 and 1932. These limiting versions have been improved in recent times; details are given later on. The Onsager–Fuoss treatment of conductance (and diffusion) incorporates a number of complex and advanced mathematical theorems. In the present account many of the conclusions reached with these are stated without giving the details.

In general, the problem consists of investigating the migration of ions in a field of forces, namely external forces $k_1, k_2, \ldots k_s$ per ion acting on ionic species 1, 2, . . . s, and a balancing force k_0 acting on the solvent molecules, i.e.,

$$-n_0k_0 = \textstyle\sum_1^s n_ik_i \tag{3.49}$$

where n_i is the number of ions of the species i, n_0 the number of molecules, per c.c.

If k_j is the external force acting on a specific j ion, which will be taken as isolated, its velocity v_j relative to the solvent is

$$v_j = k_j/\rho_j = k_j\omega_i \tag{3.50}$$

where ρ_j is its coefficient of friction and ω_j is the ion mobility or velocity under unit applied force. In practice, this equation is an approximate one since ions cannot be isolated, and they affect each other's motions. As already mentioned, there are two factors to consider, the direct transfer of electric forces between ions, and an electrophoretic effect resulting from an ion having an ionic atmosphere which moves in the opposite direction. An approximate estimate of the "force" or "relaxation" effect can be obtained as follows. When an ion moves, at first, its atmosphere is left behind, but the displaced ion soon exerts an electric pull on the lagging atmosphere which will continue to form around the moving ion. This ion will lead its atmosphere by a time τ (the time of relaxation of the atmosphere) corresponding to a distance $v_j\tau$. The ratio of this to $1/\kappa$, (see 3.12), the radius of the ionic atmosphere, gives, from (3.50), the relative dissymmetry of the latter, namely $\kappa k_j\tau/\rho_j$. The directed component Δk_j of the total force between ion and atmosphere, namely $e_j^2\kappa^2/\epsilon$ is given (approximately) by

$$\Delta k_j = - e_j^2\kappa^2(\kappa k_j\tau)/\epsilon \tag{3.51}$$

and, according to Onsager and Fuoss, τ is given by

$$\tau = \rho_j/\kappa^2 kT \tag{3.52}$$

whence

$$\Delta k_j = - e_j^2\kappa k_j/\epsilon kT \tag{3.53}$$

This agrees with the more exact treatment given later on, except for a numerical factor. The difference is that account is taken of the movement of the ions of the ionic atmosphere due to the external forces, resulting in the central j ion becoming part of the atmospheres of these other migrating ions. The relationship of the forces is thus a reciprocal one; an ion contributes to the relaxation of its own atmosphere.

Considering now the electrophoretic effect, in the case of conductance, k_0 of (3.49) is zero, and $k_j = Xe_j$ for an ion j, where X is the applied field. The ionic atmosphere of j will have a charge of $- e_j$ (i.e., equal and opposite to that of the j ion), so it will be subjected to a force of $- Xe_j$. If it is assumed that the entire charge of the atmosphere is situated on a spherical shell of radius $1/\kappa$, then by Stokes's law (velocity = applied force/$6\kappa\eta r$, where r = radius), the interior of this shell will move with a velocity of

$$\Delta v_j = - Xe_j\kappa/6\pi\eta \tag{3.54}$$

where η is the viscosity of the solution. The liquid which surrounds the central ion thus moves with this velocity, and the central ion has to move against it. Now if this ion had no atmosphere, its velocity would be, from (3.50),

$$v_j = k_j/\rho_j = Xe_j/\rho_j = Xe_j\omega_j \tag{3.55}$$

but owing to the atmosphere, the ion is subjected to a force $k_j - \Delta k_j$, and not k_j. The corresponding velocity is, however, relative to a neighbourhood of velocity Δv_j, so the net velocity is

$$v_j = (k_j - \Delta k_j)/\rho_j + \Delta v_j \tag{3.56}$$

In terms of electrical conduction this becomes, from (3.54, 3.55)

$$v_j = (Xe_j - \Delta k_j)\omega_j - Xe_j\kappa/6\pi\eta \tag{3.57}$$

Since both Δk_j and Δv_j depend on κ, both are proportional to $\sqrt{c}$.

2. General Treatment of Electrophoresis and the Relaxation Effect

Equation (3.49) will now be written in a more convenient form, namely

$$n_\sigma k_\sigma + n_0 k_0 = 0 \tag{3.58}$$

where σ signifies summation. The force acting on the ions in an element of volume dV near a j ion is $n_{j\sigma}k_\sigma dV$, where $n_{j\sigma}$ is the concentration of other ions per unit volume near the j ion. The force acting on the solvent molecules is n_0k_0dV. The net force acting on dV is thus $(n_{j\sigma}k_\sigma + n_0k_0)dV$, and if dV is a shell of radius r and thickness dr, while $d\mathscr{F}$ is the force, then from (3.58),

$$d\mathscr{F} = (n_{j\sigma} + n_0k_0)dV = (n_{j\sigma} - n_\sigma)k_\sigma dV = (n_{j\sigma} - n_\sigma)k_\sigma 4\pi r^2 dr \qquad (3.59)$$

This force is distributed over the shell and since this is the same as that acting on a solid sphere moving through a liquid of viscosity η, Stokes's law may be applied;

$$dv = d\mathscr{F}/6\pi\eta r \qquad (3.60)$$

The force $d\mathscr{F}$ will cause all the points inside a shell of radius r to move with a velocity dv.

The Boltzmann distribution law and expansion by the exponential series can be used as they were in (3.6 to 3.10) to give

$$n_{j\sigma} = n_\sigma\{1 - (e_\sigma\psi_j/kT) + \tfrac{1}{2}(e_\sigma\psi_j/kT)^2\} \qquad (3.61)$$

The 3rd term of the expansion is included because it is significant in the theory of diffusion. The potential ψ_j is expressed by (3.21). Substitution in (3.59) gives

$$d\mathscr{F} = 4\pi r^2\left[\frac{-\ e_je_\sigma \exp \kappa a \,.\, \exp -\ \kappa r}{\epsilon kT(1 + \kappa a)r} + \frac{1}{2}\left(\frac{e_je_\sigma \exp \kappa a}{kT(1 + \kappa a)}\right)^2 \frac{\exp -\ 2\kappa r}{r^2}\right] n_\sigma k_\sigma dr \qquad (3.62)$$

which will be abbreviated to

$$d\mathscr{F} = 4\pi[-\ A_1 r \exp -\ \kappa r + (A_2/r) \exp -\ 2\kappa r]dr \qquad (3.63)$$

so from (3.60),

$$dv_j = (2/3\eta)[-\ A_1 \exp -\ \kappa r + (A_2/r) \exp -\ 2\kappa r]dr \qquad (3.64)$$

To obtain the velocity effect due to the entire atmosphere, integration is effected from $r = a$ to $r = \infty$, giving

$$\Delta v_j = (2/3\eta)[-\ (A_1/\kappa) \exp -\ \kappa a + A_2 \mathrm{Ei}(2\kappa a)] \qquad (3.65)$$

where $\mathrm{Ei}(2\kappa a)$ is obtained by means of the exponential integral function

$$\mathrm{Ei}(x) = \int_x^\infty (\exp -\ t)dt/t = -\ 0.57722 - \ln x + x - x^2/2.2! + x^3/3.3! \ldots \qquad (3.66)$$

Resubstituting the values of A_1 and A_2 establishes the electrophoretic correction,

$$\Delta v_j = \frac{-\ 2e_je_\sigma n_\sigma k_\sigma}{3\eta\epsilon kT\kappa(1 + \kappa a)} + \frac{1}{3\eta}\left(\frac{e_je_\sigma \exp \kappa a}{\epsilon kT(1 + \kappa a)}\right)^2 n_\sigma k_\sigma \mathrm{Ei}(2\kappa a) \qquad (3.67)$$

At low concentrations, κa is small so the second term on the right is negligible, and since, with the present nomenclature (3.12) is

$$\kappa^2 = 4\pi e_j e_\sigma n_\sigma / \epsilon kT \tag{3.68}$$

then with Xe_σ for k_σ, (3.54) is once more derived.

The relaxation or force transfer effect which has been expressed as Δk_j in (3.56) involves mathematical discussions which will be omitted because of their complexity. If ΔX_j is the added field caused by the finite time of relaxation of the ionic atmosphere, then for a simple electrolyte, where s of (3.49) is 2, e_σ is e_i,

$$\Delta X_j = e_i e_j q^* \kappa X / 3\epsilon kT(1 + \sqrt{q^*}) \tag{3.69}$$

where

$$q^* = (e_j \omega_j - e_i \omega_i)/(e_j - e_i)(\omega_j + \omega_i) \tag{3.70}$$

For a symmetrical electrolyte, $e_j = - e_i$, and $q^* = 0.5$.

The total field acting on a j ion is $X + \Delta X_j$, and the corresponding velocity would be $e_j\omega_j(X + \Delta X_j)$, but it is reduced by the electrophoretic effect. Accordingly, from (3.57) and (3.67), within the conditions laid down for the latter,

$$v_j = (X + \Delta X_j)e_j\omega_j - (X + \Delta X_j)e_j\kappa/6\pi\eta \tag{3.71}$$

The symbol ω has been used for mobility in general, and if $\bar{u}_j$ is the velocity of a j ion in an applied field of 1V per cm, if c^* is the velocity of light in cm/sec,

$$\bar{u}_j = v_j/10^{-8}Xc^* = [e_j\omega_j - (e_j\kappa/6\pi\eta)][1 + \Delta X_j/X]/10^{-8}c^* \tag{3.72}$$

Also,

$$e_j\omega_j/10^{-8}c^* = \bar{u}_j^0, \quad F\bar{u}_j = \lambda_j \tag{3.73}$$

where $\bar{u}_j^0$ is the mobility at zero concentration (from 3.72, with $\kappa = 0$, when $\Delta X_j = 0$, also), so (3.72) becomes

$$\lambda_j = [\lambda_j^0 - Fe_j\kappa/6\pi\eta 10^{-8}c^*](1 + \Delta X_j/X) \tag{3.74}$$

On substituting for κ from (3.12), and for n_i by (3.29), since $e_j = z_j e$,

$$Fe_j\kappa/6\pi\eta 10^{-8}c^* = 29.167 z_j \Gamma^{\frac{1}{2}}/\eta(\epsilon T)^{\frac{1}{2}} \tag{3.75}$$

Equation (3.70) likewise becomes (regarding z as always positive)

$$q^* = z_i z_j(\lambda_i^0 + \lambda_j^0)/(z_i + z_j)(z_j\lambda_i^0 + z_i\lambda_j^0) = z_i z_j/(z_i + z_j)(z_j T_i^0 + z_i T_j^0) \tag{3.76}$$

where T^0 represents transference numbers at zero concentration. Also, from (3.69), since $e_i e_j = - \mid z_i z_j \mid e^2$,

$$\Delta X_j/X = -\frac{\mid z_i z_j \mid e^2}{3\epsilon kT}\left(\frac{4\pi e^2 N}{1{,}000\epsilon kT}\right)^{\frac{1}{2}} \frac{q^*\Gamma^{\frac{1}{2}}}{(1 + \sqrt{q^*})} = -\frac{1.981 \times 10^6 \mid z_i z_j \mid q^*}{(\epsilon T)^{\frac{3}{2}}(1 + \sqrt{q^*})} \tag{3.77}$$

and since $\Lambda = \lambda_i + \lambda_j$, the limiting conductance equations for a single ionic species and for a single electrolyte are (on ignoring the small product $\Delta X_j F e_j \kappa / X$),

$$\lambda_j = \lambda_j^0 - \left(\frac{1.9806 \times 10^6\, q^* \mid z_i z_j \mid \lambda_j^0}{(\epsilon T)^{\frac{3}{2}}(1 + \sqrt{q^*})} + \frac{29.167 \mid z_j \mid}{\eta(\epsilon T)^{\frac{1}{2}}} \right) \Gamma^{\frac{1}{2}} \tag{3.78}$$

$$\Lambda = \Lambda^\circ - \left(\frac{1.9806 \times 10^6\, q^* \mid z_i z_j \mid \Lambda^\circ}{(\epsilon T)^{\frac{3}{2}}(1 + \sqrt{q^*})} + \frac{29.167(\mid z_i \mid + \mid z_j \mid)}{\eta(\epsilon T)^{\frac{1}{2}}} \right) \Gamma^{\frac{1}{2}} \tag{3.79}$$

Example 3.4. The following figures [22] refer to KCl in water at 25°C (c = equiv./l.)

10^4c	1.9825	3.0907	4.1381	4.9013	5.2151	6.0093	8.1535	9.8869
(exptl.)	148.61	148.29	148.10	147.88	147.81	147.58	147.30	147.09
(calc.)	148.62	148.29	148.03	147.86	147.79	147.63	147.25	146.98

A plot of Λ (exptl.) against $\sqrt{c}$ gives $\Lambda^0 = 149.58 \pm 0.01$. Since $\epsilon = 78.54$, $T = 298.16$, $\eta = 0.008949$ poise, $\Gamma = 2c$, from (3.79), Λ (calc.) $= 149.85 - (0.2289\, \Lambda^0 + 60.24)\sqrt{c} = 149.85 - 94.55\sqrt{c}$. To obtain a more exact estimate of Λ^0, Λ (exptl.) $+ 94.55\sqrt{c} = \Lambda_0'$ was plotted against c. The values of Λ_0' are almost constant, varying by not more than 0.01 over the whole range. A fair estimate of Λ^0 by this method is 149.95, and this in (3.79) gives $\Lambda = 149.95 - 94.56\sqrt{c}$, which was used to obtain Λ (calc.) in the last row of the Table. Up to 0.0008 N, the agreement is good; beyond this the difference increases steadily. Although the range is restricted, it is enough to show that the evidence, from conductance theory, is that KCl dissociates completely in aqueous solution. A similar conclusion is reached [23] about the other alkali metal halides (excluding fluorides), perchlorates and the corresponding acids. The conductances of the alkaline earth metal and lanthanide salts of the same anions point to the same conclusion also, but the range of validity of the theoretical equation (3.79) is even more restricted than with the 1 : 1 salts. On the other hand there are many instances [23] where the experimental conductances of a number of 1 : 1, 1 : 2 and 2 : 1 electrolytes in water are lower than those calculated by (3.79), although the difference may be small, as it is for instance with a number of nitrates. The quantitative discussion of this in terms of ion-association is dealt with in Chapter 8.

3. Extended Forms of the Onsager–Fuoss Limiting Conductance Equation

Some of the approximations in the Onsager–Fuoss limiting conductance treatment have been pointed out in the previous section; to summarize, they are to regard ions as point charges, to use the Debye–Hückel approximation for potentials, to use an approximate form of the electrophoretic correction (3.67), and since the total field acting on an ion is $X + \Delta X_j$, this should be used instead of X alone in both correction terms. One way of at least partly allowing for ion sizes is [11] to replace κ by $\kappa/(1 + \kappa a)$ in (3.69) and (3.77). The result of doing this is that in equations (3.78) and (3.79), $\Gamma^{\frac{1}{2}}/(1 + \kappa a)$ replaces $\Gamma^{\frac{1}{2}}$, and by selecting a suitable value of a, the concentration range in which calculated and experimental conductances of fully dissociated electrolytes are identical is increased considerably.

Falkenhagen [24] made allowance for ion sizes by means of Eigen and Wicke's distribution function [25] which takes account of the limit that ion size imposes on the number of ions which can group around a central ion to form its atmosphere. Robinson and Stokes [11] have reviewed and discussed this treatment and they have shown that by making minor approximations, κ in the relaxation effect equation, (3.69), becomes $\kappa/(1 + \kappa a)$. The full expression for electrophoresis, (3.67), already contains this factor so there is justification on theoretical grounds for altering the limiting conductance equations in the manner mentioned in the preceding paragraph.

Onsager and Fuoss [21] used the second approximation of Debye and Hückel to express the potential (see 3.61). Pitts [25] has endeavoured to improve on this by means of the fuller expression of Gronwall;[5] the resulting conductance equation is rather elaborate, but by choosing a suitable value of the parameter a, there is good agreement between the calculated and experimental conductances of some 1 : 1 salts up to about 0.1 N. Pitts's treatment has been criticized [11] on two points, firstly some of the required values of a are suspect, and secondly the potential formula does not satisfy the requirement of self-consistency. Stokes [26] has also discussed the consequences of using a more complete form of the Boltzmann distribution formula. He concludes that the approximate form as applied in (3.61) is adequate for 1 : 1 electrolytes in water, but not for higher valent types. On the other hand, the inclusion of higher terms is not satisfactory since they contradict the consistency condition just mentioned.

Example 3.5. On introducing the ion size parameter, (3.79) becomes

$$\Lambda = \Lambda^0 - \left[\frac{1.9806 \times 10^6 q^* \,|z_i z_j|\, \Lambda^0}{(\epsilon T)^{\frac{3}{2}}(1 + \sqrt{q^*})} + \frac{29.167(|z_i| + |z_j|)}{\eta(\epsilon T)^{\frac{1}{2}}}\right] \frac{\Gamma^{\frac{1}{2}}}{[1 + 35.56\, \Gamma^{\frac{1}{2}} \mathring{a}/(\epsilon T)^{\frac{1}{2}}]} \qquad (3.81)$$

where $\mathring{a}$ is in Å units. This may be tested by comparison with the conductances of aqueous NaCl, $CaCl_2$ and $LaCl_3$ at 25°C. For [27] NaCl, $\Lambda^0 = 126.45$, and $\mathring{a} = 3.80$ are required for agreement at $c = 0.01$, i.e., $\Lambda = 126.45 - 89.20 \sqrt{c}/(1 + 1.25\sqrt{c})$. For [28] $CaCl_2$, $\Lambda^0 = 135.84$, $\lambda^0_{Cl} = 76.35$, $\Gamma = 3c$, and from (3.76), $q^* = 0.4268$. For agreement at $c = 0.01$, $\mathring{a} = 3.70$ is required, i.e., $\Lambda = 135.84 - 177.8\sqrt{c}/(1 + 1.49\sqrt{c})$. For [29] $LaCl_3$, $\Lambda^0 = 145.95$, $\Gamma = 4c$, $q^* = 0.3665$, $\mathring{a} = 3.60$ for agreement at $c = 0.01$, so $\Lambda = 145.95 - 280.9\sqrt{c}/(1 + 1.67\sqrt{c})$. Some figures are given in Table 3.2.

TABLE 3.2.

Conductances[19] of NaCl, $CaCl_2$ and $LaCl_3$ at 25°C.

c	0.0005	0.001	0.005	0.01	0.02	0.05	0.10
NaCl(exptl.)	124.50	123.74	120.65	118.51	115.76	111.06	106.74
NaCl(calc.)	124.51	123.74	120.70	118.52	115.73	110.86	106.23
$CaCl_2$(exptl.)	131.93	130.36	124.25	120.36	115.65	108.47	102.46
$CaCl_2$(calc.)	131.99	130.47	124.45	120.37	115.07	106.01	97.61
$LaCl_3$(exptl.)	139.6	137.0	127.5	121.8	115.3	106.2	99.1
$LaCl_3$(calc.)	139.9	137.7	128.2	121.9	113.8	100.2	87.8

Equation (3.81) is clearly successful for NaCl up to at least 0.02 N, is quite satisfactory for $CaCl_2$ up to 0.01 N, but is rather poor when applied to $LaCl_3$. The selected parameters for $CaCl_2$ and $LaCl_3$ are somewhat smaller than those derived by other methods (see Chapter 14). Robinson and Stokes [11] have given other examples illustrating the success of (3.81) when applied to 1 : 1 electrolytes.

4. The Extended Conductance Theory of Fuoss and Onsager [30]

The limiting conductance equation (3.79) for dilute solutions can be represented by

$$\Lambda = \Lambda_0 - (\alpha\Lambda_0 + \beta)\sqrt{c} \tag{3.82}$$

for 1 : 1 electrolytes, where from (3.79),

$$\alpha = 1.9806 \times 10^6 \times 0.2929\sqrt{2}/(\epsilon T)^{\frac{3}{2}} = 8.204 \times 10^5/(\epsilon T)^{\frac{3}{2}} \tag{3.83}$$

$$\beta = 29.167 \times 2\sqrt{2}/\eta(\epsilon T)^{\frac{1}{2}} = 82.49/\eta(\epsilon T)^{\frac{1}{2}} \tag{3.84}$$

The extension of (3.82) described in the preceding section included the replacement of $\beta\sqrt{c}$ by $\beta\sqrt{c}/(1 + \kappa a)$; this is inherent in the limiting treatment. Its extension to the relaxation term α was justified by Robinson and Stokes [11] on the basis of Falkenhagen's treatment [24] of the relaxation effect, but it has been shown by Fuoss and Onsager [30] that Falkenhagen's equation neglects terms of the order c and $c \ln c$, and they have re-examined their limiting treatment with this in mind. This time, ion sizes have been taken into account, and the number of approximations reduced. The Debye–Hückel equation for the potential is retained but the new equation for the relaxation field contains extra inhomogeneous terms. The new relaxation term which replaces (3.69) is

$$\Delta X = \Delta X_0\{1 + 0.5\,\kappa a(1 + q)[1 - (1 + \kappa a)/b]\}/[(1 + \kappa a)(1 + q\kappa a + q^2\kappa^2a^2/3)] \quad (3.85)$$

$$= \Delta X_0(1 + \alpha_1\kappa a + \alpha_2\kappa^2a^2 + \ldots) \quad (3.86)$$

where

$$\Delta X_0 = e_1e_2q^2\kappa X/(3\epsilon kT)(1 + q) \quad (3.87)$$

Equation (3.87) is identical with the point charge model expression, (3.69), with a slight change of symbols; e_1e_2 for e_ie_j, q^2 for q^*, ΔX_0 for ΔX_j. In (3.85), b is the Bjerrum parameter (see Chapter 14),

$$b = e_1e_2/a\epsilon kT \quad (3.88)$$

Since the treatment is confined to moderate concentrations of symmetrical electrolytes, (3.86) is replaced by

$$\Delta X = \Delta X_0(1 - \Delta_1) \quad (3.89)$$

where

$$\Delta_1 = (1 + q)(b + 1)\kappa a/2bp_3; \ (\kappa a < 0.2) \quad (3.90)$$

and

$$p_3 = 1 + q\kappa a + \kappa^2a^2/6 \quad (3.91)$$

The inclusion of higher field terms and cross-products is summarized by writing for the full relaxation term:

$$-\Delta X/X = \alpha\sqrt{c}(1 - \Delta_1 + \Delta_2) + \beta\Delta_3'\sqrt{c}/\Lambda_0 \quad (3.92)$$

where Δ_2 and Δ_3' are these extra contributions. The conductance equation is now

$$\Lambda = [\Lambda_0 - (\beta\sqrt{c})/(1 + \kappa a)][1 + \Delta X/X] \quad (3.93)$$

but if the Δ terms are replaced by their definitions, terms of the order c and higher powers would appear, and since they have been omitted earlier on, they are also left out now, so (3.93) becomes

$$\Lambda = \Lambda_0 - (\alpha\Lambda_0 + \beta)\sqrt{c} + cB(\kappa a) \quad (3.94)$$

in which

$$B(\kappa a) = (\kappa^2a^2b^2/c)[(1 + 2b)\phi_1/b^2 - \phi_2]\Lambda_0 + (\beta\kappa ab/\sqrt{c})(\phi_3 + \phi_4/6) + \alpha\beta = B_1\Lambda_0 + B_2 \quad (3.95)$$

where

$$\phi_1 = 1/12p_3 \quad (3.96)$$

$$\phi_2 = (11\sqrt{2} - 3)/24p_2p_3 + 1/4p_2 + F''(\kappa a)/6(1 + \kappa a)^2 \quad (3.97)$$

$$\phi_3 = [(\sqrt{2} + 5)/32p_2 + F''(\kappa a)/4]/(1 + \kappa a)^2 \quad (3.98)$$

$$\phi_4 = 8/9(1 + \kappa a) \tag{3.99}$$

$$p_1 = 1 + \kappa a + \kappa^2a^2/2 \simeq \exp \kappa a \tag{3.100}$$

$$p_2 = 1 + q\kappa a + \kappa^2a^2/4 \simeq \exp q\kappa a \tag{3.101}$$

Using the approximate forms of p_1 and p_2, and putting $(1 + \kappa a)^2 \simeq \exp 2\kappa a$,

$$F''(\kappa a)/(1 + \kappa a)^2 = -\ 1.2269 - 0.5 \ln \kappa a \tag{3.102}$$

Values of the ϕs for different values of κa have been calculated and tabulated by Fuoss and Onsager,[30] and may be used in calculating values of Λ after Λ_0 and a have been found as follows (an alternative somewhat simpler way is given later). Since $B(\kappa a)$ has an involved form, (3.94) is expanded, and with neglect of higher terms, approaches the limiting form

$$\Lambda''' = \Lambda'' - J_1c + J_2c^{\frac{3}{2}} \tag{3.103}$$

$$\Lambda''(1 - \alpha\sqrt{c}) = \Lambda + \beta\sqrt{c} - (\alpha'\Lambda_0 - \beta')c \ln c \tag{3.104}$$

where

$$\alpha' = \kappa^2a^2b^2/24c,\ \beta' = ab\beta/16\sqrt{c} \tag{3.105}$$

$$J_1 = \theta_1\Lambda_0 + \theta_2,\ J_2 = \theta_3\Lambda_0 + 8\beta\kappa^2a^2/9c \tag{3.106}$$

$$\theta_1 = 2\alpha'[g(b) + 0.9074 + \ln(\kappa a/\sqrt{c})];\ g(b) = (1 + 2b)/b^2 \tag{3.107}$$

$$\theta_2 = \alpha\beta + 8\beta\kappa a/9\sqrt{c} - 2\beta'[0.8504 + \ln(\kappa a/\sqrt{c})] \tag{3.108}$$

$$\theta_3 = (\alpha'\kappa a\sqrt{2}/\sqrt{c})(g(b) - 1.0774) \tag{3.109}$$

When the correct value of a is used, $\Lambda''' = \Lambda_0$ for small values of c, so Λ'' is computed by (3.103) to (3.109) at a series of concentrations, trying several values of a. Next, Λ''' is plotted against c to give a set of curves which intersect at Λ_0 on the ordinate. The correct value of a gives a horizontal plot; since J_1 is almost linear over the range $a = 3$ to 5 Å, this correct value is easily interpolated from a plot of the slopes against a.

Example 3.6. Fuoss and Onsager [30] have applied their method to some KBr data [31] in water at 25°C. Plots of the kind mentioned above indicate $a = 3.58$ Å. A first estimate of Λ_0 by the methods described earlier on (see example 3.4) is 151.68. From (3.88), $ab = 16.708 \times 10^{-4}/\epsilon T$; from (3.12, 3.29), $\kappa^2/c = 8\pi e^2N/1{,}000\epsilon kT = 2529.5 \times 10^{16}/\epsilon T$, whence from (3.105), $\alpha' = 2.9421 \times 10^{12}/(\epsilon T)^3 = 0.22911$ in water at 25°C. From (3.84, 3.105) and ab above, $\beta' = 4.3323 \times 10^7/\eta\epsilon^2T^2 = 8.828$ in water at 25°C. From (3.83, 3.84, 3.104) and the values of α' and β', $\Lambda''(1 - 0.2289\sqrt{c}) = \Lambda + 60.24\sqrt{c} - (0.22911 \times 151.68 - 8.828)(2.3026\, c \log c)$ and with $a = 3.58$ Å, $b = 1.993$.

From (3.106 to 3.109) and the above figures, $\theta_1 = 0.4582\,(1.2552 + 0.9074 + 0.1626) = 1.0655$, $\theta_2 = 13.79 + 63.00 - 17.89 = 58.90$, $\theta_3 = 0.06776$, $J_1 = 220.5$ and $J_2 = 84.4$.

[A minor amendment has resulted [32] from the consideration of a small term due to interionic collisions. Owing to the asymmetry of the moving ionic atmosphere, if the central ion is, say, a cation, it will be struck more often from behind by anions than in front, giving a slight encouragement to the velocity of the cation. The outcome is that θ_1, i.e., (3.107) is replaced by

$$\sigma_1 = 2\alpha'[h(b) + 0.9074 + \ln (\kappa a/\sqrt{c})] \text{ where } h(b) = (2b^2 + 2b - 1)/b^3 \qquad (3.107a)$$

The effect of replacing $g(b)$ by $h(b)$ is small.]

At $10^4c = 13.949$, Λ (exptl.) = 148.27, $\Lambda'' = 152.058$, $J_1c = 0.308$, $J_2c^{\frac{3}{2}} = 0.004$, $\Lambda''' = 151.754$. Fuoss and Onsager, taking $a = 3.6$ Å, obtained $J_1 = 221.7$, $J_2 = 85.7$, and the figures shown in Table 3.3; the last five rows are discussed below.

TABLE 3.3.

Conductance of KBr in Water at 25°C.

10^4c	13.949	27.881	42.183	59.269	71.696
Λ(exptl.)	148.27	146.91	145.88	144.90	144.30
$\beta\sqrt{c}$	2.248	3.179	3.909	4.634	5.097
$(\alpha'\Lambda_0 - \beta')c \ln c$	0.238	0.425	0.598	0.788	0.918
Λ''	152.056	152.356	152.657	153.018	153.286
J_1c	0.309	0.618	0.935	1.314	1.590
$J_2c^{\frac{3}{2}}$	0.004	0.013	0.023	0.039	0.052
Λ'''	151.751	151.751	151.745	151.743	151.748
$B_1\Lambda_0$(see 3.95)	− 67.34	− 45.65	− 32.21	− 21.28	− 15.27
B_2 (see 3.95)	114.31	107.20	102.89	99.20	97.13
$B(\kappa a)$ (see 3.95)	46.97	61.55	70.68	77.92	81.86
$(\alpha\Lambda_0 + \beta)\sqrt{c}$	3.545	5.013	6.165	7.308	8.038
Λ(calc.)	148.27	146.91	145.89	144.91	144.30

The row of Λ''' figures indicates that Λ_0 is 151.75; this is 0.07 unit higher than the original estimate [31] by the older methods.

To calculate Λ, values of $B(\kappa a)$ are found by interpolating values of the ϕs for each value of κa, which, from (3.37), is given by $1.176\sqrt{c}$. At $10^4c = 13.949$, $\kappa å = 0.0439$, and by linear interpolation between 0.04 and 0.05, $\phi_1 = 0.0808$, $\phi_2 = 0.183$, $\phi_3 = 0.285$, $\phi_4 = 0.849$. From (3.95) and (3.105), $B_1 = 24\,\alpha'[\phi_1(1 + 2b)/b^2 - \phi_2] = -0.4443$, $B_2 = 16\beta'(\phi_3 + \phi_4/b) + \alpha\beta = 114.7$, whence $B(\kappa a) = 47.4$, so from (3.94), Λ (calc.) = 148.27_1, in perfect agreement with the experimental figure. The last row of Table 3.3 shows all the calculated conductances; in only two instances do these differ from the experimental ones by as much as 0.01 units. It is of interest to compare with these calculations those based on the extended form of the limiting equation, i.e., (3.81). Using the method described in example (3.4), $å = 3.3$ Å gave constant values of Λ_0, namely 151.67 ± 0.01_5, which is 0.08 unit lower than that derived from Table 3.3. The calculated conductances are 148.25, 146.92, 145.90, 144.91, 144.29, which are very close to the experimental figures but show slightly more differences than do the last row of the Table.

Fuoss and Onsager [30] have applied the extended treatment to several potassium salts in water at various temperatures. The corresponding Λ_0 and $\mathring{a}$ values, together with those from other sources,[31,32] and those calculated by Robinson and Stokes by means of (3.81) are shown in Table 3.4. On the whole, the two sets of $\mathring{a}$ values are in fair agreement, but considering accurate conductances are usually quoted to 0.01 units, the new values are markedly higher.

TABLE 3.4.

Λ_0 and $\mathring{a}$ Values for KCl, KBr and KI in Water; (*a*) Ref. (30), eqn. (3.103);(*b*) Refs. (31, 33); (*c*) Ref. (11), eqn. (3.81).

Salt	°C.	(*a*)Λ_0	(*b*)Λ_0	(*c*)Λ_0	(*a*) $\mathring{a}$	(*c*) $\mathring{a}$
KCl	5	94.27	94.26	94.21	3.29	3.3
KCl	25	149.89	149.88	149.82	3.50	3.3
KCl	55	245.97	245.69	245.73	3.39	3.3
KCl (33)	25	149.90	149.88	149.80	3.60	3.3
KBr	5	96.01	96.00	95.92	3.58	3.8
KBr	25	151.75	151.68	151.60	3.58	3.8
KBr	55	247.32	247.15	247.04	3.61	3.8
KI	5	95.33	95.32	95.25	3.94	4.5
KI	25	150.49	150.34	150.32	3.90	4.5
KI	55	244.86	244.73	244.65	4.15	4.5

5. MODIFICATIONS BY FUOSS TO THE FUOSS–ONSAGER CONDUCTANCE THEORY

At low concentrations when $\kappa a < 0.2$, Fuoss [34] has pointed out that (3.103) may be written as

$$\Lambda = \Lambda_0 - \mathrm{S}\sqrt{c} + Ec \log c + J_1 c(1 - \alpha\sqrt{c})(1 - J_2\sqrt{c}/J_1) \quad (3.110)$$

where

$$S = \alpha\Lambda_0 + \beta;\ E/2.3026 = (\kappa^2 a^2 b^2/24c)\Lambda_0 - \kappa ab\beta/16\sqrt{c} \quad (3.111)$$

From (3.111), E for KBr in water at 25°C, using Λ_0 derived as described in example (3.6) is 60.64, and using this and J_1 and J_2 derived in this example, values of Λ can be calculated by (3.110) and compared with those given in the last row of Table 3.3.

Fuoss [35] has made further simplifications and improvements to (3.110). These consist of replacing J_1 and J_2 by J, incorporating the osmotic term,

(3.107a), and a term which takes into account any increase in the viscosity due to the ions. The equation produced is

$$\Lambda_\eta = \Lambda(1 + F'c) = \Lambda_0 - S\sqrt{c} + Ec \log c + Jc \quad (3.112)$$

where S and E are expressed by (3.111), and

$$J = \sigma_1 \Lambda_0 + \sigma_2; \text{ (see 3.107a for } \sigma_1); \ \sigma_2 = \alpha\beta + (11\beta\kappa a/12\sqrt{c}) - (\kappa a b\beta/8\sqrt{c})[1.0170 + \ln(\kappa a/\sqrt{c})] \quad (3.113)$$

For simple inorganic electrolytes, the viscosity correction is negligible, but it becomes significant if either or both ions are large, e.g., are of the quaternary ammonium type. If viscosity data are available, F' is derived from

$$F' = [(\eta - \eta_0)/\eta_0 - S_\eta \sqrt{c}]/c \quad (3.114)$$

where [36]

$$S_\eta = (\beta/320)(\Lambda_0/\lambda_1^0\lambda_2^0)\{1 - 0.6863[(\lambda_1^0 - \lambda_2^0)/\Lambda_0]^2\} \quad (3.115)$$

If viscosity data are not available, $F'c$ is taken as 5/2 times the volume fraction of solute,[37] and

$$F' = \pi N R_a^3/300 = 6.308 \times 10^{21} R_a'^3 \text{ if one large}, = 12.62 \times 10^{21} R_a''^3 \text{ if both ions large.} \quad (3.116)$$

where R_a' is the radius of the large ion, and R_a'' is the average radius of both ions.

Fuoss, Kraus and their associates have utilized the extended Fuoss–Onsager conductance theory to investigate electrolytic dissociation equilibria. This is dealt with in Chapter 8.

6. Conductances in Concentrated Solutions

For aqueous solutions of 1 : 1 electrolytes, the working range of the extended Fuoss–Onsager treatment is up to about 0.02 N. Some of the most convincing attempts to devise a conductance equation for higher concentrations are those of Stokes and his associates, although these too are confined to the 1 : 1 types. Thus Wishaw and Stokes,[38] by means of a modified form of Falkenhagen *et al.*'s expression [24] for the relaxation effect, $\Delta X/X$, and Walden and Ulich's "rule," [39]

$$\Lambda_0\eta_0 = \text{constant}; \quad \lambda_j^0\eta_0 = \text{constant} \quad (3.117)$$

formulated the following expression:

$$\Lambda = [\Lambda_0 - (\beta/c)(1 + B''a\sqrt{c})(1 - F_1\alpha)/(1 + B''a\sqrt{c})]\eta/\eta_0 \quad (3.118)$$

where $F_1 = [(\exp.\ 0.2929\ \kappa a) - 1]/0.2929\ \kappa a$, in which $0.2929 = q^2/(1 + q)$. Values of F_1 for various values of κa have been tabulated.[33] From (3.37), in water at 25°C, $B''å = 0.3282\ å$, where $å$ is the adjustable ion size parameter in Å. The viscosities η and η_0 refer to solutions and solvent respectively. By taking $\mathring{a} = 4.35$, they obtain very good agreement between the experimental conductances of NH_4Cl from 0.1 N to 5.0 N in water at 25°C, and those calculated by (3.118). Similar tests with KBr and LiCl (up to 9 N) are equally effective.[11]

A viscosity correction of the above type is an approximate one. Walden and Ulich's rule is fairly true when the ions are large, but small ions show considerable deviations when tested by conductances in a variety of solvents.[6] Robinson and Stokes [11] have reported some valuable information on the effect of viscosities on the conductances of inorganic ions, and they have also given an exhaustive review and discussion on this subject.

Transference Numbers

The relationship between transference numbers and conductances expressed by (1.11) clearly indicates that the dependence of the former on concentrations should be capable of interpretation by the conductance treatments described in the preceding sections. Thus, from the limiting Onsager–Fuoss equations (3.72 to 3.79),

$$T_j = \lambda_j/\Lambda = [\lambda_j^0 - 29.167\,|\,z_j\,|\,\Gamma^{\frac{1}{2}}/\eta_0(\epsilon T)^{\frac{1}{2}}]/[\Lambda_0 - 29.167(\,|\,z_i + z_j\,|\,)\Gamma^{\frac{1}{2}}/\eta_0(\epsilon T)^{\frac{1}{2}}] \qquad (3.119)$$

(the term $1 + \Delta X/X$ due to the relaxation effect cancels out). Unfortunately this is of little practical use since transference numbers cannot be determined with sufficient precision in the region where (3.119) is applicable, but Robinson and Stokes [11] have shown that if the extension applied in (3.81) is adopted, so that (3.119) is converted into

$$T_j = \frac{\lambda_j^0 - 29.167\,|\,z_j\,|\,\Gamma^{\frac{1}{2}}/(1 + \kappa a)\eta_0(\epsilon T)^{\frac{1}{2}}}{\Lambda_0 - 29.167\,(\,|\,z_i + z_j\,|\,)\Gamma^{\frac{1}{2}}/(1 + \kappa a)\eta_0(\epsilon T)^{\frac{1}{2}}} \qquad (3.120)$$

then this gives an excellent quantitative account of the actual transference numbers of, e.g., HCl to 3 N, LiCl to 0.5 N, NaCl to 0.2 N, KCl to 1 N.

Substituting for 1 : 1 electrolytes by means of (3.84) leads to

$$T_j^0 = T_j + \beta\sqrt{c}(0.5 - T_j)/\Lambda_0(1 + B''a\sqrt{c}) \qquad (3.121)$$

This provides a convenient way of deriving T_j^0 and hence λ_j^0. The parameter a is adjusted until T_j^0 is constant over the range of concentrations investigated.

Example 3.7. At 25°C in water, $\beta = 60.24$, $B'' = 0.3283$. The moving boundary method was used to obtain the T_+ values [40] shown in Table 3.5. $\Lambda_0 = 126.45$ and the most suitable value of a was found to be 5.5 Å. The average T_+^0 value is $0.3962_4 \pm 0.0001_6$, whence $\lambda_{\mathrm{Na}}^0 = 50.10_5 \pm 0.020$.

This method fails when applied to higher valent electrolytes. Several empirical procedures have been devised for these; Shedlovsky [41] has given one of the most useful ones, namely

$$1/T_j^0 = 1/T_j - A^*\sqrt{c} + B^*c \qquad (3.122)$$

This can be made to fit the data up to about 0.2 N by adjustment of the parameters, A^* and B^*.

TABLE 3.5.

Transference Numbers of NaCl in Water at 25°C. Estimation of T_+^0.

10^3c	15.024	19.965	29.132	59.853	99.75	100.28
T_+	0.3910_5	0.3900_5	0.3893	0.3873_7	0.3853_5	0.3853_5
T_+^0	0.3961_1	0.3959_5	0.3961_8	0.3965_3	0.3963_3	0.3963_4

Conductances and Transference Numbers in Non-Aqueous Solvents

The dielectric constants of most organic solvents, and their mixtures with water are lower than those of water itself, but there are a few such as liquid HCN, concentrated H_2SO_4, formamide, N-methyl formamide and similar materials, which have higher values. The conductances of a number of 1 : 1 electrolytes have been measured in these solvents [42] and in others such as [43] methanol, ethanol, acetone, etc., and their mixtures with water. In general, the Onsager–Fuoss conductance expressions point to the dissociation of 1 : 1 electrolytes being largely complete in methanol and solvents of higher dielectric constants, but in ethanol and other solvents of lower dielectric constant, ion-association occurs.

Very few transference numbers in non-aqueous and mixed solvents have been determined with accuracy up to the present time.[44] Robinson and Stokes [11] have discussed some of these, and also the conductances of a number of electrolytes in these media.

REFERENCES

1. Debye, P. and Hückel, E. *Phys. Z.* **24,** 185, 305 (1923). See "The Collected Papers of P. J. Debye." Interscience Publishers, Inc., New York (1954), for English translations.
2. Guntelberg, E. *Z. phys. Chem.* **123,** 199 (1926).
3. Guggenheim, E. A. *Phil. Mag.* **19,** 588 (1935).
4. Davies, C. W. *J. chem. Soc.* 2093 (1938).
5. Gronwall, T. H., La Mer, V. K. and Sanved, K., *Phys. Z.* **29,** 358 (1928); La Mer, V. K., Gronwall, T. H. and Grieff, L. T. *J. phys. Chem.* **35,** 2245 (1931).

6. Ref. 2, Ch. 1.
7. Müller, H. *Phys. Z.* **29**, 78 (1928).
8. Fowler, R. H. and Guggenheim, E. A. "Statistical Thermodynamics." Cambridge University Press, Cambridge, U.K., 2nd edn. (1939, 1956).
9. Guggenheim, E. A. *Disc. Faraday Soc.* **24**, 53 (1957).
10. Bagchi, S. N. *J. Indian chem. Soc.* **27**, 199 (1950). Wicke, E. and Eigen, M. *Naturwissenschaften* **38**, 456 (1951); **39**, 545 (1952); *Z. Elektrochem.* **56**, 551 (1952); **57**, 319 (1953).
11. Ref. 11, Ch. 1.
12. Kirkwood, J. G. *J. chem. Phys.* **2**, 767 (1934).
13. Falkenhagen, H. and Kelbg, G. *Disc. Faraday Soc.* **24**, 20 (1957).
14. Bell, G. M. and Levine, S. *ibid.* **24**, 69 (1957).
15. Mayer, J. E. *J. chem. Phys.* **18**, 1426 (1950).
16. Poirier, J. C. *ibid.* **21**, 965, 972 (1953).
17. Mason, C. M. *J. Amer. chem. Soc.* **60**, 1638 (1938).
18. Shedlovsky, T. *ibid.* **72**, 3680 (1950). Stokes, R. H. and Levien, B. J. *ibid.* **68**, 333 (1946).
19. Ref. 2, Ch. 1.
20. Onsager, L. *Phys. Z.* **27**, 388 (1926); **28**, 277 (1927).
21. Onsager, L. and Fuoss, R. M. *J. phys. Chem.* **36**, 2689 (1932).
22. Davies, C. W. *J. chem. Soc.* 432 (1937).
23. Ref. 1, 2 and 11, Ch. 1.
24. Falkenhagen, H., Leist, M. and Kelbg, G. *Ann. Phys. Lpz.* **11**, 51 (1952).
25. Pitts, E. *Proc. roy. Soc.* **217A**, 43 (1953).
26. Stokes, R. H. *J. Amer. chem. Soc.* **75**, 4563 (1953).
27. Shedlovsky, T., Brown, A. S. and MacInnes, D. A. *Trans. electrochem. Soc.* **66**, 165 (1934).
28. Shedlovsky, T. and Brown, A. S. *J. Amer. chem. Soc.* **56**, 1066 (1934).
29. James, J. C. and Monk, C. B. *Trans. Faraday Soc.* **46**, 1041 (1950).
30. Fuoss, R. M. and Onsager, L. *J. phys. Chem.* **61**, 668 (1957).
31. Owen, B. B. and Zeldes, H. *J. chem. Phys.* **18**, 1083 (1950). See also Chambers, J. F. *J. phys. Chem.* **62**, 1136 (1958).
32. Fuoss, R. M. and Onsager, L. *J. phys. Chem.* **62**, 1339 (1958).
33. Gunning, H. E. and Gordon, A. R. *J. chem. Phys.* **10**, 126 (1942).
34. Fuoss, R. M. *J. Amer. chem. Soc.* **79**, 3301 (1957).
35. Fuoss, R. M. *ibid.* **80**, 3163 (1958); **81**, 2659 (1959); *J. phys. Chem.* **63**, 633 (1959). Fuoss, R. M. and Accascina, F. "Electrolytic Conductance." Interscience Publishers, Inc., New York (1959).
36. Falkenhagen, H. and Dole, M. *Z. phys. Chem.* **6**, 159 (1929); *Phys. Z.* **30**, 611 (1929).
37. Einstein, A. *Ann. Phys.* **19**, 289 (1906); **34**, 591 (1911).
38. Wishaw, B. F. and Stokes, R. H. *J. Amer. chem. Soc.* **26**, 2065 (1954).
39. Walden, P. and Ulich, H. *Z. phys. Chem.* **107**, 219 (1923).
40. Longsworth, L. G. *J. Amer. chem. Soc.* **54**, 2741 (1942).
41. Shedlovsky, T. *J. chem. Phys.* **6**, 845 (1938).
42. (a) HCN: Coates, J. E. and Taylor, E. G. *J. chem. Soc.* 1245, 1495 (1936). Lange, J. Berga, J. and Konopik, N. *Mh. Chem.*, **80**, 708 (1949). (b) substituted amides: Dawson, L. R., Sears, P. G. *et al. J. Amer. chem. Soc.* **76**, 6024 (1954); **77**, 1986 (1955); **78**, 1569 (1956); **79**, 298, 3004, 4269, 5906 (1957); **80**, 4233 (1958); *J. phys. Chem.* **59**, 16, 373 (1955); **60**, 1076, (1956). (c), aq. glycine: Monk, C. B. *Trans. Faraday Soc.* **46**, 645 (1950).

43. (a) MeOH: Gordon, A. R., *et al. J. chem. Phys.* **16,** 336 (1948); **19,** 749, 752 (1951); *J. Amer. chem. Soc.* **75,** 2855 (1953); **79,** 2347, 2350, 2352 (1957). Hartley, H., *et al. J. chem. Soc.* 2488 (1930); *Phil. Mag.* **15,** 610 (1933). (b) EtOH: Hartley, H., *et al. J. chem. Soc.* 2492 (1930). Gordon, A. R., *et al. J. Amer. chem. Soc.* **79,** 2350, 2352 (1957). Butler, J. A. V., *et al. Proc. roy. Soc.* **147A,** 418 (1934). (c) Reviews: Lange, J. *Z. phys. Chem.* **188A,** 284 (1941). Ref. 1, Ch. 1. Fischer, V. L., Winkler, G. and Gander, G. *Z. phys. Chem.* **62,** 11 (1958).

44. (a) MeOH: Gordon, A. R. *J. chem. Phys.* **19,** 749 (1951). (b) Gordon, A. R., *et al. J. Amer. chem. Soc.* **79,** 2350, 2352 (1957). (c) Substituted amides: Dawson, L. R., *et al. loc. cit.* 298, 3004, 4269; *J. phys. Chem.* **60,** 1076 (1956).

CHAPTER 4

REVERSIBLE E.M.F. CELLS

An electromotive force (e.m.f.) cell is a means of converting the energy of a chemical reaction, or the energy released on altering the concentration of an electrolyte, into electrical energy. If, as so often is possible, an applied e.m.f. from an outside source can inject electrical energy into the cell and thereby reverse the chemical or physical processes that produce the e.m.f. of the cell, then the cell is said to be "reversible." Such cells may be classified as:

(a) "Galvanic" cells, or cells without liquid junction. One of the electrodes of such cells is reversible to an anion, and the other to a cation.

(b) Concentration cells without transference. The two electrodes across which the potential is measured (or calculated in some cases) are identical, and are reversible to the same anion or cation. The two electrodes are in solutions of the same electrolyte which differ in concentration. The two solutions are connected either in practice or in theory by a middle electrode which is reversible to the other ionic species. Very often the potentials of such cells are obtained by simply subtracting the potentials of two cells of type (a).

(c) Concentration cells with transference. In these, two identical electrodes are immersed in two different concentrations of the same electrolyte. These two solutions are in direct contact at an interface across which ions migrate.

The potentials of these cells can, with the aid of thermodynamic principles, provide information on electrolytes in solution providing the cells operate under true reversible conditions. By this is meant that if the potential of the cell is balanced exactly by an opposing potential from another source, no reactions or changes of concentration should take place in the cell, and furthermore, if there is an infinitesimal decrease or increase in the opposing potential, the cell reaction will proceed or be reversed by an infinitesimal amount. Under these reversible conditions, the cell system remains in a state of equilibrium, which is the necessary condition for the application of thermo dynamic theorems.

Reversible potentials are measured by potentiometers. The chief principles of this can be explained by reference to Fig. 4.1. A battery B is connected across a uniform resistance R_1. C is the reversible cell; it is connected in opposition to B. The contact on R_1 is varied until the galvanometer G is at its null-point. In most instruments, R_1 is made in sections, each with stepped contact connections. In the first section, the contacts may be in 0.01 V steps, in the second section in 0.0001 V steps, but other variations are also adopted.

For research purposes, the potential is recorded to the nearest 0.001 mV; this may be read directly or estimated from the deflection of the galvanometer per 0.01 mV change. This means that G must be highly sensitive, e.g., about 10 microV (μV) per mm of scale deflection. To calibrate R_1, S is switched to W, which is a standard reference "Weston" cell. The potential of this is known accurately, it is reproducible, has a small known temperature coefficient and changes very little with age. To make the calibration, R_1 is set to the e.m.f. of the standard and R_2 is adjusted until G is at its null-point. There are other variations such as switching G to R_2, part of which is calibrated in small voltage steps, which cover that of the standard cell over a range of temperatures.

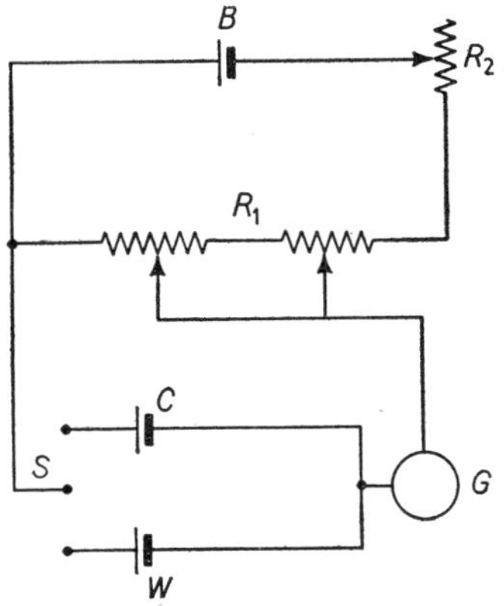

FIG. 4.1. Potentiometer circuit.

The relationships between electrode potentials and the activities of electrolyte solutions can be obtained by considering a reaction in which x_1 moles of X_1 and x_2 of X_2 react to form y_1 moles of Y_1 and y_2 moles of Y_2, i.e., $x_1X_1 + x_2X_2 \rightleftharpoons y_1Y_1 + y_2Y_2$.

For component X_1, from (1.15),

$$\bar{G}_{X_1} - \bar{G}^0_{X_1} = x_1RT \ln a_{X_1} \tag{4.1}$$

where $\bar{G}_{X_1}$ is the partial free energy of x_1 moles of X_1, and $\bar{G}^0_{X_1}$ is the corresponding free energy in the standard state. Similar expressions can be written for the other components. If now

$$\Delta\bar{G} = \bar{G}_{Y_1} + \bar{G}_{Y_2} - \bar{G}_{X_1} - \bar{G}_{X_2} \tag{4.2}$$

$$\Delta\bar{G}^0 = \bar{G}^0_{Y_1} + \bar{G}^0_{Y_2} - \bar{G}^0_{X_1} - \bar{G}^0_{X_2} \tag{4.3}$$

then for the reaction,

$$\Delta\bar{G} - \Delta\bar{G}^0 = RT \ln a^{y_1}_{Y_1}a^{y_2}_{Y_2}/a^{x_1}_{X_1}a^{x_2}_{X_2} \tag{4.4}$$

This can easily be extended to more complex systems.

When the reaction is at equilibrium, the composition of each phase is fixed, so $\Delta\bar{G} = 0$. Hence from (4.4),

$$\Delta\bar{G}^0 = -RT \ln K \tag{4.5}$$

where K is the equilibrium constant of the reaction, and is defined by

$$K = (a^{y_1}_{Y_1})_e(a^{y_2}_{Y_2})_e/(a^{x_1}_{X_1})_e(a^{x_2}_{X_2})_e \tag{4.6}$$

where $(a)_e$ signifies activities at equilibrium.

The electrical work generated by a cell under reversible conditions corresponds to the decrease in the free energy of the cell as the result of the reaction that takes place. Accordingly, if the cell is allowed to produce an infinitesimally small quantity of electricity of df coulombs, then if E is the potential difference (in absolute volts) between the two electrodes, and dG is the corresponding decrease in free energy at constant temperature, pressure and composition, then $-dG = Edf$. In terms of mole quantities this will be written as

$$-\Delta\bar{G} = nEF \tag{4.7}$$

where nF is the number of coulombs passing when one mole of electrode material or of electrolyte is allowed to react, and F is the faraday.

From (4.4), if all the activities are unity, $\Delta\bar{G} = \Delta\bar{G}^0$. Under such conditions $E = E^0$, the "standard" potential of the cell, and (4.7) becomes

$$-\Delta\bar{G}^0 = nE^0F \tag{4.8}$$

and from (4.5) and (4.8),

$$nE^0F = RT \ln K \tag{4.9}$$

and from (4.4) as well,

$$E = E^0 - (RT/nF) \ln (a^{y_1}_{Y_1}a^{y_2}_{Y_2}/a^{x_1}_{X_1}a^{x_2}_{X_2}) \tag{4.10}$$

Example 4.1. The reaction that takes place in the cell

$$\text{Pt} \mid H_2 \mid \text{HCl in aqueous solution} \mid \text{AgCl} \mid \text{Ag} \tag{4.11}$$

where the vertical lines are used to separate different phases is, in terms of one mole of electrode material,

$$\text{AgCl} + \tfrac{1}{2}H_2 \rightleftharpoons \text{Ag} + \text{HCl} \tag{4.12}$$

Here, $n = 1$, and putting 2.3026 $RT/F = k'$, from (4.10),

$$E = E^0 - k' \log (a_{\text{Ag}}a_{\text{HCl}}/a_{\frac{1}{2}H_2}a_{\text{AgCl}}) \tag{4.13}$$

The activity of a solid phase is taken as unity. This may be seen from (1.15), in which $\bar{G} = \bar{G}^0$ for a pure component, so $RT \ln a = 0$, $a = 1$. Thus

$$E = E^0 - k' \log (a_{\text{HCl}}/a_{\frac{1}{2}H_2}) \tag{4.14}$$

Electrode and Cell Conventions

Lewis and Randall[2] initiated certain conventions with regard to e.m.f. cells, electrode reactions and the signs of potentials. These have been adopted in the United States, and are also used elsewhere except in one important respect which is discussed further on. Attempts were made at a Commission in Stockholm in 1953 to remove this difference of convention. Latimer[3] has given a lucid account of the recommendations and how they differ from the Lewis system. It is intended to use the Stockholm system in this book, but since it has so much in common with that of Lewis, the following summary is intended to cover both versions.

(i) The cell is written so that when E is positive, the tendency is for negative electricity to pass spontaneously through the cell from right to left. The right-hand electrode is clearly the positive one. This is common to both conventions. The cell described by (4.11) is written accordingly. Electrons enter the Ag | AgCl electrode from the outer circuit, and the reaction $AgCl + e^- \rightarrow Ag + Cl^-$ liberates Cl^- ions which migrate from right to left through the solution. On the Pt surface of the left-hand electrode, the overall reaction is $\frac{1}{2}H_2 \rightleftharpoons H^+ + e^-$, and the released electrons pass through the outer circuit to the Ag | AgCl electrode.

(ii) A cell may be regarded as being composed of two half-cells. The compositions of these, and the corresponding electrode reactions are represented thus:

$$\mathrm{Pt} \mid \mathrm{H_2} \mid \mathrm{H^+} \qquad \mathrm{H^+} + e^- \rightleftharpoons \tfrac{1}{2}\mathrm{H_2} \tag{4.15}$$

$$\mathrm{Ag} \mid \mathrm{AgCl} \mid \mathrm{Cl^-} \qquad \mathrm{AgCl} + e^- \rightleftharpoons \mathrm{Ag} + \mathrm{Cl^-} \tag{4.16}$$

The electrode reactions are written so that the associated electrons are on the left-hand side of the equation. The corresponding potential of the half-cell is called the electrode potential, e.g., E_{H}, E_{AgCl}.

The Lewis convention is the opposite to the above. The released electrons are placed on the right-hand side of the equation, that is to say, the equations are written the other way round.

(iii) A potential equation can be written for a single electrode. From (4.10), (4.15) and (4.16),

$$E_{\mathrm{H}} = E^0_{\mathrm{H}} - k' \log (a_{\frac{1}{2}\mathrm{H_2}}/a_{\mathrm{H^+}}) \tag{4.17}$$

$$E_{\mathrm{AgCl}} = E^0_{\mathrm{AgCl}} - k' \log a_{\mathrm{Cl^-}} \tag{4.18}$$

If the Lewis convention is used, because of (ii), the two log terms would be the reciprocals of those which are written.

(iv) The potential of the whole cell, E, is the difference between the potentials of the two half-cells, i.e. between those of the two electrodes. The correct order of subtraction can be established by separating the overall cell reaction into its constituent electrode reaction. Thus (4.12) becomes (a) $\frac{1}{2}H_2 \rightleftharpoons H^+ + e^-$, (b) $AgCl + e^- \rightleftharpoons Ag + Cl^-$; (b) — (a) gives $\frac{1}{2}H_2 + AgCl = Ag + HCl$ (or $H^+ + Cl^-$). This decides that the order is such that the potential of the left-hand side is subtracted from that on the right:

$$E = E_{AgCl} - E_H \tag{4.19}$$

This is the converse to the order which applies when the Lewis convention is used. From (4.17 to 19), the potential of the cell is given by

$$E = E^0_{AgCl} - E^0_H - k' \log (a_{H^+} a_{Cl^-}) + k' \log a_{\frac{1}{2}H_2} \tag{4.20}$$

The Conventional Zero of Potential. The Hydrogen Electrode

In order to assign numerical values to single electrode potentials, it is necessary to have an arbitrary standard, since measured cell potentials are the difference of those of the two electrodes. It was suggested by Nernst [4] in 1900 that the potential of a hydrogen electrode, when the H_2 pressure is 760 mm of Hg, which is in reversible equilibrium with a solution containing one g ion per litre of H^+ ions, be taken as the zero of potential. This was later modified by Lewis,[5] and this electrode is now taken to have zero potential when the activity of the H^+ ions is unity. This convention applies at all temperatures and for all solvents. From (4.17), this convention leads to

$$E_H = 0 - k' \log (a_{\frac{1}{2}H_2}/a_{H^+}) = 0 + k' \log a_{H^+} - 0.5k' \log \{(P - p_w)/760\} \tag{4.21}$$

where P is the total gas pressure and p_w is the vapour pressure of the solution, both in mm. For dilute solutions, p_w can be taken as that of the solvent; The taking of the H_2 pressure to represent its activity has been tested over a wide range of pressures,[6] and found satisfactory.

Example 4.2. At 25°C, p_w for water = 23.7 mm so if $P = 770$ mm, with $k' = 0.059156$, (4.20) gives

$$E = E_{measured} + 0.00023 \text{ V} = E^0_{AgCl} - k' \log a_{H^+} a_{Cl^-} \tag{4.22}$$

The Cell Pt | H_2 | HCl Solution | AgCl | Ag

A conventional design of cell [7,8,22] for this and similar systems is shown in Fig. 4.2. The cell is built of pyrex glass; the height from X to the base is about 8 cm, and the diameter of the electrode compartments is about 4.5 cm. X and X' connect to capillary tube gas exits; these are shaped to prevent the back-diffusion of air. C and C' are bubblers for the saturation of the gases. N_2 is used to sweep out O_2 since this causes a side-reaction, $2Ag + \frac{1}{2}O_2 + 2HCl = 2AgCl + H_2O$. An alternative procedure is to boil out gases from solution and to fill the cell in the absence of air, but this is scarcely necessary provided the H_2 and N_2 are passed for a sufficient time before taking readings, e.g. overnight. A convenient rate of bubbling after saturation is about one bubble per sec. Once constant readings are obtained at one temperature, the thermostat of the bath can be altered in 5° or 10°C steps and the new readings obtained in an hour or so.

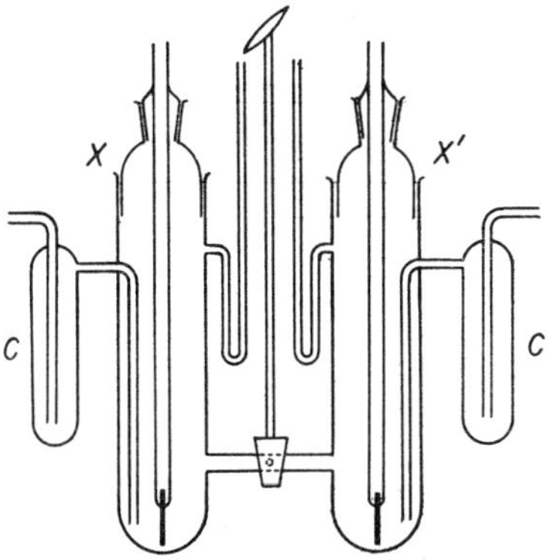

FIG. 4.2. The cell Pt | H_2 | HCl solution | AgCl | Ag.

The H_2 electrode consists of a Pt—Pt black electrode; the deposit of the Pt black should be light. Hills and Ives,[9] who have examined and discussed some of the errors to which this electrode is susceptible, recommend electrolysis in 2 % platinic chloride in 2 N HCl with a current of 10 to 20 mA/cm^2 for about 15 min. The H_2 gas must be free of other gases and since O_2 is the commonest impurity, it is passes over a Pt or Pd catalyst before entering the cell. Hills and Ives have shown that there is a minor correction for the depth of the gas tube below the surface of the solution; e.g., 20 μV for 3.3 cm, 42 μV for 7.8 cm, etc.

Reliable and reproducible potentials cannot be obtained with the Ag | AgCl electrode unless it is made under specified conditions. A comprehensive review of the different methods of manufacture and of the analogous AgBr and AgI electrodes has been given by Janz and Taniguchi,[10] and many details are to be found in the numerous papers of Harned and his associates.[11]

One of the most reliable methods for forming Ag | AgCl electrodes, which is termed the thermal-electrolytic process, is as follows.[12,13] Silver oxide is made by adding hot dilute NaOH to hot dilute $AgNO_3$, and the product washed about forty times by stirring it into a large volume of conductivity water [14] to remove all impurities. The silver oxide paste is spread on to a Pt wire spiral fused at one end into a glass tube. The coated spiral is heated to 400°C so that a spongy mass of silver forms. The Pt wire should be completely covered by this. The electrode is then electrolyzed in 1 N HCl free of bromide [7] until it has been partly converted to AgCl. After it has aged for about 30 h,[15] the electrode is then ready. This ageing process is essential for all the silver halide electrodes, no matter what the method of preparation, otherwise the potentials are on the high side. To ensure that the Pt wire is completely coated, a layer of Ag is sometimes deposited by electrolysis before treating with silver oxide paste.

Brown [16] has made a thorough study of the electrolytic type. For this, a straight Pt wire is fixed into the end of a glass tube, the other end of the wire (about 2 to 3 cm long) is fused to smooth off rough edges. Batches of six are cleaned by hot HNO_3, washed, and then plated in 1 % potassium silver cyanide which has been freed of excess cyanide by a trace of $AgNO_3$. The bright Pt anode used for the electrolysis is kept in a compartment with a porous end; this keeps the main solution free of silver cyanide deposits. The plated electrodes are washed with NH_4OH, stored in water for at least 16 h to remove all traces of cyanide, then electrolyzed in 0.1 N HCl with a current of about 0.2 mA for 30 min. Smith and Taylor [15] have confirmed that this method gives very reproducible electrodes.

A third way of forming Ag | AgCl electrodes has been developed by Purlee and Grunwald.[17] About 2 cm^2 of Pt foil is cleaned after sealing its wire to a glass tube, and Ag deposited by a Rochelle salts silvering process.[18] Sodium tartrate can also be used. About 0.8 mg/cm^2 of Ag is deposited, and three separate deposits are desirable. The process also appears to be satisfactory with thick Pt wire. The AgCl layer is produced by electrolysis in 0.05 N HCl for 2 to 3 min with a current of about 1 $mA/cm.^2$ These electrodes have so far been used only in solutions containing a high proportion of alcohol or dioxan and very dilute acid chlorides; they gave very reproducible readings, and quickly reached equilibrium.

The Standard Potential of the Ag | AgCl | Cl^- Electrode

Assuming that E has been corrected to 760 mm, (4.22) can be rearranged to

$$E + 2k' \log m + 2k' \log \gamma_{\pm} = E^0_{\text{AgCl}} \qquad (4.23)$$

by means of (1.27); m is the molality of HCl. The activity coefficients can be calculated within close limits in dilute solutions by a generalized version of the D.H. equation such as (3.42, 3.43). A plot of the left-hand side of (4.23) against some convenient function such as m, should, on extrapolation, give E^0. In some of the earlier extrapolations of this kind, a limiting form of the D.H. equation, (3.39), with the addition of an extension term, Qm, was used. This produced curved plots which cannot be extrapolated with much certainty; Harned and Ehlers [12] used this way to assess E^0 in aqueous HCl, and at 25°C for instance, their answer is 0.22239 int. V., = 0.22243 abs. V. Guggenheim and Prue,[19] by using (3.42) obtained a straight line plot up to about 0.1 m, the slope of which gives Q''. In fact, by putting $AQ'' = 0.234$, E^0 between $m = 0.003$ and 0.097 averages 222.5 ± 0.1 mV (abs.). In other words, an extrapolation is not needed. A similar analysis of Carmody's data [20] gave $E^0 = 222.3 \pm 0.1$ mV.

The most recent investigation of E^0_{AgCl} in aqueous HCl is that of Bates and Bower.[13] They used the thermal-electrolytic type of electrode, and to

TABLE 4.1.

E^0_{AgCl} in Aqueous Chloride Solution (abs. mV). (a) = obs., (b) = calc.

°C	0	5	10	15	20	25	30	35	40
$E°(a)$	236.55	234.13	231.42	228.57	225.57	222.34	219.04	215.65	212.08
$E°(b)$	236.59	234.08	231.40	228.56	225.57	222.40	219.10	215.66	212.07

°C	45	50	55	60	70	80	90	95
$E°(a)$	208.35	204.49	200.56	196.49	187.82	178.73	169.52	165.01
$E°(b)$	208.34	204.49	200.51	196.41	187.85	178.85	169.46	—

minimize uncertainties due to slight differences between individual electrodes, they prepared and obtained measurements from large numbers. Their average results, together with those calculated from their empirical formula $E^0 = 0.23659 - (4.8564 \times 10^{-4})t - (3.4205 \times 10^{-6})t^2 + (5.869 \times 10^{-9})t^3$, where $t =$ °C, are given in Table 4.1.

Example 4.3. Bates and Bower [13] have given smoothed values of E at 25°C. Taking $AQ'' = 0.245$, i.e., $-\log \gamma_\pm = 0.5083 \sqrt{m}/(1 + \sqrt{m}) - 0.245\, m$,

TABLE 4.2.

$E°$ of the AgCl Electrode in Aqueous Solution at 25° C (mV).

m	0.001	0.002	0.005	0.01	0.02	0.05	0.07
E	579.09	544.18	498.40	464.12	430.19	385.79	369.57
E^0_a	222.34	222.33	222.34	222.32	222.31	222.32	222.40
E^0_b	222.35	222.36	222.37	222.38	222.40	222.36	222.35

the E^0_a values are as shown above. They average 222.34 ± 0.02 mV. Bates and Bower have pointed out that E^0 depends slightly on the ion size parameter. Now if $\sqrt{m}$ is replaced by 1.5 $\sqrt{m}$ in the denominator of the above expression (equivalent to a being changed from 3.0 Å to 4.5 Å), this alteration requires 0.245 being replaced by 0.095 for constant E_b^0 values (average 222.37 $\pm$ 0.015 mV).

The E^0 values of the Ag | AgCl electrode in a large number of other solvents have been derived. Most of the references and results have been collected and tabulated by Harned and Owen, and by Robinson and Stokes.[21] In certain solutions, such as 70 % and 82 % dioxan, and in ethanol, which have low dielectric constants, HCl is not completely dissociated, and the extrapolation procedures and the derivation of the activity coefficients must take this into account. This is illustrated by [22] the evaluation of E^0_{AgCl} in ethanol. Assuming complete dissociation of HCl, 0.08138 V at 25°C was derived, while allowing for HCl ion-pairs, $E^0 = 0.09383$ V. A similar recalculation for E^0 in 82 % dioxan [23] is not too satisfactory.

The AgBr and AgI Electrodes

These have not been studied in such detail as has the AgCl electrode. Some of the more recent investigations of the Ag | AgBr electrode are those of Owen and Foering [24, 26] and of Harned and Donelson.[25] In both cases thermal electrodes were made by coating a Pt wire spiral with a paste consisting of 9 parts of silver oxide and 1 part of $AgBrO_3$ moistened with water, and heating at 650°C for 7 min. Smith and Taylor [15] have undertaken a careful intercomparison of this type with electrodes formed by the thermal-electrolytic and electrolytic methods that have been described for the Ag | AgCl electrode. They aged the electrodes for a week in O_2-free 0.05 N KBr before testing, and found little difference between all three types.

Owen formed Ag | AgI electrodes by the thermal method (using $AgIO_3$), and the thermal-electrolytic process. In the opinion of Taylor and Smith,[15] the latter is the more satisfactory. Since HI solutions are unstable, buffered borax KI was used for the E^0 in this instance.[26] All these E^0 data have been tabulated together.[21]

The Hg | HgCl, Hg | HgBr, Hg | HgI Electrodes

Hills and Ives [27] have shown that the cell Pt | H_2 | aq. HCl | HgCl | Hg is capable of giving highly accurate and reproducible potentials. This has been confirmed by Grzybowski [28] who has extended the temperature range of the measurements (0°C to 60°C). The calomel electrode is prepared in one compartment of a cell similar to that shown in Fig. 4.2; the essential modifications are shown in Fig. 4.3a. The calomel is made either electrically by means of a bright Pt cathode, a mercury pool anode, stirred 2 N HCl and a current of about 2 V and 0.5 amp, or 2 N HCl is added rapidly with stirring to pure mercurous nitrate in dilute HNO_3 standing over a pool of Hg. The HgCl is washed free of excess chloride, dried in vacuum and stored in the dark over P_2O_5. To form the electrode, a little of the preparation is shaken with dry Hg for several minutes, and a little of the grey skin which results is added to dry

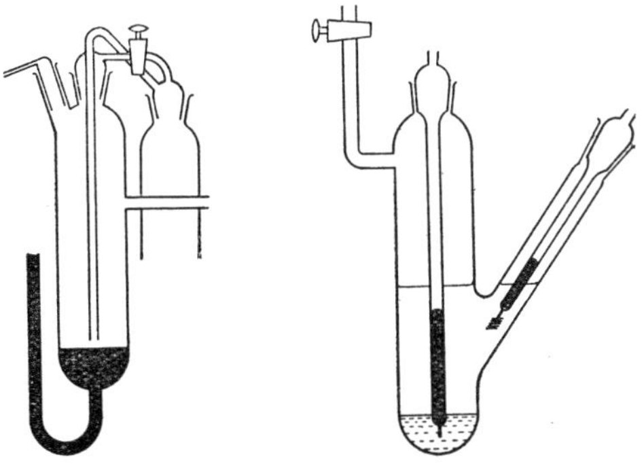

FIG. 4.3. Calomel and zinc bromide cells.

Hg in the cell. The cell is swept out with N_2 before filling with solution. The dry conditions and the absence of O_2 are essential. At 25°C and potentials of 0.0016 to 0.119 *m* HCl, E^0_{HgCl} was calculated by the methods described for AgCl to be [27, 19] 267.96 mV, although Grzybowski has derived a value of 268.13 mV; he has discussed the reasons and possible explanations for the difference. Studies of the analogous HgBr and HgI electrodes are less extensive.[29] A few studies of the HgCl electrode in mixed solvents have been reported by Schwabe.[30]

Some Other Galvanic Cells

Numerous series of measurements of the cell type

$$M\text{—}Hg \mid MX_2 \mid AgX \mid Ag \tag{4.24}$$

in which M = Zn^{2+}, Cd^{2+} or Pb^{2+}, and X = Cl^-, Br^- or I^- have been made available.[31] The reason for employing an amalgam electrode, two phase M—Hg, is that it often happens that reliable and reproducible potentials cannot be obtained with the pure metal owing to the presence of oxide films, crystal strains, etc. On forming an amalgam, the potential changes until a certain proportion of M is present, after which it remains constant at the true potential of M. The cell reaction is

$$M + 2AgX \rightleftharpoons MX_2 + 2Ag \tag{4.25}$$

The electrode reactions and potential equations are those of (4.16, 4.18) together with

$$M^{2+} + 2e^- \rightleftharpoons M;\ E_M = E^0_M - (k'/2)\log(1/a_{M^{2+}}) \qquad (4.26)$$

By analogy with (4.20), the potential of the cell is expressed by

$$E = E_{AgX} - E_M = E^0_{AgX} - E^0_M - (k'/2)\log a_{M^{2+}}a^2_{X^-} \qquad (4.27)$$

and if m is the molality of MX_2, then from (1.23, 1.27),

$$a_{M^{2+}}a^2_{X^-} = m\gamma_{M^{2+}}(2m\gamma_{X^-})^2 = 4m^3\gamma^3_\pm \qquad (4.28)$$

so that

$$E = E^0_{AgX} - E^0_M - (3k'/2)\log m\gamma_\pm(4)^{\frac{1}{3}} \qquad (4.29)$$

In order to establish E^0_M, this equation is expanded as in (4.23), and $\gamma_\pm$ calculated by a suitable form of the D.H. expression (3.41). Also, E^0_{AgCl} is needed.

Example 4.4. Stokes [31] has obtained potentials of cell (4.25) with $ZnBr_2$ of $m = 0.002$ to 0.99, and from 20° to 40°C. 5 % Zn amalgam was made by heating the metals together, washing the product with dilute acid, and filtering while hot. To prevent oxidation, the cell and solution were swept out with H_2. Fig. 4.3b shows the design of the cell for the more dilute solutions. Some of the measured potentials are given below. By putting

$$-\log\gamma_\pm = 0.5083 \times 2\{\sqrt{3m}/(1 \times 1.807\sqrt{3m}) - 0.15(3m)\} = \phi(I) \qquad (4.30)$$

where 1.807 corresponds to $\mathring{a} = 5.5$ Å, the left-hand side of $E + (3k'/2)\log m\sqrt{4} - \phi(I) = E^0_{AgBr} - E^0_{Zn}$ gave the last row of figures shown below:

10^4m	20.26	32.50	42.50	76.22	82.16
E(mV, 25°C)	1061.08	1044.20	1034.73	1014.36	1011.72
$E^0_{AgBr} - E^0_{Zn}$	833.82	833.80	833.81	833.83	833.79

The average of these is 833.81 ± 0.01 mV, and taking the mean value [21] of $E^0_{AgBr} = 71.33$ mV, $E^0_{Zn} = -862.48 \pm 0.04$ mV. The sign here is opposite to that given by the Lewis convention.

The calculation of E^0_{Cd} and E^0_{Pb} by the above process is not so straightforward owing to ion-pair formation. This is returned to in Chapter 9.

Other examples are:

(a) [32] Cd—Hg | $CdSO_4$ | $PbSO_4$ | Pb—Hg: (b) [33] Zn—Hg | $ZnSO_4$ | Hg_2SO_4 | Hg for which

$$E = E^0_{PbSO_4}\ (\text{or } E^0_{Hg_2SO_4}) - E^0_{Cd}\ (\text{or } E^0_{Zn}) - k'\log m\gamma_\pm \qquad (4.31)$$

and (c) [34] M—Hg | M_2SO_4 | $PbSO_4$ | Pb—Hg, where M = Li, Na, K. The potential expression is similar to (4.29). Similar cells where M = H are discussed in Chapter 5.

Experimental Values of Activity Coefficients

Once the standard electrode potentials of the cells have been obtained it is a simple matter to obtain values of $\gamma_\pm$ at various concentrations, by simple substitution in the potential equations. These activity coefficients are stoichiometric values unless allowance is made for any ion-association that takes place.

The mean activity coefficients of HCl in aqueous solution at 25°C have been obtained by means of cell (4.11), the analogous calomel electrode cell and from concentration cells with transference (described further on). The values for dilute solutions are listed below:

m	Ref. 12	Ref. 13	Ref. 26	Ref. 28	Refs. 46, 49
0.001	0.9656	0.9650	0.9650	0.9655	0.9653
0.002	0.9521	0.9520	0.9519	0.9526	0.9525
0.005	0.9285	0.9283	0.9280	0.9291	0.9287
0.01	0.9048	0.9045	0.9040	0.9053	0.9049
0.02	0.8755	0.8753	0.8747	0.8760	0.8757

It has been suggested [35] that owing to the slight scatter between the potentials of Ag | AgCl electrodes when measured in the same solution, E^0 for each electrode should be found by measuring it in 0.01 M HCl at 25°C and using $\gamma_\pm = 0.904$ in (4.23); (judging by the above list, 0.905 is better).

Concentration Cells without Transference

If two cells such as that depicted by (4.11) containing different molalities of HCl, m_1 and m_2, are connected in opposition, a concentration cell without transference would be formed, represented by

$$\text{Ag} \mid \text{AgCl} \mid \text{HCl}, m_1 \mid \text{H}_2 \mid \text{Pt} \mid \text{H}_2 \mid \text{HCl}, m_2 \mid \text{AgCl} \mid \text{Ag} \qquad (4.32)$$

the potential of which is given by

$$E = E_2 - E_1 = k' \log (a_\text{H} a_\text{Cl})_1/(a_\text{H} a_\text{Cl})_2 = 2k' \log (m_1\gamma_1/m_2\gamma_2) \qquad (4.33)$$

where 1 and 2 refer to m_1 and m_2, and γ_1 and γ_2 are mean activity coefficients. This equation follows directly from (4.23). The order of subtraction is $E_2 - E_1$ if $m_1 > m_2$, so that E is positive. In practice such a cell as (4.32) is not set up, for its potential is obtained by subtraction of the appropriate values of (4.11). There are cells of this class which are constructed as one unit, and which are used to obtain activity coefficients and thermodynamic data, e.g.,[36]

$$\text{Ag} \mid \text{AgX} \mid \text{MX}, m_1 \mid \text{M—Hg} \mid \text{MX}, m_2 \mid \text{AgX} \mid \text{Ag} \qquad (4.34)$$

The main features are shown in Fig. 4.4a. The Ag | AgX electrodes are sited in the outer compartments while the two compartments are fed by a stream of amalgam from the reservoir *A*. The amalgam is made by electrolyzing a solution of MOH with a Pt anode and a mercury pool cathode until the latter contains about 0.01 % of M. The flow rate of amalgam from *A* into the cell is about 1 ml. per 15 to 25 sec. The halide solutions MX are freed of O_2 and introduced into the cell via *B* and *B′* under vacuum conditions. The lowest concentration for reliable, reproducible readings is about 0.1 M. This is taken as the reference (m_2) while m_1 is varied in steps from about 0.2 M upwards.

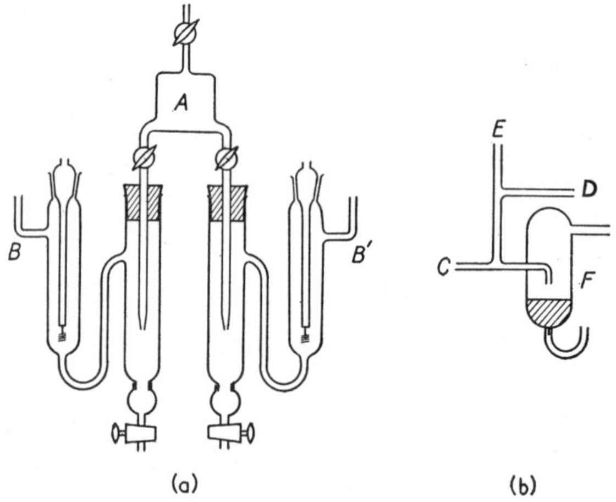

FIG. 4.4. Amalgam concentration cells.

Example 4.5. Harned and Crawford used cell (4.34) to ascertain the activity coefficients of NaBr in aqueous solution from 0° to 40°C, and from 0.1 M to 4 M. At 0.1 M and 25°C, from the isopiestic vapour pressure method (see Chapter 5), $\gamma_{\pm} = 0.782$. By using this in (4.33), and the data given below (the potentials have been corrected to abs. mV), $\gamma_{\pm}$ from 0.2 M to 4 M was derived.

E(mV)	32.75	52.00	76.54	93.24	111.59	133.69	150.64	165.06	177.51	188.59	198.64
M	0.2	0.3	0.5	0.7	1.0	1.5	2.0	2.5	3.0	3.5	4.0
$\gamma_{\pm}$	0.739_5	0.717	0.694	0.686	0.686	0.703	0.733_5	0.777	0.825	0.877	0.933_5

Some further examples are:

(a) [37] Hg | Hg_2SO_4 | M_2SO_4, m_1 | M—Hg | M_2SO_4, m_2 | Hg_2SO_4 | Hg
(M = Li, Na, K)

(b) [38] Pb—Hg | $PbSO_4$ | Na_2SO_4, m_1 | Na—Hg | Na_2SO_4, m_2 | $PbSO_4$ | Pb—Hg
(6 to 10 % Pb)

(c) [39] Ag | AgCl | MCl_2, m_1 | Na—Hg | MCl_2, m_2 | AgCl | Ag (M = Ca, Sr, Ba)

For all of these,

$$E = (3k'/2) \log (m_1\gamma_1/m_2\gamma_2) \quad m_1 > m_2 \tag{4.35}$$

Instead of a flowing amalgam, Tippetts and Newton [39] made use of a stationary two-phase amalgam for studying $BaCl_2$ in cell (c) above. The cell was similar to that devised by Danner;[40] the amalgam compartment is shown in Fig. 4.4b, C connects to the electrolyte compartment and D to a calomel electrode, i.e., the cell is Ba—Hg | $BaCl_2$, m | HgCl | Hg. E contained a plunger for cutting off D while fresh solution was allowed to flow over the amalgam in F. To prevent oxidation, since no precautions were taken to exclude air, the amalgam surface was covered with light mineral oil (cyclohexane has proved suitable for such purposes [41]). In some instances H_2 was evolved at the amalgam surface; these measurements were rejected.

Harned, Akerlöf and their associates [42] have made several studies of the cell

Pt | H_2 | MOH, m_1 | M—Hg | MOH, m_2 | H_2 | Pt (M = Li, Na, K, Cs)

Following Harned's treatment,[43] the expression relating the potentials to the activities of MOH can be derived in this way. In the cell, M | MOH | H_2 | Pt, the reaction is $M + H_2O \rightleftharpoons \frac{1}{2}H_2 + M^+ + OH^-$. The electrode reactions and potentials, are, as in (4.14, 4.15),

$$M^+ + e^- \rightleftharpoons M; \quad E_M = E^0_M - k' \log (1/a_{M^+}) \tag{4.36}$$

$$H_2O + e^- = \tfrac{1}{2}H_2 + OH^-; \quad E_{OH} = E^0_{OH} - k' \log (a_{\frac{1}{2}H_2} a_{OH^-}/a_{H_2O}) \tag{4.37}$$

To conform to the cell reaction, (4.36) is subtracted from (4.37), so

$$E_1 \text{ and } E_2 = E_{OH} - E_M = E^0_{OH} - E^0_M - k' \log (a_{\frac{1}{2}H_2} a_{OH^-} a_{M^+}/a_{H_2O}) \tag{4.38}$$

Accordingly, for the concentration cell,

$$\begin{aligned} E = E_2 - E_1 &= k' \log\{(a_{OH}a_M)_1/(a_{OH}a_M)_2\} + k' \log \{(a_{H_2O})_2/(a_{H_2O})_1\} \\ &\qquad + \tfrac{1}{2}k' \log \{(a_{H_2})_1/(a_{H_2})_2\} \\ &= 2k' \log (m_1/m_2) + 2k' \log (\gamma_1/\gamma_2) + k' \log \{(a_{H_2O})_2/(a_{H_2O})_1\} \\ &\qquad + \tfrac{1}{2}k' \log \{(P - p_1)/(P - p_2)\} \end{aligned} \tag{4.39}$$

where p_1 and p_2 are the vapour pressures of solution m_1 and m_2, and P is the barometric pressure, When m_1 and m_2 are of the same order, the last two terms are negligible, but since the reference m_2 is kept at 0.1 M while m_1 extends to high concentrations, the two terms are often significant. The evaluation of the third right-hand term has been discussed by Harned and Cook [42] and by Akerlöf and Kegeles.[42] The latter treatment has been generalized by Stokes;[44] the free energy change, from (4.7), is given by

$$\Delta G = -nFE = \bar{G}_{m_2} - \bar{G}_{m_1} - r(\bar{G}_{w_2} - \bar{G}_{w_1}) \tag{4.40}$$

where $\bar{G}_m$ refers to the electrolyte and w to the solvent, and r is an integer expressing the moles of solvent transferred per mole of electrolyte transferred. If m_2 is kept constant, differentiation of (4.40) gives

$$-nFdE = -d\bar{G}_{m_1} + rd\bar{G}_{w_1} \tag{4.41}$$

The thermodynamic expression known as the Gibbs–Duhem equation states that for a two component system,

$$N_1 d_1\bar{G} = -N_2 d\bar{G}_2 \tag{4.42}$$

where N_1 and N_2 are the mole fractions of the two components. For aqueous electrolyte solutions, this becomes, from (4.1),

$$(1{,}000/M_1)RT \text{ d ln } a_1 = -mRT \text{ d ln } a_2 \tag{4.43}$$

where M_1 is the molecular weight of water, a_1 is its activity. Thus, from (4.41),

$$-nFdE/RT = (55.51/m) \text{ d ln } a_1 + r \text{ d ln } a_1 \tag{4.44}$$

If $nm/(55.51 + rm) = m'$, and integration between m_2' and m_1' is effected,

$$-k' \log (a_{w_2}/a_{w_1}) = \int_{m_1'}^{m_2'} m'dE \tag{4.45}$$

Example 4.6. A plot of E against m' can be used to evaluate the integral of (4.45) by graphical means. The following [42] figures for KOH at 25°C can be used for this purpose. The reference, $m_2 = 0.05$ M, for which $\gamma_\pm = 0.824$. Also, in (4.41), $r = n = 1$. The given potentials have been corrected for the last term of (4.39).

m_1	0.10	0.15	0.25	0.35	0.50
E(abs. mV)	34.27	53.32	78.37	94.54	112.41
$10^3\, m'$	1.80	2.70	4.48	6.27	8.92

The area under the curve of the plot of E against m' between $10^3\, m' = 0.90$ (its value at 0.05 M) and $10^3\, m' = 8.92$ is 0.60 mV. Thus, from (4.39), $\gamma_\pm$ for 0.5 M = 0.720.

Harned and Cook [42] have given another way. Since $\ln a_2 = \ln m\gamma_\pm$, for a 1 : 1 electrolyte, from (4.43),

$$-0.5 \text{ d ln } a_1 = (m/55.51)\text{d ln } m\gamma_\pm = (\text{d}m/55.51) + (m/55.51)\text{d ln } \gamma_\pm \tag{4.46}$$

Integrating,

$$0.5 \log (a_{w_1}/a_{w_2}) = (m_1 - m_2)/(2.303 \times 55.51) + \int_{m_1}^{m_2} m \text{ d} \log (\gamma_1/\gamma_2) \tag{4.47}$$

For a first approximation, the second right-hand term is ignored, and the resulting first value put into (4.39) to obtain provisional values of log (γ_1/γ_2), which are used to obtain the second right-hand term of (4.47) by graphical integration.

Concentration Cells with Transference

There is an important series of concentration cells which permit activity coefficients to be derived with a high order of accuracy, especially in dilute solutions. These cells consist of identical electrodes in different concentrations of the same electrolyte which meet at a common interface; so far, these investigations have been confined to the type

$$Ag \mid AgX \mid MX_n, m_1 : MX_n, m_2 \mid AgX \mid Ag \quad (4.48)$$

where $X = Cl^-$, Br^-. Considering a univalent chloride, for the passage of $1F$, if the right-hand electrode is the positive one, the electrode process at its surface is $AgCl + e^- \rightleftharpoons Ag + Cl^-$, and the reverse process occurs at the left-hand electrode. There is a stream of Cl^- ions from right to left through the cell, and an electron stream from left to right through the outer circuit. Ion migration takes place at the liquid junction between the two solutions; $T_{Cl^-} = 1 - T_{M^+}$ of Cl^- ions migrate from left to right across the interface, i.e., from m_2 to m_1. At the same time, T_{M^+} of M^+ ions migrate in the opposite direction. The net result is that solution m_1 loses one equivalent of Cl^- ions by the electrode reaction, and gains T_{Cl^-} of Cl^- ions by migration, so the total loss is $1 - T_{Cl^-} = T_{M^+}$ of Cl^- ions. Also, T_{M^+} of M^+ ions are lost to m_2 by ion migration. Hence T_{M^+} of MCl is lost by m_1 and gained by m_2.

It follows from (4.33) that the potential, E_T, is given by

$$E_T = k'T_{M^+} \log (a_1/a_2) = 2k'T_{M^+} \log (m_1\gamma_1/m_2\gamma_2) \quad (4.49)$$

For E_T to be positive, $m_1 > m_2$; this also follows from the tendency for concentrations to become equal as a result of the cell reaction.

Equation (4.50) assumes that T_{M^+} is constant; this is only true when $a_1 = a_2$. If a and $a + da$ are the activities of the two solutions, where da is infinitesimally small, and dE_T is the corresponding potential, then

$$dE_T = k'T_{M^+} \log \{(a + da)/a\} = k'T_{M^+} d \log a \quad (4.50)$$

since $\log (1 + da/a) \simeq da/a = d \log a$ when da is very small. Accordingly,

$$\int_0^E dE_T = E_T = 2k' \int_{m_2}^{m_1} T_{M^+} d \log m\gamma_\pm \quad (4.51)$$

The first application of cell (4.48) and the associated development of (4.51) is due to Brown and MacInnes.[45] They determined the potentials for NaCl solutions at 25°C, keeping the reference at $m_1 = 0.1$ M, and varying m_2 from 0.005 to 0.08. The liquid junction was formed by the sheared boundary device that was described in Chapter 2 for moving boundary transference numbers. A detailed sketch is given in the original article. In subsequent studies,[46,48] a different type of cell was used which contained no moving parts. A somewhat similar type, used by Gordon and others [47,48] is shown in Fig. 4.5.

The Ag | AgX electrodes were formed on thick Pt discs, the solutions m_1 and m_2 were placed in the two side compartments and the centre section was filled with the more concentrated of these. The liquid junction forms around A or A'.

Spedding used cell (4.48) to derive the activity coefficients of a number of lanthanide chlorides and bromides.[44,49] The cell consisted of two electrode compartments joined by a stopcock of the hollow high-vacuum type. The more dilute solution was placed in the compartment joined to the centre opening of the stopcock, and this compartment contained a small trap as a safeguard against the entry of the stronger solution.

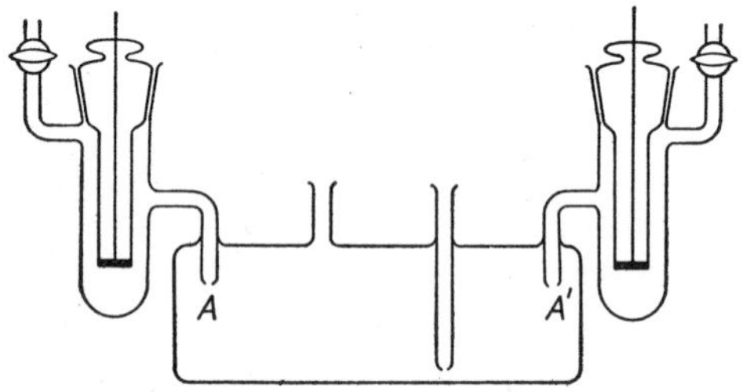

FIG. 4.5. Concentration cell with transference.

The first method of computation was devised by Longsworth (private communication to Brown and MacInnes[45]) and consists of firstly replacing T_M by $T_1 + \Delta T$, where T_1 is the transference number of the reference concentration. Putting this in (4.51), and rearranging,

$$\Delta \log \gamma = \log (\gamma_1/\gamma_2) = E_T/2k'T_1 - \log (m_1/m_2) - (1/T_1)\int_{m_2}^{m_1} \Delta T \, d \log (m_1/m_2) - (1/T_1)\int_{\gamma_2}^{\gamma_1} \Delta T \, d \log (\gamma_1/\gamma_2) \quad (4.52)$$

The first calculation ignores the last term, and the third right-hand term is obtained by graphical integration. In this way, a preliminary value of $\log (\gamma_1/\gamma_2)$ is obtained which is used to obtain the last term, also by graphical integration. Thus the true values of $\log (\gamma_1/\gamma_2)$ are derived. This method was used in the earlier papers of MacInnes and Shedlovsky [45] and of Gordon,[47] but a simpler way was later adopted by the former authors [48] and also by Spedding.[50] For this, $(1/T_M)$ is replaced by $\delta + (1/T_1)$ in the differential form of (4.51), i.e.,

$$\delta dE + dE/T_1 = k' d \log m\gamma \quad (4.53)$$

Integration and rearrangement results in

$$\log (\gamma_1/\gamma_2) = E_T/\nu k'T_1 + (1/\nu k')\int_0^E \delta dE_T - \log (m_1/m_2) \quad (4.54)$$

The integral is obtained by graphical integration of the area under the curve of the plot of E_T against δ.

Guggenheim and Prue [51] have devised another way. Writing

$$\Delta E_T = \nu k' \bar{T} \Delta \log m\gamma \tag{4.55}$$

where $\bar{T}$ denotes the average value of T_M between two consecutive concentrations m' and m'', and ΔE is the corresponding difference in potentials, values of $\Delta \log \gamma$ corresponding to these intervals are obtained. If the intervals are such that the change in T_M is considerable, $\bar{T}$ should be taken at the midpoint of $\log m'$ and $\log m''$.

The next stage is to devise ways of calculating values of γ from the $\log (\gamma_1/\gamma_2)$ data. Shedlovsky and MacInnes [48, 49] did this by plotting $\log (\gamma_1/\gamma_2) - A''\sqrt{m}$ against $B - \log (\gamma_1/\gamma_2)$, where A'' is the theoretical constant of the D.H. equation (see 3.36), and B is an empirical factor obtained by a short series of approximations in which its final value agrees with the intercept of the plot. Values of γ are then obtained from $\log \gamma = B - \log (\gamma_1/\gamma_2)$. Substitution of these in (3.35) decides the value of $B''a$. With 1 : 1 electrolytes, this is satisfactory up to about 0.04 N, and by including the parameter Q as in (3.40, 3.41), activity coefficients could be obtained up to 0.1 N.

Gordon's way [47] is to plot $\log (\gamma_1/\gamma_2) + (A''\sqrt{m}/(1 + B''a\sqrt{m})$ against m, $B''a$ being adjusted until the plot is linear. The slope of the line gives Q'', the intercept gives $\log \gamma_1$ and thus γ can be deduced at each concentration.

Example 4.7. The potentials listed in Table 4.3 were obtained by Brown and

TABLE 4.3.

Activity Coefficients of NaCl in Water at 25°C from Concentration Cells with Transference

m_2	E_T(abs.mV.)	T_M	$-10^2\delta$	$-\int\delta dE_T$	$-\log(\gamma_1/\gamma_2)$	$-\log \gamma_1'$	γ	γ(calc.)
0.005	56.469	0.3930	4.99	1.223	(0.0731)	(0.1067)	0.9260	0.9270
0.007	49.923	0.3925	4.66	1.048	(0.0690)	(0.1082)	0.9149	0.9157
0.01	43.043	0.3918	4.21	0.839	(0.0632)	(0.1094)	0.9009	0.9020
0.02	29.814	0.3902	3.16	0.456	0.0491	0.1121	0.8706	0.8706
0.03	22.188	0.3891	2.44	0.271	0.0386	0.1137	0.8498	0.8493
0.04	16.824	0.3883	1.91	0.161	0.0304	0.1151	0.8339	0.8333
p.05	12.699	0.3876	1.44	0.091	0.0233	0.1162	0.8204	0.8204
0.06	9.320	0.3870	1.04	0.049	0.0179	0.1179	0.8102	0.8095
0.08	4.057	0.3861	0.44	0.009	0.0080	0.1200	0.7920	0.7923
0.10	0	0.3854_5	0	0	0	—	0.7779	0.7791

MacInnes [45] for aqueous NaCl at 25°C (corrected to abs. mV). The molality of the reference solution $m_1 = 0.10$ M. The transference numbers for this and the varying concentration, m_2, were obtained by the moving boundary method and are shown in the Table. Values of $\log (\gamma_1/\gamma_2)$ have been calculated by means of (4.54) and are given in the 6th column. A value of $\log \gamma_2$ was calculated by using (3.41) with $Q'' = 0$, whence the values of $\log \gamma_1'$ in the 7th column were derived. These were plotted against m and gave a straight line providing the points corresponding to $m_2 = 0.005$, 0.007 and 0.01 are omitted. Extrapolation to $m = 0$ gives $-\log \gamma_1 = 0.109_2$ for 0.1 M NaCl.

The figures for the three most dilute concentrations just given were recalculated by use of (4.55). Thus for $m'' = 0.02$, $m' = 0.01$, $E = 13.229$ mV, $\bar{T} = 0.3910$, whence $E_T/2k'\bar{T} = 0.2860 = 0.3010 + \Delta \log \gamma$, so $\Delta \log \gamma = -0.0150$. Assuming that $\log (\gamma_1/\gamma_2)$ is correct at 0.02 M, then at $m_2 = 0.01$, $-\log (\gamma_1/\gamma_2) = 0.0491 + 0.0150$, whence $-\log \gamma_1' = 0.1103$. Similarly, at $m_2 = 0.007$, $-\log (\gamma_1/\gamma_2) = 0.0707$, $-\log \gamma_1' = 0.1099$, and at $m_2 = 0.005$, $-\log (\gamma_1/\gamma_2) = 0.0759$, $-\log \gamma_1' = 0.1095$. These fresh figures for γ_1' fit the drawn straight line; (4.54) breaks down when $T_M - T_1 > 0.015$.

Taking the final extrapolation of $-\log \gamma_1 = 0.109_3 \pm 0.0003$, and the final values of $\log (\gamma_1/\gamma_2)$, the values of γ shown in column 8 were derived. Also, from the slope of the line, Q'' of (3.41) is 0.138/0.5083, and using this, the figures of column 9 are found.

REFERENCES

1. "International Critical Tables." McGraw-Hill Book Co. Inc., New York **6**, 312 (1929).
2. Ref. 10, Ch. 1.
3. Latimer, W. M. *J. Amer. chem. Soc.* **76**, 1200 (1954).
4. Nernst, W. *Z. Elektrochem.* **5**, 253 (1900).
5. Lewis, G. N. *J. Amer. chem. Soc.* **35**, 1 (1913).
6. Lewis, G. N. and Randall, M. *ibid.* **36**, 1969 (1914).
7. Bates, R. G. and Acree, S. F. *J. Res. nat. Bur. Stand.* **30**, 129 (1943).
8. Nair, V. S. K. and Nancollas, G. H. *J. chem. Soc.* 4144 (1958).
9. Hills, G. J. and Ives, D. J. G. *ibid.* 305 (1951).
10. Janz, T. G. and Taniguchi, H. *Chem. Rev.* **53**, 397 (1953).
11. Ref. 2, Ch. 1.
12. Harned, H. S. and Ehlers, R. W. *J. Amer. chem. Soc.* **54**, 1350 (1932).
13. Bates, R. G. and Bower, V. E. *J. Res. nat. Bur. Stand.* **53**, 283 (1954).
14. Harned, H. S. *J. Amer. chem. Soc.* **51**, 416 (1929).
15. Smith, E. R. and Taylor, J. K. *J. Res. nat. Bur. Stand.* **20**, 837 (1938); **22**, 307 (1939).
16. Brown, A. S. *J. Amer. chem. Soc.* **56**, 646 (1934).
17. Purlee, E. L. and Grunwald, E. *J. phys. Chem.* **59**, 1112 (1955).
18. "Handbook of Chemistry and Physical Chemistry." Chemical Rubber Publishing Co., Cleveland, Ohio, 37th edn., 2,999 (1955–6).
19. Guggenheim, E. A. and Prue, J. E. *Trans. Faraday Soc.* **50**, 231 (1954).
20. Carmody, W. R. *J. Amer. chem. Soc.* **51**, 2905 (1929).
21. Ref. 11, Ch. 1., Ref. 2, Ch. 1. Aq. fructose—Crockford, H. D., Little, W. F. and Wood, W. A. *J. phys. Chem.* **61**, 1674 (1957). Aq.MeOH, Aq. EtOH, Formamide—I.T. Oiwa, *Sci. Rep., Tokohu Univ.* **41**, 47 (1957).
22. Taniguchi, H. and Janz, G. J. *J. phys. Chem.* **61**, 688 (1957).

23. Danyluk, S. S. Tanniguchi, H. and Janz, G. J. *ibid.* **61,** 688 (1957).
24. Owen, B. B. and Foering, L. *J. Amer. chem. Soc.* **58,** 1575 (1936).
25. Harned, H. S. and Donelson, J. G. *ibid.* **60,** 339, 2128 (1938).
26. Owen, B. B. *ibid.* **57,** 1526 (1935).
27. Hills, G. J. and Ives, D. J. G. *J. chem. Soc.* 311, 318 (1951).
28. Grzybowski, A. K. *J. phys. Chem.* **62,** 550 (1958).
29. (a) HgBr: Larson, W. D. *J. Amer. chem. Soc.* **62,** 765 (1940); Dakin, T. W. and Ewing, D. T. *ibid.* **62,** 2230 (1940). (b) HgI: Bates, R. G. and Vosburgh, W. C. *ibid.* **59,** 1188 (1937).
30. Schwabe, K. and others, *Z. Elektrochem.* (a) Aq.MeOH: **62,** 172 (1958). (b) Aq. dioxan, aq. glycol: **63,** 441, 445 (1959).
31. (a) $ZnCl_2$: Robinson, R. A. and Stokes, R. H. *Trans. Faraday Soc.* **36,** 740 (1940). (b) $ZnBr_2$: Stokes, R. H. and Stokes, J. M. *ibid.* **41,** 688 (1945). (c) ZnI_2: Bates, R. G. *J. Amer. chem. Soc.* **60,** 2983 (1938). (d) $CdCl_2$: Harned H. S. and Fitzgerald, M. E. *ibid.* **58,** 2624 (1936). (e) $CdBr_2$: Bates, R. G. *ibid.* **60,** 2983 (1938). (*f*) CdI_2: Bates, R. G. *ibid.* **63,** 399 (1941). (g) $PbCl_2$: Carmody, W. R. *ibid.* **51,** 2905 (1929).
32. La Mer, V. K. and Parks, W. G. *ibid.* **53,** 2040 (1931); **55,** 4343 (1933).
33. Bray, U. B. *ibid.* **49,** 2372 (1927).
34. Akerlöf, G. *ibid.* **48,** 1160 (1926); Harned, H. S. and Hecker, J. C. *ibid.* **56,** 650 (1934).
35. Bates, R. G., Prue, J. E. *et al. J. chem. Phys.* **25,** 361 (1956); **26,** 222 (1957).
36. Harned, H. S. and co-workers *J. Amer. chem. Soc.* (a) KCl: **59,** 1290 (1937). (b) NaCl: **54,** 423 (1932). (c) NaBr: **59,** 1903 (1937). (d) CsCl: **52,** 3886 (1930). (e) NaI,KI: **48,** 3095 (1926). (f) KBr,LiCl,LiBr: **51,** 416 (1929).
37. Akerlöf, G. *J. Amer. chem. Soc.,* **48,** 1160 (1926). Harned, H. S. and Akerlöf, G. *Phys. Z.* **27,** 411 (1926).
38. Harned, H. S. and Hecker, J. C. *J. Amer. chem. Soc.* **56,** 650 (1934).
39. Lucasse, W. W. *ibid.* **47,** 743 (1925). Scatchard, G. and Teft, R. F. *ibid.* **52,** 2265 (1930). Tippetts, E. A. and Newton, R. F. *ibid.* **56,** 1675 (1934).
40. Danner, P. S. *ibid.* **46,** 2385 (1924).
41. Truemann, W. B. and Ferris, L. M. *ibid.* **80,** 5048 (1958).
42. (a) LiOH: Harned, H. S. and Schwindells, F. E. *ibid.* **48,** 126 (1926). (b) NaOH: Akerlöf, G. and Kegeles, G. *ibid.* **62,** 620 (1940). (c) KOH: Akerlöf, G. and Bender, P. *ibid.* **70,** 2366, (1948); Harned, H. S. and Cook, M. A. *ibid.* **59,** 496 (1937). (d) CsOH: Harned, H. S. and Schupp, O. E. *ibid.* **52,** 3886 (1930).
43. Harned, H. S. *J. Amer. chem. Soc.* **47,** 676 (1925).
44. Stokes, R. H. *ibid.* **67,** 1686 (1945).
45. Brown, A. S. and MacInnes, D. A. *ibid.* **57,** 1356 (1936).
46. Shedlovsky, T. and MacInnes, D. A. *ibid.* **58,** 1970 (1936).
47. Hornibrook, W. J., Janz, G. J. and Gordon, A. R. *ibid.* **64,** 513 (1942).
48. (a) KCl: Shedlovsky, T. and MacInnes, D. A. *ibid.* **59,** 503, (1937); Janz, G. J. and Gordon, A. R. *ibid.* **65,** 218 (1943). (b) $CaCl_2$: Shedlovsky, T. and MacInnes, D. A. *ibid.* **59,** 503 (1937); McLeod, H. G. and Gordon, A. R. *ibid.* **68,** 58 (1946). (c) $LaCl_3$: Shedlovsky, T. and MacInnes, D. A. *ibid.* **61,** 200 (1939).
49. Shedlovsky, T. *ibid.* **72,** 3680 (1950).
50. Spedding, F. H., Porter, P. E. and Wright, J. M. *ibid.* **74,** 2781 (1952); Spedding, F. H. and Yaffe, I. S. *ibid.* **74,** 4751 (1952).
51. Guggenheim, E. A. and Prue, J. E. "Physico-Chemical Calculations." North-Holland Publishing Co., Amsterdam (1955).

CHAPTER 5

THE DETERMINATION OF ACTIVITY COEFFICIENTS FROM SOLUBILITY, FREEZING-POINT AND VAPOUR-PRESSURE MEASUREMENTS

IN ADDITION to the e.m.f. methods of deriving activity coefficients that have been described in Chapter 4, there are several other ways of obtaining them. Three of these form the subject matter of the present chapter.

Solubility Measurements

In a saturated solution, an equilibrium is established between the solute that is in solution, and the solute that remains undissolved. For an electrolyte giving ν_i cations of valency z_i and ν_j anions of valency z_j, the equilibrium constant between dissolved and undissolved electrolyte, K_s, is described by

$$K_s = a_{\mathrm{M}}a_{\mathrm{L}}/a_{\mathrm{ML}} = [\mathrm{M}^{z_i}]^{\nu_i}[\mathrm{L}^{z_j}]^{\nu_j}\, y_{\mathrm{M}}y_{\mathrm{L}} \qquad (5.1)$$

where M^{z_i} and L^{z_j} are the cation and anion. The activity of the undissolved electrolyte, a_{ML}, is taken as unity since it is independent of the amount of undissolved material. Under given conditions of temperature, K_s is constant and is termed the activity solubility-product. In the presence of another soluble salt with a common ion, the concentration of either M^{z_i} or L^{z_j} is increased, but since K_s is constant, the concentration of the other ion is decreased, so that the solubility of ML diminishes. On the other hand, in the presence of non-common ions, the solubility of ML is increased since the increased ionic strength results in lower values of the activity coefficients y_{M} and y_{L}. It should be observed that these remarks assume that ion-association does not take place; if it does, then the solubility will rise under certain conditions even in the presence of a common ion. This topic is reserved for Chapter 11.

Bronsted, La Mer and others [1] measured the solubilities of some sparingly soluble salts in solutions of electrolytes in order to test the validity of the Debye-Hückel activity coefficient expressions. It is perhaps more informative to show how, in certain cases, solubilities may be used to obtain mean experimental activity coefficients with the aid of the D.H. treatment, and how these may be used to find the activity coefficient parameters. The following calculations illustrate one procedure. At the same time, a convenient form of saturator is described (a comprehensive review on the determination of solubilities has been given by Zimmermann [2]).

Example 5.1. MacDougall and Davies [3] measured the solubility of barium iodate in water, KCl, KNO_3 and some other salt solutions. The saturator that they used has been described briefly by Money and Davies,[4] and since it has proved to be such a convenient device, some account of it will be given. It consists of a bulb 200 to 400 ml capacity attached to a tube containing the sparingly soluble salt. This is about 15 cm long and 1.5 cm diameter, and is connected at its base by a capillary tube to a collecting tube about 20 cm long and 3 cm diameter into which the saturated solution percolates; the whole is U-shaped. Convenient sized crystals are packed into the saturating tube above a plug of cotton wool. A general way of making the crystals is exemplified by barium iodate, namely by allowing solutions of iodic acid and of barium hydroxide to drip slowly into about a litre of water without stirring. The solvent is allowed to percolate through the crystals at about 100 ml./h. Several small quantities of the solvent are allowed to pass, pipetted out and rejected before a convenient quantity is collected for analysis. The saturation is easily checked by recycling a portion. The pipette is kept in an empty tube placed in the same thermostat bath.

The solubility of barium iodate in water and in KCl solutions at 25°C is shown by figures in Tables 5.1. From (5.1),

$$K_s = [Ba^{2+}][IO_3^-]^2 y_{Ba} y^2_{IO_3} \tag{5.2}$$

This is converted to

$$\log K_s = \log C_1(2C_1)^2 + 3 \log y_{\pm} \tag{5.3}$$

where C_1 is the molarity of the barium iodate and $y_{\pm}$ is its mean activity coefficient at $I = 3C_1 + C_2$, C_2 being the molarity of the KCl. A general form of the D.H. equation is used to calculate $y_{\pm}$ (at low ionic strengths, $y_{\pm} = f_{\pm}$), namely

$$-\log f_{\pm} = 0.5091 \times 2\sqrt{I}/(1 + \sqrt{I}) - Q_1 I \tag{5.4}$$

As a start, Q_1 is taken as zero, and from the solubility in water, $\log K_s = \bar{9}.1840$. The 3rd row of the Table has been calculated from

$$\bar{9}.1840 - \log 4C_1^3 + 6 \times 0.5091\sqrt{I}/(1 + \sqrt{I}) = 3Q_1 I \tag{5.5}$$

and a plot of these figures against I has a slope of 0.2, so $Q_1 = 0.067$. This could be used in (5.4) and (5.3) to recalculate $\log K_s$, but in this instance the solubility in water is so small that it makes no difference. The 4th row gives the experimental values of $y_{\pm}$ derived from (5.3) and the derived value of K_s, and the last row shows the values of $f_{\pm}$ obtained with (5.4) by putting $Q_1 = 0.067$.

The solubilities of higher valent salts, particularly in solutions of salts which themselves contain higher valence ions, cannot be interpreted in the manner just described. Bronsted [5] attributed the apparent disharmonies with the

D.H. theory to what are called the specific effects of ions on the activity coefficients of other electrolytes. There is evidence from various sources that in mixtures of strong electrolytes the mean activity coefficient of an electrolyte

TABLE 5.1

Mean Activity Coefficients of Barium Iodate in KCl Solutions at 25°C from Solubilities

C_2	0	0.001	0.002	0.0035	0.005	0.0075	0.010	0.020	0.050
10^4C_1	8.10	8.27	8.40	8.59	8.74	8.99	9.18	9.85	11.17
$3Q_1I$	0	− 0.0009	0.0015	0.0008	0.0027	0.0006	0.0033	0.0034	0.0110
$y_\pm$	0.8958	0.8772	0.8638	0.8447	0.8303	0.8069	0.7903	0.7365	0.6495
$f_\pm$	0.8960	0.8780	0.8634	0.8448	0.8294	0.8078	0.7900	0.7370	0.6495

is influenced in a specific manner by the concentration and nature of the other ions that are present (see Chapter 5); these are not significant at the low concentrations concerned in solubility studies, and the marked effects found can be attributed to ion-association [5] as shown in Chapter 11.

Freezing-point Measurements

The determination of activity coefficients from freezing-point depressions calls for extremely refined measurements, especially at low concentrations. An error of 0.0002°C, which is about the limit of precision, corresponds to an error of about 1 % in a 0.001 M solution,[7] although it is much less at higher concentrations.

A full description of the necessary apparatus, the sources of error and their elimination has been given by Scatchard, Jones and Prentiss [7] and also by Lange.[8] Differential measurements between ice-water and ice-solution are obtained by means of multiple thermocouples. The two liquids are contained in plated vessels fixed into Dewar flasks which are contained in a thermostat containing ice and water.

As an illustration, a brief account of the apparatus constructed by Brown and Prue [9] will be given with the aid of Fig. 5.1. Two silvered Dewar flasks, one of which is shown, were fitted into a lagged tank containing about equal amounts of ice and water stirred by four air-jet circulators; these kept the tank to within ± 0.01°C. Each Dewar was fitted with a rubber bung carrying five tubes for (a) drainage, (b) sampling, (c) air inlet, (d) internal circular support and (e) thermo-element. The air for (c) was passed from a compressor through a solution of NaOH, water, and then three wash-bottles in the tank. The thermocouple consisted of twenty-four varnish insulated chromel-constantan couples and one chromel-copper couple. The optimum wire cross-sections and number of junctions were determined by equations.[10] The whole of this unit was fitted into a U-shaped tube fitted at its ends with thin glass tubes filled with liquid paraffin. To determine the potential developed across the thermocouple, which is marked as T in the circuit shown in Fig. 5.1b, the potential drop across R_3 (one ohm) was adjusted via R until it equalled that of T. The potential drop across R_2 (100 ohms) was then measured by potentiometer (F), thus obtaining a reading 100 times that of T. The galvanometer G_1 was so sensitive that a mm deflection on its scale corresponded to 10^{-5}°C. By substituting R_3 for T, a check could be made on parasitic currents.

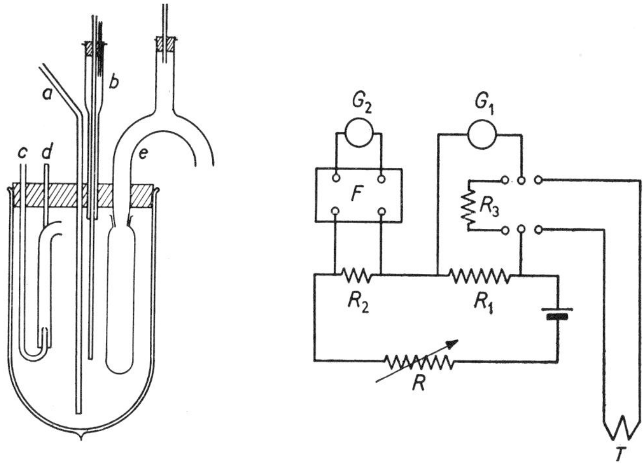

FIG. 5.1. Brown and Prue's [9] freezing-point apparatus and electrical circuit.

The Dewar flasks were one-third filled with ice made from conductivity water, then precooled water added until the flasks were three-quarters full. All connections were made overnight so that equilibrium was reached the next morning and nothing need be touched. The bath was filled, the circulators started and equilibrium observed when there was no potential in T. Then 20 ml. was withdrawn and replaced by 20 ml. of precooled stock solution of electrolyte. Equilibrium was again reached in about 30 min, the potential measured, after which 40 ml. was withdrawn into a 20 ml. conductance cell mounted in the thermostat, the first 20 ml. being used to rinse out the cell; this conductance determined the concentration. 40 ml. of more stock solution was added to the flask and the operations repeated. In this way, about ten solutions between 0.005 M and 0.1 M were measured.

Dissolved gases influence the freezing-point of a solution. This has been investigated,[7] and since N_2 is about half as soluble as O_2, it is preferable to saturate the solution with the former. Although the solubility increases about

2 % per degree drop, the gas is salted out as the electrolyte concentration increases; the two effects almost cancel. It is essential, however, that the degree of saturation in the two Dewars is the same to within about 1 %.

The thermocouple was calibrated against the data for aqueous NaCl obtained by Scatchard [11] and by Harkins and Roberts [12] (these are in excellent agreement). Another way [13] is to use a four-leads platinum resistance thermometer and measure the freezing-points of $NaNO_3$ solutions down to − 10°C.

Equations for calculating the activities of solutes from freezing-points were derived by thermodynamic principles by Lewis and Randall.[14] The first step is effected by means of the Gibbs–Helmholtz equation,

$$\mathrm{d}(\Delta G/T)_P/\mathrm{d}T = -\Delta H/RT^2 \tag{5.6}$$

The free energy change is that for the isothermal transfer of a constituent from a solution to another of different concentration, and from (1.15) is given by

$$\Delta G = RT \ln (a/a') \tag{5.7}$$

ΔH is the change in heat content during the transfer and since $\Delta H = \bar{L} - \bar{L}'$, where $\bar{L}$ and $\bar{L}'$ are the relative partial molal heat contents of the two solutions (see Chapter 6), then

$$\mathrm{d} \ln (a/a')/\mathrm{d}T = -(\bar{L} - \bar{L}')/RT^2 \tag{5.8}$$

If the transfer is from a standard state, i.e., $a' = 1$, $\bar{L}' = \bar{L}_0 = 0$, then

$$\mathrm{d} \ln a = -\bar{L}\mathrm{d}T/RT^2 \tag{5.9}$$

$\bar{L}$ can be expressed as a function of the temperature by $\mathrm{d}\bar{L}/\mathrm{d}T = \bar{C}_p - \bar{C}_p^0$, where $\bar{C}_p$, $\bar{C}_p^0$ are partial molal heat capacities (see Chapter 6), the latter referring to the standard state. Thus

$$\bar{L} = \bar{L}_0 + (\bar{C}_p - \bar{C}_p^0)T \tag{5.10}$$

For the transfer of a substance from the pure liquid to the pure solid phase, the former can be taken as the standard state. If the activity of the solid is a_s and its heat content is L_s, then for the liquid solvent, $a_1 = 1$, and $L_1 = 0$, whence from (5.9),

$$\mathrm{d} \ln a_s = -L_s\mathrm{d}T/RT^2 = -\Delta H_s\mathrm{d}T/RT^2 \tag{5.11}$$

in which ΔH_s will be the molal heat of solidification. Also, from (5.10),

$$L_s = \Delta H_0 - (C_{ps} - C_{pl})\theta = \Delta H_0 - \Delta C_p\theta \tag{5.12}$$

where ΔH_0 is the heat of solidification at T_0, the freezing-point of the pure solvent, and $\theta = T_0 - T$ where T is the freezing-point of the solution. Substituting (5.12) and θ in (5.11), since $\mathrm{d}\theta = -\mathrm{d}T$,

$$\mathrm{d} \ln a_s = [(\Delta H_0 - \Delta C_p\theta)\mathrm{d}\theta]/R(T_0 - \theta)^2 \tag{5.13}$$

This can be expanded as a series; Young [15] has shown that the expression can be written as

$$\mathrm{d}\ln a_s = (A_1 + B_1\theta + C_1\theta^2 + D_1\theta^3 + E_1\theta^4)\mathrm{d}\theta \tag{5.14}$$

where

$$A_1 = \Delta H_0(RT_0^2)^{-1},\ B_1 = 2\Delta H_0(RT_0^3)^{-1} - b_1(RT_0^2)^{-1},\ C_1 = 3\Delta H_0(RT_0^4)^{-1} - 2b_1(RT_0^3)^{-1} + b_2(2RT_0^2)^{-1} \tag{5.15}$$

The term ΔC_p varies with the temperature and may be expressed by

$$\Delta C_p = b_1 + b_2 t \quad (t = {}^\circ\mathrm{C}) \tag{5.16}$$

From the data collected by Guggenheim and Turgeon,[16] $\Delta H_0 = 1436.3$ cal/mole deg., $b_1 = -9.109$ c/m.d. while Young [15] gives $b_2 = 0.057$ c/m.d, The values of A_1, etc., are

$$A_1 = -9.685 \times {}^{-}10^3, \quad B_1 = -9.49 \times 10^{-6}, \quad C_1 = 1.5 \times 10^{-7}, \quad D_1 = 2.0 \times 10^{-9} \tag{5.17}$$

By means of the Gibbs–Duhem equation (4.42), it is possible to convert (5.13) into an expression for the activity of the solute, a_2, for

$$\mathrm{d}\ln a_2 = -(N_1\,\mathrm{d}\ln a_1)/N_2 = -(1{,}000\,\mathrm{d}\ln a_s)/M_1 m \tag{5.18}$$

whence from (5.14),

$$\mathrm{d}\ln a_2 = -1{,}000(A_1 + B_1\theta + \ldots)\mathrm{d}\theta/M_1 m \tag{5.19}$$

Now when $\theta \to 0$, $a_2 \to m_2$, and $B_1\theta$, etc., tend to disappear. Also, $\ln m_2 = \ln m = -1{,}000\,A_1\,\mathrm{d}\theta/M_1 m$, so from (5.15),

$$\mathrm{d}\theta/\mathrm{d}m = -\nu M_1 R T_0^2/1{,}000\,\Delta H_0 = \nu\lambda \tag{5.20}$$

where λ is termed the molal lowering of the freezing-point and has the value of [16] 1.860 ± 0.001 in water.

So far as dilute solutions are concerned, it is not easy to derive activity coefficients from freezing-point data. A useful procedure, which is given here, was advised by Guggenheim and Turgeon.[16] From (5.14), after integration and substitution from (5.17) and (5.20), since C_1 and D_1 are negligible at low concentrations,

$$\ln a_s = A_1\theta + B_1\theta^2/2 = -M_1\theta(1 + b\theta)/1{,}000\lambda \tag{5.21}$$

where $b = 1/T_0 - b_1/2\Delta H_0 = 0.00048$. A "practical" osmotic coefficient is defined by

$$\phi = -1{,}000\ln a_1/\nu m M_1 \tag{5.22}$$

Since $a_1 = a_s$, and $\mathrm{d}\ln a_2 = \nu\mathrm{d}\ln m + \nu\mathrm{d}\ln\gamma_\pm$, from (5.20, 5.22),

$$\mathrm{d}\ln m + \mathrm{d}\ln\gamma_\pm = -\mathrm{d}\phi, \text{ whence } \mathrm{d}[m(1-\phi)]/m = -\mathrm{d}\ln\gamma_\pm \tag{5.23}$$

Now $d \ln \gamma_\pm$ can be expressed in terms of the D.H. equation, namely

$$-\log \gamma_\pm = A'' \mid z_i z_j \mid \sqrt{I}/(1 + B''a\sqrt{I}) \quad (5.24)$$

by putting $x = B''a\sqrt{I}$ and $k'' = 2.303\ A'' \mid z_i z_j \mid /B''a$, whence

$$-d \ln \gamma_\pm = k''dx/(1 + x)^2 \quad (5.25)$$

Since m (i.e., $n'I$ where n' is an integer) is proportional to x^2 when m is small, substitution in (5.23) gives

$$1 - \phi = (k''/x^2) \int_0^x x^2 dx/(1 + x)^2 \quad (5.26)$$

$$= (k''x/x^3)\{(1 + x) - 2 \ln (1 + x) - 1/(1 + x)\} \quad (5.27)$$

$$= k''x\sigma/3 = 2.303\ A'' \mid z_i z_i \mid \sigma\sqrt{I}/3 \quad (5.28)$$

where

$$\sigma = \sum_1^\infty 3n(-x)^{n-1}/(n + 2) = 3(\tfrac{1}{3} - \tfrac{2}{4}x + \tfrac{3}{5}x^2 - \tfrac{4}{6}x^3 + \tfrac{5}{7}x^4 - \ldots) \quad (5.29)$$

Values of σ have been tabulated [17] for various values of x.

It is assumed that (5.28) defines a standard value, ϕ_{st}, and to allow for deviations from theory, ϕ is expressed as

$$1 - \phi = 2.303\ A'' \mid z_i z_j \mid \sigma\sqrt{I}/3 - \beta m \quad (5.30)$$

Also, from (5.21) and (5.22),

$$\theta(1 + b\theta) = \phi\nu m\lambda \quad (5.31)$$

whence

$$\theta(1 + b\theta) - \nu\lambda m\phi_{st} = \nu\lambda m(\phi - \phi_{st}) = \beta\nu\lambda m^2 \quad (5.32)$$

so a plot of the left-hand expression against m^2 should give a straight line of slope $\nu\lambda\beta$, from which the extension parameter to (5.24) can be derived, or alternately, the value of a in $B''a$ can be found.

Example 5.2. Brown and Prue [9] measured the freezing-points of some dilute KCl solutions. The figures and associated calculations are shown in Table 5.2. For these, $B''a$ is taken as 1.0, so $x = \sqrt{I} = \sqrt{m}$. A line of slope $0.21 = 2\lambda\beta$ can be drawn up to almost 0.1 M so the activity coefficients can be calculated by means of A'' from (3.36):

$$\log \gamma_\pm = -0.4879\sqrt{m}/(1 + \sqrt{m}) + 0.049\ m.$$

For the evaluation of activity coefficients in more concentrated solutions, a function introduced by Lewis and Randall [14] is helpful; this is

$$j = 1 - (\theta/\nu m\lambda) \quad (5.33)$$

Differentiating this,

$$d\theta/\nu m\lambda = (1 - j)d \ln m - dj \quad (5.34)$$

Also, by differentiating (5.19) and substituting for a_2,

$$-\,\mathrm{d}\ln\gamma_{\pm} = 1{,}000\ A_1\,\mathrm{d}\theta/M_1 m + 1{,}000(B_1\theta + C_1\theta^2 + D_1\theta^3)\mathrm{d}\theta/M_1 m + \mathrm{d}\ln m \qquad (5.35)$$

and from (5.17, 5.20, 5.34)

$$1{,}000\ A_1\mathrm{d}\theta/M_1 m\nu = 1{,}000\ \varDelta H_0\mathrm{d}\theta/RT_0^2 M_1\nu m = -\ \mathrm{d}\theta/\nu m\lambda \qquad (5.36)$$

whence, from (5.34), and (5.35),

$$-\,\mathrm{d}\ln\gamma_{\pm} = \mathrm{d}j + j\mathrm{d}\ln m + 1{,}000\ \lambda(1-j)\mathrm{d}\theta(B_1 + C_1\theta + D_1\theta^2)/M_1 \qquad (5.37)$$

so

$$-\ln\gamma_{\pm} = j + 2\int_0^m j\mathrm{d}\sqrt{m}/\sqrt{m} + (1{,}000\ \lambda/M_1)\int_0^\theta (1-j)(B_1 + C_1\theta + D_1\theta^2)\mathrm{d}\theta \qquad (5.38)$$

(since $\mathrm{d}\ln m = 2\ \mathrm{d}\sqrt{m}/\sqrt{m}$).

TABLE 5.2

Freezing-Point Data for Dilute KCl Solutions[9]

m	θ	σ	ϕ_{st}	$\theta(1 + b\theta) - 2\lambda m\phi_{\mathrm{st}}$
0.01555	0.0558	0.8372	0.9609	0.0002
0.03927	0.1381	0.7611	0.9435	0.0003
0.06737	0.2343	0.7065	0.9313	0.0010
0.08095	0.2806	0.6857	0.9269	0.0016
0.08814	0.3048	0.6726	0.9252	0.0015
0.1050	0.3615	0.6527	0.9208	0.0020

The first integral on the right-hand side is solved by plotting $j/\sqrt{m}$ against $\sqrt{m}$, and the second integral by plotting $1 - j$ against θ, θ^2 and θ^3. Up to a fairly high concentration the last two of these are unimportant. Young[15] has estimated that in 4 N NaCl they affect $\gamma_{\pm}$ by 0.28 % and 0.02 % respectively.

It is difficult to obtain a reliable plot for the first integral on the right-hand side below 0.1 M. Scatchard[7] has shown that the errors may be minimized by the following procedure. Ignoring the small $\mathrm{d}\theta$ term in (5.31), j' can be defined as $1 - \phi$, and this can be converted from a limiting into an extended form by the same process that is applied to the D.H. equation, i.e., instead of defining j' by (5.28), j'' is defined by

$$j'' = 2.303\ A''\sigma\ |z_i z_j|\ \sqrt{I}/3(1 + B''a\sqrt{I}) + Q_1 I \qquad (5.39)$$

whence, for 1 : 1 electrolytes, putting $\delta j = j - j''$,

$$2\int j\,\mathrm{d}\sqrt{m}/\sqrt{m} = \int j\,\mathrm{d}\ln m = Q_1 m + 4.606\,A''\int \sigma\,\mathrm{d}\sqrt{m}/3(1 + B''a\sqrt{m})$$
$$+ \int \delta j\,\mathrm{d}\ln m = Q_1 m + 4.606\,A''\sigma \ln(1 + B''a\sqrt{m})/3B''a + \int \delta j\,\mathrm{d}\,ln\,m \quad (5.40)$$

Values are assigned to $B''a$ and Q_1 to make the δj values as small as possible (generally about 0.02 maximum). Spencer [18] has given Tables of $j/\sqrt{m}$ using more complex formulae then (5.40). The difficult region of 0 to 0.1 M can also be dealt with by the method associated with Example 5.3 and suitably modifying (5.38) into

$$\log \gamma_{\pm} = \log \gamma_{0\cdot 1} - \frac{(j_m - j_{0\cdot 1})}{2.303} - \frac{2\int_{0\cdot 1}^{m} j\mathrm{d}\sqrt{m}/\sqrt{m}}{2.303}$$
$$- \frac{10^3}{M_1}\int_{\theta_m}^{\theta_{0\cdot 1}} (1 - j)(B_1 + C_1\theta + D_1\theta^2)\mathrm{d}\theta \quad (5.41)$$

Example 5.3. Scatchard *et al.* [7] measured the freezing-points of a number of alkali metal halides up to 1 M. Some of the figures for KCl are given in Table 5.3. The original calculations have been changed from those based on $\lambda = 1.858$ to $\lambda = 1.860$. Since the functions from which the integrals

TABLE 5.3

Freezing-Point Data for Concentrated KCl Solutions

m	θ	j	1st integral	2nd integral	$-\log\gamma_{\pm}$	$-\log\gamma_{\pm}$[S.P.]
0.1	0.3343	0.0744	—	—	0.1123	0.1121
0.2	0.6771	0.0900	0.0260	0.0001	0.1440	0.1430
0.3	1.0058	0.0988	0.0415	0.0002	0.1642	0.1632
0.4	1.3321	0.1048	0.0543	0.0003	0.1795	0.1782
0.5	1.6570	0.1091	0.0646	0.0004	0.1916	0.1903

are evaluated change rather slowly after 0.1 M, the integrals may be evaluated after 0.1 M by tabular methods; the mean value for two $j/\sqrt{m}$ consecutive concentrations is multiplied by the corresponding difference in $\sqrt{m}$ and the products added from 0.1 M to M. The second integral is likewise derived from mean $(1 - j)$ values and θ differences (and θ^2 and θ^3 differences if these are significant). From the equation derived in Example 5.2, $-\log\gamma_{\pm}$ at 0.1 M is 0.1123. For comparison, the figures of Scatchard *et al.* [7] are included.

CONVERSION TO 25°C

Activity coefficients obtained by means of (5.38) and (5.41) refer to the actual freezing-points of the solutions. In order to obtain them at a fixed temperature, a further term is needed. If a_1' is the activity of the solvent at $T_0 - \theta = T'$, and a_1'' is its value in the same solution at a fixed temperature T'', then if $\ln (a_1''/a_1') = x_1$, from (5.9),

$$R \ln (a_1''/a_1') = Rx_1 = - \int_{T'}^{T''} \bar{L}_1 \mathrm{d}T/T^2 \qquad (5.42)$$

where $\bar{L}_1$ is the relative partial molal heat content of the solvent. Also (see Chapter 6), $\bar{L}_1 = \bar{L}_{1(T'')} + \bar{J}_1(T - T'')$, and $\bar{J}_1 = \bar{C}_{pl} - \bar{C}_{pl}^0$ is the relative partial molal heat capacity of the solvent. By substitution and integration,

$$\log (a_1''/a_1') = x_1/2.303 = - y(\bar{L}_1)_{T''} + z\bar{J}_1 \qquad (5.43)$$

where $y = (T'' - T')/2.303\, RT'T''$ and $z = T''y - R^{-1} \log (T''/T')$. Tables of y and z at various values of T' when $T'' = 25°\mathrm{C}$ are available,[14] so if $\bar{L}_1$ and $\bar{J}_1$ data are also available, x_1 can be found. From (5.18) and (5.42),

$$\mathrm{d} \ln a_2'' = - 1{,}000 \,\mathrm{d} \ln a_1''/M_1 m = - 1{,}000(\mathrm{d} \ln a_1' - \mathrm{d}x_1)/M_1 m \qquad (5.44)$$

so by analogy with (5.38),

$$- \ln \gamma_{\pm T''} = - \ln \gamma_{\pm T'} + (1{,}000/M_1 \nu) \int_0^{x_1} \mathrm{d}x_1/m \qquad (5.45)$$

An alternative procedure is to use

$$R \ln (a_2''/a_1') = Rx_2 = - \int_{T'}^{T''} \bar{L}_2 \mathrm{d}T/T^2 \qquad (5.46)$$

where a_2 and $\bar{L}_2$ refer to the electrolyte. Since $\bar{L}_2 = \bar{L}_{2(T'')} + (T - T'')\bar{J}_2$, and $\bar{J}_2 = \bar{C}_{p_2} - \bar{C}_{p_2}^0 = S_c\sqrt{m}$ (see Chapter 6), then

$$- \ln \gamma_{\pm T''} = - \ln \gamma_{\pm T'} + y\bar{L}_{2(T'')}/\nu - z\bar{J}_2/\nu \qquad (5.47)$$

Values of $\bar{L}_2$ and $\bar{J}_2$ are available for many electrolytes.[19] These, as explained in Chapter 6 have been obtained from calorimetric and e.m.f. measurements at various temperatures.

Example 5.4. The activity coefficients for KCl in Table 5.3 are converted to values at 25°C, using available values of $\bar{L}_2$, and $\bar{J}_2$ by using $S_c = 16.8$. The figures in the 7th and 8th columns of Table 5.4 (see page 80) are those obtained from vapour pressures (as described in the next section), and from e.m.f. measurements,[20] both at 25°C.

Vapour Pressure Measurements. The Isopiestic Method.

One of the most useful ways of deriving activity coefficients from about 0.1 M up to very high concentrations depends on equilibrating the vapour pressure of an electrolyte solution with that of known references. In its present form, due to the refinements introduced by Robinson and Sinclair,[21] the "isopiestic" method can provide data as precise as other methods. The apparatus consists of four gold-plated silver or copper [59] dishes (platinum or stainless steel are used for corrosive substances) fitted with flap-lids which are kept open by a wire during the equilibration. About 2 ml. of solution of known concentration is weighed into each dish, two of the same unknown,

TABLE 5.4

Activity Coefficients of KCl at 25°C from Freezing-Point Data

m	$\bar{L}_2$	$\bar{J}_2$	$0.5\bar{L}_2y$	$0.5\bar{J}_2z$	$-\log\gamma_\pm$	$-\log\gamma_\pm$ (V.P.)	$-\log\gamma_\pm$ (E.M.F.)
0.1	91	5.3	0.0031	0.0023	0.1131	0.1136	0.1141
0.2	72	7.5	0.0025	0.0034	0.1440	0.1438	0.1433
0.3	44	9.2	0.0016	0.0043	0.1615	0.1627	0.1624
0.4	12	10.6	0.0004	0.0051	0.1748	0.1767	—
0.5	− 21	11.9	− 0.0008	0.0060	0.1848	0.1876	0.1864

and two of the same reference. The dishes are placed on a thick copper block in a desiccator. Both the upper surface of the block and the bases of the dishes should be as flat as possible to ensure good thermal contact. The desiccator is evacuated, placed in a thermostat bath and gently rocked. Distillation among the solutions proceeds until they have equal vapour pressures. Above about 1 M, 24 h are sufficient, but below this concentration, several days may be required and 0.1 M is about the lowest practical limit. By using a more elaborate procedure, Gordon [22] has obtained satisfactory answers down to 0.03 M although this took about 18 days. He found the most serious difficulty to be the removal of air. In his opinion, the extreme slowness in attaining equilibrium with dilute solutions is due to the rate of transport of solvent in the vapour phase and not to the rate of heat transfer between solutions. Davies and Thomas [23] have devised a volumetric form of equilibrium based on the design of Lassettre and Dickenson.[24] It consists of a wide inverted U-tube, the arms of which hold the test and reference solutions. Near to the bottoms of these arms, and at right angles to them, narrow tubes are joined; the solutions can be drained into these and their volume changes measured. The whole apparatus is rocked in a precision thermostat bath.

At equilibrium,

$$(a_1)_m = (a_1)_R \tag{5.48}$$

where $(a_1)_m$ is the activity of the solvent of the test solution of molality m, and $(a_1)_R$ is that of the reference solution of molality m_R. From the definition of ϕ given in (5.22), it follows that

$$m\nu\phi = m_R\nu_R\phi_R \tag{5.49}$$

(this is not exact [25] since ϕ depends on the pressure, but the error is negligible). The references in common usage are NaCl, KCl, $CaCl_2$, sucrose and H_2SO_4 solutions. The ϕ_R values of these have been obtained by several ways:

(a) Direct Vapour Pressures. Those of KCl at 20°C were measured by a static method.[26] The data give p and p_0, the vapour pressures of solution and solvent, whence $(a_1)_R = p/p_0$. The answers were converted to the 25°C basis by (5.42), and (5.22) used to derive ϕ_R. By plotting ϕ_R against $\sqrt{m}$, the isopiestic values can be read off.

Shankman and Gordon [27] measured the vapour pressures of H_2SO_4 solutions by means of a manometer containing a special pump oil. The removal of residual air from the solutions was effected by boiling them under vacuum followed by freezing; this was repeated several times. The range covered was 1.9 M to 22.6 M and up to 3 M the derived activities agreed with those calculated from the e.m.f. measurements of Hamer [32] (described below) but disagreed from 3 M to 7 M.

Stokes [28] has obtained the a_1 values of NaOH solutions from 5 M to 13.8 M by a bithermal vapour pressure method in which solutions at 25°C were equilibrated with water at a lower temperature. Thence, from the isopiestic comparisons of NaOH and H_2SO_4 obtained by Olynyk and Gordon,[29] he obtained the a_1 values of H_2SO_4 from 2 M to 27 M. These are in excellent agreement with those of Shankman and Gordon [27] up to 7 M, and within 0.25 % up to 11 M. Hornung and Giauque [30] made fresh studies of 14, 19 and 27 M H_2SO_4. They are within 0.2 % of the first set.[27]

(b) E.M.F. Measurements. Harned and Hamer [31] obtained potentials of the cell

(A) $\quad$ $Pt \mid H_2 \mid H_2SO_4, m \mid Hg_2SO_4 \mid Hg$ $\quad$ 0.05 M to 17.5 M

where the reaction is

$$H_2 + HgSO_4 \rightleftharpoons 2Hg + H_2SO_4$$

while Hamer,[32] and Wynne-Jones and his colleagues [33] have studied the cell

(B) $\quad$ $Pt \mid H_2 \mid H_2SO_4, m \mid PbSO_4 \mid PbO_2 \mid Pt$ $\quad$ 0.0005 M to 7 M

in which the reaction

$$H_2 + PbO_2 + H_2SO_4 \rightleftharpoons PbSO_4 + 2H_2O$$

occurs.

Using the principles described in Chapter 4,

$$E_A = E^0_{Hg_2SO_4} - 0.5\,k' \log m_H^2 m_{SO_4} - 1.5\,k' \log \gamma_\pm \tag{5.50}$$

where $E^0_{Hg_2SO_4}$ refers to the process $Hg_2SO_4 + 2e^- \rightleftharpoons 2Hg + SO_4^{2-}$, and

$$E_B = E^0_{PbO_2} - 0.5\,k' \log m_H^2 m_{SO_4} - 1.5\,k' \log \gamma_\pm - k' \log a_w \quad (5.51)$$

where $E^0_{PbO_2}$ refers to the process $PbO_2 + H_2SO_4 + 2e^- \rightleftharpoons PbSO_4 + 2OH^-$ and a_w is written for $a_H a_{OH}$. Since $K_{HSO_4^-}$ is of the order of 0.01 (see Chapters 8, 9 and 10), allowance should be made for incomplete dissociation of the acid in the formulae for deriving E^0. One way is to estimate $x = m_{HSO_4}$ of $K/\gamma^3_\pm = 2m^2 - 3mx + x^2$ by taking a generalized form of the D.H. equation such as (3.43), with $I = 3m - 2x$. For a first approximation, x^2 is ignored and I taken as $3m$. Approximations are continued until x is constant. The method of extrapolating for E^0 is described in Chapter 4. These calculations can only be applied if $I \not> 0.1$, the useful limit of (3.43). This is true of Hamer's measurements for cell B, and at 25°C leads to $E^0_{PbO_2} = 1685.9$ mV. Hamer has give two answers (a) 1684.88 mV by ignoring x, and (b) 1685.97 mV by taking x into account. Wynne-Jones [33] used another way, namely by inserting published values [34] of $\gamma_\pm$ and a_w into (5.51); his answer is 1687.05 mV.

At the same molality, combination of cells A and B gives

$$E_B + E_A = E^0_{PbO_2} + E^0_{Hg_2SO_4} - k' \log a_w \quad (5.52)$$

Cell A cannot be studied much below 0.05 M owing to hydrolysis, but since the last term of (5.52) is small at low concentrations, $E^0_{Hg_2SO_4}$ can be derived by extrapolating the plot of $E_B + E_A - E^0_{PbO_2}$ against m, and hence a_w calculated.[31]

A more general procedure for obtaining a_w can be applied whenever activity coefficients have been computed from the measurements. This is to integrate a suitable form of the Gibbs–Duhem equation (5.18). For aqueous solutions, the expression is

$$-\,55.51 \times 2.303 \log a_w = m + \int_0^m m \,\mathrm{d} \log \gamma_\pm \quad (5.53)$$

Reference data have been obtained in this way for NaCl, KCl and NaBr, the activity coefficients having been obtained [34] by means of cell (4.48).

Example 5.5. The following figures refer to KCl solutions at 25°C:

m	0.005	0.01	0.05	0.10	0.20	0.50
$\gamma_\pm$	0.9275	0.9021	0.8172	0.7701	0.7191	0.6516

From a plot of $\log \gamma_\pm$ against m, graphical integration gives 0.0216 for 0.5 M KCl, whence $-\log a_w = 0.007044$, so from (5.22), $\phi = 0.9005$.

(c) Freezing-Points. The activity coefficients derived from freezing-points are converted to the 25°C scale by means of (5.42), and ϕ data calculated as described in Example 5.5.

Robinson has collected together all the available calculations on the ϕ_R

values of reference solutions, and has also made isopiestic intercomparisons. The figures have been tabulated.[35, 36]

In order to establish activity coefficients of "unknowns" from a set of isopiestic comparisons, the first step is to plot R against m where from (5.49),

$$R = \nu_R m_R / \nu m = \phi / \phi_R \tag{5.54}$$

The object of this is to obtain a means of reading off values of R at round molalities of the test solution, and hence get ϕ. Activity coefficients are then derived from an integrated form of (5.23), namely,

$$-\ln \gamma_{\pm} = (1 - \phi) + \int_0^m (1 - \phi) \, \mathrm{d} \ln m \tag{5.55}$$

Since measurements are not usually made below 0.1 M, it is necessary to extrapolate the integration curve that is constructed to solve the integral. A plot of $1 - \phi$ against $\ln m$ is unsatisfactory since it tends to infinity as m tends to zero. A better way is to use the fact that $\mathrm{d} \ln m = 2\mathrm{d}\sqrt{m}/\sqrt{m}$, and to plot $2(1 - \phi)/\sqrt{m}$ against $\sqrt{m}$, for this has a limiting value at $\sqrt{m} = 0$, which can be calculated. In very dilute solutions, $1 - \phi$ can be taken as constant over a very small range, so from (5.55), the limit of $\log \gamma_{\pm}$ as $m \to 0 = -3(1 - \phi)/2.303$. Also, from the limiting form of the D.H. equation, at 25°C, $-\log \gamma_{\pm} = 0.5083 \sqrt{m}$ for 1 : 1 electrolytes in water, so the limiting value of $(1 - \phi)/\sqrt{m} = 0.5083 \times 2.303/3 = 0.3903$.

Since the activity coefficients of the reference solutions are known, an alternative procedure is available. From the Gibbs–Duhem equation (5.18), and from the isopiestic condition (5.49),

$$\nu_R m_R \, \mathrm{d} \ln m_R \gamma_R = \nu m \, \mathrm{d} \ln m\gamma \tag{5.56}$$

and from (5.54), on rearrangement,

$$\mathrm{d} \ln \gamma = \mathrm{d} \ln (m_R/m) + \mathrm{d} \ln \gamma_R + (R - 1) \, \mathrm{d} \ln m_R \gamma_R \tag{5.57}$$

Now as $m \to 0$, m_R also $\to 0$, while ϕ and $\phi_R \to 1$, so from (5.54), the integrated form of $\mathrm{d} \ln (m_R/m)$ between 0 and m is $\ln R$, whence

$$\ln \gamma = \ln R + \ln \gamma_R + \int_0^{m_R} (R - 1) \, \mathrm{d} \ln a_R \tag{5.58}$$

The last term is best obtained by graphical integration of $2(R - 1)/\sqrt{a_R}$ against $\sqrt{a_R}$.

If the activity coefficient of the test electrolyte at 0.1 M is known by other means, the difficulty of extrapolating from the lowest practical limit of 0.1 M to 0 can be overcome by dividing the integral into two parts, from 0 up to the molality of the reference which is isopiestic with the 0.1 M test solution, then by substitution in (5.58), the "residual" integral is obtained by difference.

Example 5.6. Wishaw and Stokes [37] found the following solutions of NaCl ($= m_R$) and NH_4NO_3 to be isopiestic:

m_R	0.1106	0.2322	0.4352	0.6170	0.7881	1.001
m	0.1135	0.2426	0.4670	0.6772	0.8833	1.148

After plotting $R = m_R/m$ against m, the values of R at the given round concentrations of m in Table 5.5 were read off. By taking known values [36] of ϕ_R at the values of m_R corresponding to these round concentrations of m, $\phi = R\phi_R$ is found. A plot of $(1 - \phi)/\sqrt{m}$ against $\sqrt{m}$ gave a smooth curve, indicating that the integral of (5.55) could be obtained with sufficient accuracy by the tabular method described in Example 5.3, taking the limit of 0.3903 at $\sqrt{m} = 0$. The activity coefficients thus derived from (5.55) are shown in Table 5.5.

The calculation of γ by means of (5.58) is also demonstrated. The first integral was evaluated by plotting $(R - 1)/\sqrt{a_R}$ against $\sqrt{a_R}$, drawing a smooth curve to 0,0 and obtaining the area under the curve up to a_R corresponding to 0.1 M NH_4NO_3. The remaining integrals were obtained by the tabular method.

TABLE 5.5

Activity Coefficients of NH_4NO_3 at 25°C by the Isopiestic Method

m	0.10	0.20	0.30	0.40	0.50	0.60	0.70	0.80	0.90	1.00
$10R$	9.78	9.63	9.50	9.39	9.28	9.19	9.09	8.99	8.90	8.82
$10\phi_R$	9.32	9.25	9.22	9.20	9.21	9.23	9.26	9.29	9.32	9.36
10ϕ	9.11	8.91	8.76	8.64	8.55	8.48	8.42	8.35	8.29	8.26
$10a_R$	7.78	7.35	7.10	6.93	6.81	6.73	6.67	6.62	6.59	6.57
$10(R - 1)/\sqrt{a_R}$	0.79	0.96	1.08	1.16	1.23	1.28	1.33	1.39	1.43	1.46
$100\int(5.58)$	1.27	2.18	2.98	3.71	4.39	5.04	5.66	6.27	6.87	7.45
$10\gamma(5.55)$	7.40	6.77	6.36	6.05	5.81	5.62	5.45	5.30	5.16	5.06
$10\gamma(5.58)$	7.42	6.78	6.35	6.04	5.79	5.59	5.41	5.25	5.11	4.99

The osmotic coefficients and activity coefficients of a large number of electrolytes, particularly those of the 1 : 1, 1 : 2 and 2 : 1 types have been determined up to very high concentrations by the isopiestic method. Robinson and Stokes have published collections of these data,[36, 38] and have propounded a treatment to account for the very large and characteristic activity coefficients of electrolytes at high concentrations in terms of ion-solvation; this is outlined below (see also Chapter 14).

Activity Coefficients in Concentrated Solutions

As the concentration is increased, the activity coefficients of many strong electrolytes fall to a minimum and then rise to values which may be considerably greater than unity, the limiting value at zero concentration. Some of the values in concentrated solutions are astonishingly high, e.g.,[39] for 6 M solutions, $HCl = 3.22$, $HClO_4 = 4.76$, $LiBr = 3.92$, $KOH = 2.2$, and for 5 M solutions, $MgCl_2 = 14.0$, $CaBr_2 = 18.5$, $MgI_2 = 115$. The main cause for these high values is attributed to ion hydration, and semi-theoretical treatments with this in mind have been given by Bjerrum,[40] Harned,[41] Scatchard[42] and Frank,[43] but the first thorough examination of this idea is due to Stokes and Robinson[44] with the aid of the extensive data that they have amassed,[39, 45] particularly by the isopiestic method. It is assumed that the total interaction between ions and solvent (where this is $> kT$) can be allowed for as a binding of h molecules of water in ion hydration shells. For a solution of molality m, there are hm molecules of "bound" water to every $(55.51 - hm)$ molecules of "free" water, so the "true" molality m' is given by

$$m' = 55.51\, m/(55.51 - hm) = m/(1 - 0.018\, hm) \qquad (5.59)$$

Applying the Gibbs–Duhem equation, (4.42), and letting γ' and γ represent the activity coefficients for molalities m' and m,

$$\mathrm{d} \ln \gamma' = -(55.51/\nu m')\mathrm{d} \ln a_w - \mathrm{d} \ln m' \qquad (5.60)$$

$$\mathrm{d} \ln \gamma = -(55.51/\nu m)\mathrm{d} \ln a_w - \mathrm{d} \ln m \qquad (5.60)$$

so from (5.59),

$$\mathrm{d} \ln \gamma' = \mathrm{d} \ln \gamma + (h/\nu)\mathrm{d} \ln a_w + \mathrm{d} \ln (1 - 0.018\, hm) \qquad (5.61)$$

whence, on integrating from 0 to m and applying (3.48) to the primed quantities,

$$\ln \gamma = \ln f' - (h/\nu) \ln a_w - \ln [1 - 0.018(h - \nu)m] \qquad (5.62)$$

where f' is the mean rational activity coefficient, so from (3.31),

$$\log \gamma = -(A \mid z_i z_j \mid \sqrt{\Gamma})/(1 + Ba\sqrt{\Gamma}) - (h/\nu) \log a_w - \log [1 - 0.018\,(h - \nu)m] \qquad (5.63)$$

The middle right-hand term can be calculated by means of experimental ϕ values and (5.22), but a and h are adjustable parameters. By suitable selection of these, the experimental γ values of strong electrolytes can be reproduced[44] within 0.001 up to molalities ranging from 0.7 M (e.g., HI, CaI_2, $Zn(ClO_4)_2$) up to 4 to 5 M (e.g., $NaClO_4$, NaX, KX). Some h and a values are shown in Table 5.6.

Stokes and Robinson[44] also derived a single parameter equation in which a of (5.63) is replaced by a', where

$$a' = [3(30h + V_+)/4\pi]^{\frac{1}{3}} + r_- - \Delta \qquad (5.64)$$

In this 30 (in Å^3) is the normal volume of a water molecule calculated from the density of water at 25°C, V^+ is the apparent ionic volume of the cation in

Å^3 and is calculated from $V_+ = V_{app} - 6.47\, z_1 r^3$ where V_{app} is the apparent molal volume at 1 M, z_1 is the cation valence, r_- is the anion crystallographic radius, and 6.47 comes from the fact that $V_{app} = 6.47(r_+^3 + r_-^3)$Å^3 per molecule for CsX (X = Cl, Br, I) and RbX. The correction term Δ is to equate a' to a of the two parameter equation, and is 0.7 Å and 1.3 Å for 1 : 1 and 2 : 1 electrolytes respectively. Some values of h and a' are shown in Table 5.6.

Glueckauf [46] has pointed out that Stokes and Robinson did not take into account the effect of ion sizes of the "co-volume" effect (short-range interionic repulsion forces), and has shown that both this and the hydration effect can be interpreted by volume fraction statistics.[47] The Gibbs free energy of a solution of one mole of hydrated electrolyte, using for the entropy contribution [48]

$$G^s = RT\sum_i n_i \ln (n_i \bar{v}_i / V) \tag{5.65}$$

where

$$V = n_i(\nu_1^+ \bar{v}_1 + \nu_1^- \bar{v}_1) + n_2 \bar{v}_w = n_1 \phi_v + n_w v_w^0 \tag{5.67}$$

gives, for an electrolyte, 1, and a solvent, 2,

$$G = n_1\mu_1^0 + n_2\mu_2^0 + G^{el} + RT\{n_1\nu_1^+ \ln (n_1\nu_1^+\bar{v}_1^+/V) + n_1\nu_1^- \ln (n_1\nu_1^-\bar{v}_1^-/V) + n_2 \ln (n_2\bar{v}_w/V)\} \tag{5.68}$$

where μ_1^0 and μ_2^0 are the standard state chemical potentials of hydrated electrolyte and of "free" water, n_1 and n_w are the numbers of electrolyte and of water molecules, $\bar{v}$ represents partial molar volume (see Chapter 6), h is the hydration number of the electrolyte, n_2 is the number of "free" water molecules and is equal to $n_w - n_1 h$, ϕ_v is the apparent molar volume of the electrolyte (see Chapter 6), and G^{el} is the electrostatic contribution to the Gibbs function,[49]

$$G^{el} = \sum N_i z_i^2 e^2 \kappa \tau(\kappa a)/3\epsilon \tag{5.69}$$

where

$$\tau(x) = 3 \log (1 + x) - x + \tfrac{1}{2}x^2 = 1 - \tfrac{3}{4}x + \tfrac{3}{5}x^2 - \tfrac{3}{6}x^3 + \tfrac{3}{7}x^4 - \ldots \tag{5.70}$$

Regarding the system as unhydrated electrolyte + total water, the activity coefficient of the "unhydrated" electrolyte can be derived via

$$\mu_1 = (\partial G/\partial n_1)n_w \tag{5.71}$$

where μ_1 is the chemical potential. So from (5.68),

$$\mu_1 = \mu_1^0 - h\mu_2^0 + (\partial G^{el}/\partial n_1)n_w + RT\{n_w(\nu - r - h)/(n_w + n_1 r) + \nu \ln n_1(r + h)/(n_w + n_1 r) + h \ln (n_w + n_1 r)/(n_w - n_1 h)\} \tag{5.72}$$

where $r = \phi_v/v_w^0$. Although the exact value of the third right-hand term follows from (5.69), the electrostatic contribution for log $\gamma_{\pm}$ is given very closely by (3.31) so

$$\log \gamma_{\pm} = -(A \mid z_i z_j \mid \sqrt{\Gamma})/(1 + Ba\sqrt{\Gamma}) + 0.018\, mr(r + h - \nu)/ 2.3\nu(1 + 0.018\, mr) + (h/\nu - 1) \log (1 + 0.018\, mr) - (h/\nu) \log (1 - 0.018\, mh) \quad (5.73)$$

(m = molality). Two other expressions, treating the ionic molar volumes as variable, and on the basis of mole-fractional statistics, are also derived. Some values of h are given in Table 5.6; see also Chapter 14.

If an exact differentiation of (5.69) is made,[57] the right-hand side of (5.73) contains a further term which approximates to $+ 0.001\, z_i z_j (h + r)c$, and as a consequence, the treatment of Stokes and Robinson is equivalent to the neglect only of r in this additional term. Stokes and Robinson [58] have themselves also made a detailed analysis of the consequences of differentiating (5.68) and taking the full Debye–Hückel equation.

Miller,[50] unaware of Glueckauf's treatment,[46] has used free volume fractions to derive an expression very similar to (5.73), and he also derived a simple version of (5.63), namely

$$\log \gamma = - A'' \mid z_i z_j \mid \sqrt{I}/(1 + B''a\sqrt{I}) - \log [1 - 0.018(h - \nu)m] \quad (5.74)$$

where I is the molal ionic strength, a and h taking care of differences between C and m. Rearranging into

$$- \mid z_i z_j \mid A''\sqrt{I}/\log \gamma[1 - 0.018(h - \nu)m] = 1 + B''a\sqrt{I} \quad (5.75)$$

and choosing suitable preliminary values of h and a, the left side is plotted against $\sqrt{I}$, and repeated with different values of h till the plot is linear; the slope gives a. Because of the simplifications, the resultant figures for a and h are higher than those given by (5.63) and (5.73).

TABLE 5.6

Values of Ion Size and Hydration Numbers

	HCl	HI	LiCl	NaCl	KCl	$MgCl_2$	$MgBr_2$	$BaCl_2$
(5.63) h	8.0	10.6	7.1	3.5	1.9	13.7	17.0	7.7
(5.63) a	4.47	5.69	4.32	3.97	3.63	5.02	5.46	4.45
(5.64) h	7.3	10.6	6.5	3.5	1.9	13.9	17.0	8.4
(5.73) h	4.7	4.9	4.4	2.5	1.7	6.5	7.1	4.2
(5.74) h	13.5	17.0	12.0	6.7	4.2	24.0	29.0	14.0
(5.74) a	5.03	7.71	4.91	3.81	3.29	5.15	5.85	4.45

The mean activity coefficients of fully ionized electrolytes in mixed solutions, particularly at medium and high concentrations, depend to some extent upon the relative proportions of the component electrolytes. Investigations of these features have been made with cells such as

$$\mathrm{Pt} \mid \mathrm{H_2} \mid \mathrm{HX(m_1)}, \mathrm{MX}_n(\mathrm{m_2}) \mid \mathrm{AgX} \mid \mathrm{Ag} \tag{5.76}$$

by solubility determinations and by the isopiestic vapour pressure method.

Harned[51] summarized a number of observations for mixtures of 1 : 1 valent mixtures at $m = m_1 + m_2 =$ constant by what are known as "Harned's rule," i.e.,

$$\log \gamma_1 = \alpha_{12} m_1 + \log \gamma_{(0)1} = \log \gamma_{1(0)} - \alpha_{12} m_2 \tag{5.77}$$

$$\log \gamma_2 = \alpha_{21} m_2 + \log \gamma_{(0)2} = \log \gamma_{2(0)} - \alpha_{21} m_1 \tag{5.78}$$

where the subscripts 1 and 2 refer to HCl and MCl respectively, $\gamma_{(0)1}$ and $\gamma_{(0)2}$ refer to the mean activity coefficients of HCl and MCl when $m_1 = 0$, $m_2 = m$ and $m_2 = 0$, $m_1 = m$ respectively while $\gamma_{1(0)}$ and $\gamma_{2(0)}$ refer to HCl and MCl at $m_1 = m$, $m_2 = 0$ and $m_1 = 0$, $m_2 = m$ respectively. The values of α_{12} at different values of m can be obtained from cell (5.76) with data at $m_1 = m$, $m_2 = 0$ (giving $\gamma_{1(0)}$) and e.g., $m_1 = 0.01$, $m_2 = m - 0.01$, while the values of $\alpha_{2,1}$ can be obtained on the basis of the following analysis.

From the Gibbs–Duhem equation (e.g., 4.43) in terms of a three component aqueous system,

$$m_1 \, \mathrm{d} \ln a_1 + m_2 \, \mathrm{d} \ln a_2 + 55.51 \, \mathrm{d} \ln a_w = 0 \tag{5.79}$$

so

$$2m_1 \, \mathrm{d} \log \gamma_1 + 2m_2 \, \mathrm{d} \log \gamma_2 = -\, 55.51 \, \mathrm{d} \log a_w \tag{5.80}$$

(since $a_n = m_n^2 \gamma_n^2$). Also, from (5.77) and (5.78), $\mathrm{d} \log \gamma_1 = \alpha_{12} \mathrm{d} m_1$, $\mathrm{d} \log \gamma_2 = \alpha_{21} \mathrm{d} m_2$ so (5.80) gives

$$2m_1 \alpha_{12} \mathrm{d} m_1 + 2m_2 \alpha_{21} \mathrm{d} m_2 = -\, 55.51 \, \mathrm{d} \log a_w \tag{5.81}$$

With $m_1 = mx$, $m_2 = m(1 - x)$, and integrating,

$$x^2(\alpha_{12} + \alpha_{21}) - 2\alpha_{21} x = -\, (55.51/m^2) \int_0^x \mathrm{d} \log a_w \tag{5.82}$$

whence, at $x = 1$,

$$\alpha_{21} = \alpha_{12} + (55.51/m^2) \log (a_{w(1)}/a_{w(2)}) \tag{5.83}$$

where $a_{w(1)}$ and $a_{w(2)}$ refer to $m_1 = m$, $m_2 = 0$ and $m_1 = 0$, $m_2 = m$. Also, from (5.80) and (5.83),

$$\alpha_{21} = \alpha_{12} + (2/m^2) \int_1^{\gamma_{2(0)}} m \, \mathrm{d} \log \gamma_2 - (2/m^2) \int_1^{\gamma_{1(0)}} m \, \mathrm{d} \log \gamma_1 \tag{5.84}$$

Alternatively, from (5.22),

$$-\, 55.51 \log a_w = 2m\phi/2.303 \tag{5.85}$$

so

$$\alpha_{21} = \alpha_{12} + (2/2.303m)(\phi_{2(0)} - \phi_{1(0)}) \tag{5.86}$$

Akerlöf and Thomas,[52] from a detailed examination of solubility data, found that

$$\log \gamma_{1(0)} - \log \gamma_{2(0)} = Bm \tag{5.87}$$

where B is a constant, holds for a number of systems, so from (5.84),

$$\alpha_{12} - \alpha_{21} = B \tag{5.88}$$

whence, from (5.77) and (5.78),

$$\log \gamma_{0(1)} = \log \gamma_{0(2)} \tag{5.89}$$

However, B often decreases as m increases.

McKay [53] has shown that the assumptions about α_{21} can be avoided (if α_{12} is independent of m_1/m_2) for 1 : 1 electrolytes by using

$$(\partial \log \gamma_1/\partial m_2)_{m_1} = (\partial \log \gamma_2/\partial m_1)_{m_2} \tag{5.90}$$

From (5.77), noting that the left side of (5.90) is equal to $(\partial \log \gamma_1/\partial m)_{m_1}$, m and m_1 can be taken as independent variables, whence

$$(\partial \log \gamma_1/\partial m)_{m_1} = (\partial \log \gamma_{(0)1}/\partial m)_{m_1} - m_2(\partial \alpha_{12})_{m_1} - \alpha_{12} \tag{5.91}$$

Also, $\gamma_{(0)1}$ and α_{12} are functions of m only so the constancy of m_1 is a redundant condition on the right-hand side. Thus

$$(\partial \log \gamma_1/\partial m)_{m_1} = \mathrm{d} \log (\gamma_{(0)1}/\mathrm{d}m - m_2\mathrm{d}\alpha_{12}/\mathrm{d}m - \alpha_{12} \tag{5.92}$$

Equation (5.88) leads to

$$(\partial \log \gamma_2/\partial m)_{m_2} = \mathrm{d} \log \gamma_{(0)2}/\mathrm{d}m - [\partial(\alpha_{21}m_1)/\partial m]_{m_2} \tag{5.93}$$

so combining (5.92) and (5.93),

$$[\partial \alpha_{21}m_1/\partial m]_{m_2} = \mathrm{d} \log (\gamma_{(0)2}/\gamma_{(0)1})/\mathrm{d}m + m_2\mathrm{d}\alpha_{12}/\mathrm{d}m + \alpha_{12} \tag{5.94}$$

Integrating at constant m_2 with limits $m = m_2$, $m = m$, $\alpha_{21}m_1$ vanishes at the lower limit since $m_1 = 0$ and α_{21} can be found from

$$\alpha_{21}m_1 = \Big[\log (\gamma_{(0)2}/\gamma_{(0)1})\Big]_{m=m_2}^{m=m} + m_2\Big[\alpha_{12}\Big]_{m=m_2}^{m=m} + \int_{m_2}^{m} \alpha_{12}\mathrm{d}m \tag{5.95}$$

When $m_2 \to 0$, $\alpha_{21} \to \alpha_{21}^t$ (t = trace) as given by

$$\alpha_{21}^t = \log (\gamma_{(0)2}/\gamma_{(0)1}) + \int_0^m \alpha_{12}\mathrm{d}m \tag{5.96}$$

Also, from (5.94), expansion of the left-hand side to $\alpha_{21} + m_1(\partial \alpha_{21}/\partial m)_{m_2}$ and letting $m_1 \to 0$, the result is

$$\alpha_{21}^0 = \mathrm{d} \log (\gamma_{(0)1}/\gamma_{(0)2})/\mathrm{d}m + m\mathrm{d}\alpha_{12}/\mathrm{d}m + \alpha_{12} \tag{5.97}$$

Expressions (5.95–7) are equivalent to the earlier ones (5.83, 4, 6) provided $\log (\gamma_{(0)1}/\gamma_{(0)2})$ is a linear function of m and α_{12} is a constant; this occurs with HCl, NaCl.

In many instances $\log \gamma_1$ and $\log \gamma_2$ do not vary linearly with m_2 or m_1. Harned and Cook [54] extended (5.77) and (5.78) to accommodate this feature by means of

$$\log \gamma_1 = \log \gamma_{1(0)} - \alpha_{12} m_2 - \beta_{12} m_2^2 \tag{5.98}$$

$$\log \gamma_2 = \log \gamma_{2(0)} - \alpha_{21} m - \beta_{21} m_1^2 \tag{5.99}$$

More extensive discussions on mixed electrolyte activity coefficients are to be found in the works of Harned and Owen,[17] Robinson and Stokes,[39] and in articles such as those of Glueckauf [55] and McKay.[56]

REFERENCES

1. Bronsted, J. N. and La Mer, V. K. *J. Amer. chem. Soc.* **46,** 555 (1924). La Mer, V. K., King, V. K. and Mason, C. F. *ibid.* **49,** 363 (1927). Baxter, W. P. *ibid.* **48,** 615 (1926). La Mer, V. K. and Goldman, F. H. *ibid.* **51,** 2632 (1929).
2. Zimmermann, H. K. *Chem. Rev.* **51,** 25 (1952).
3. MacDougall, G. and Davies, C. W. *J. chem. Soc.* 1416 (1935).
4. Money, R. W. and Davies, C. W. *J. chem. Soc.* 400 (1934).
5. Bronsted, J. N. *J. Amer. chem. Soc.* **45,** 2898 (1923).
6. Blayden, H. E. and Davies, C. W. *J. chem. Soc.* 949 (1930).
7. Scatchard, G., Jones, P. T. and Prentiss, S. S. *J. Amer. chem. Soc.* **54,** 2676 (1932).
8. Lange, J. *Z. phys. Chem.* **168A,** 147 (1934).
9. Brown, P. G. M. and Prue, J. E. *Proc. roy. Soc.* **232A,** 320 (1955).
10. White, W. P. *J. Amer. chem. Soc.* **36,** 2292 (1914).
11. Scatchard, G. and Prentiss, S. S. *ibid.* **55,** 4355 (1933).
12. Harkins, W. D. and Roberts, W. A. *ibid.* **38,** 2676 (1916).
13. Beattie, J. A., Jacobus, D. D. and Gaines, J. M. *Proc. Acad. Arts. Sci.* **66,** 167 (1930).
14. Ref. 10, Ch. 1.
15. Young, T. F. *Chem. Rev.* **13,** 103 (1933).
16. Guggenheim, E. A. and Turgeon, J. C. *Trans. Faraday Soc.* **51,** 747 (1955).
17. Ref. 2, Ch. 1.
18. Spencer, H. M. *J. Amer. chem. Soc.* **54,** 4490 (1932).
19. Bichowsky, F. R. and Rossini, F. D. "Thermochemistry of Chemical Substances." Reinhold Publishing Corp., New York (1936). Ref. 2, Ch. 1.
20. Harned, H. S. and Cook, M. A. *J. Amer. chem. Soc.* **59,** 1290 (1937).
21. Robinson, R. A. and Sinclair, D. A. *ibid.* **56,** 1830 (1934).
22. Gordon, A. R. *ibid.* **65,** 221 (1943).
23. Davies, M. and Thomas, D. K. *J. phys. Chem.* **60,** 41 (1956).
24. Lassettre, E. N. and Dickenson, R. G. *J. Amer. chem Soc.* **61,** 54 (1939).
25. Ref. 2, Ch. 1.
26. Lovelace, B. F., Frazer, J. C. W. and Sease, V. R. *J. Amer. chem. Soc.* **43,** 102 (1921).
27. Shankman, S. and Gordon, A. R. *ibid.* **61,** 2370 (1939).
28. Stokes, R. H. *ibid.* **69,** 1291 (1947).
29. Olynyk, P. and Gordon, A. R. *ibid.* **65,** 204 (1943).
30. Hornung, E. W. and Giauque, W. F. *ibid.* **77,** 2744 (1955).
31. Harned, H. S. and Hamer, W. J. *ibid.* **57,** 27 (1935).

32. Hamer, W. J. *ibid.* **57,** 9 (1935).
33. Beck, W. H., Singh, K. P. and Wynne-Jones, W. F. K. *Trans. Faraday Soc.* **55,** 331 (1955).
34. (a) NaCl: Harned, H. S. and Nims, L. F. *J. Amer. chem. Soc.* **54,** 423 (1932); Harned, H. S. and Cook, M. A. *ibid.* **61,** 495 (1939). (b) KCl: Harned, H. S. and Cook, M. A. *ibid.* **59,** 1290 (1937). (c) NaBr: Harned, H. S. and Crawford, C. C. *ibid.* **59,** 1903 (1937).
35. Ref. 2, Ch. 1. Appendix to Stokes, R. H. and Levien, B. J. *J. Amer. chem. Soc.* **68,** 333, (1946). Robinson, R. A. *Trans. roy. Soc., N.Z.* **75,** 203 (1945).
36. Robinson, R. A. and Stokes, R. H. *Trans. Faraday Soc.* **45,** 612 (1949).
37. Wishaw, B. F. and Stokes, R. H. *ibid.* **49,** 27 (1953).
38. Stokes, R. H. *ibid.* **44,** 295 (1948); Ref. 11, Ch. 1.
39. Ref. 11, Ch. 1.
40. Bjerrum, N. *Z. anorg. Chem.* **109,** 275 (1920).
41. Harned, H. S. *In* "Treatise on Physical Chemistry." by H. S. Taylor; D. Van Nostrand and Co., New York, II (1924).
42. Scatchard, G. *J. Amer. chem. Soc.* **47,** 2098 (1925).
43. Frank, H. S. *ibid.* **63,** 1789 (1941).
44. Stokes, R. H. and Robinson, R. A. *ibid.* **70,** 1870 (1948). *Ann. N.Y. Acad. Sci.* **51,** 593 (1949).
45. Stokes, R. H. *Trans. Faraday Soc.* **44,** 295 (1948); Robinson, R. A. and Stokes, R. H. *ibid.* **45,** 612 (1949).
46. Glueckauf, E. *ibid.* **51,** 1235 (1955).
47. Flory, P. J. *J. chem. Phys.* **10,** 51 (1942).
48. Longuet-Higgins, L. C. *Disc. Faraday Soc.* **15,** 73 (1953).
49. Fowler, R. H. and Guggenheim, E. A. "Statistical Thermodynamics." Cambridge University Press (1956).
50. Miller, D. G. *J. phys. Chem.* **60,** 1296 (1956).
51. Harned, H. S. *J. Amer. chem. Soc.* **57,** 1865 (1935).
52. Akerlöf, G. and Thomas, H. C. *ibid.* **56,** 593 (1934).
53. McKay, H. A. C. *Nature, Lond.* **169,** 464 (1952); *Trans. Faraday Soc.* **51,** 903 (1955).
54. Harned, H. S. and Cook, M. A. *J. Amer. chem. Soc.* **59,** 2890 (1937).
55. Glueckauf, E., McKay, H. A. C. and Mathieson, A. R. *J. chem. Soc.* S 299, (1949). Glueckauf, E. *Nature, Lond.* **163,** 414 (1949). *Trans. Faraday Soc.* **47,** 428 (1951).
56. McKay, H. A. C. *ibid.* **48,** 1103 (1952). **49,** 237 (1953). McKay, H. A. C. and Mathieson A. R. *ibid.* **47,** 428 (1951). McKay, H. A. C. and Jenkins, I. L. *ibid.* **50,** 107 (1954).
57. Glueckauf, E. *Trans. Faraday Soc.* **53,** 305 (1957).
58. Stokes, R. H. and Robinson, R. A. *loc. cit.* 301.
59. Brubaker, C. H., Johnson, C. E., Knop, C. P. and Betts, F. *J. chem. Educ.* **34,** 42, (1957).

CHAPTER 6

PARTIAL MOLAL QUANTITIES

THERE are a number of properties of electrolyte solutions such as the heat content, heat capacity, free energy, volume and density which, under certain fixed conditions such as temperature and pressure, depend on the concentration, or rather on the activities of the solute and solvent. It follows that functions can be derived which express the relationships between these properties and activities by application of the laws of thermodynamics and the Debye–Hückel theory of interionic forces. However, the application of these can be troublesome since, for practical reasons, it may not be possible to obtain accurate data at low concentrations, and in addition derivatives such as $d\epsilon/dT$, $d\epsilon/dP$ and second differentials may lead to complex expressions; furthermore, the values of these terms may not be known with sufficient accuracy.

Partly because of these difficulties, but also because many of these thermodynamic properties were studied before the advent and development of the interionic attraction theory, a number of empirical formulae have been and still are used to express the properties in terms of concentrations. Some of these will be described at suitable places.

If[1] Y is any property of a solution which is a function of T, P, and amounts of constituents (i.e., it is an "extensive" property), then the derivative $(\partial Y/\partial n_1)_{P,\,T,\,n_2,\,n_3\ldots} = \bar{Y}_1$ is termed a partial molal quantity. In this, n_1 refers to n_1 moles of constituent 1, etc., and P, T, n_2, etc., are kept constant. In general,

$$dY = \bar{Y}_1 dn_1 + \bar{Y}_2 dn_2 + \ldots$$

$$Y = \bar{Y}_1 n_1 + \bar{Y}_2 n_2 + \ldots = n_i \bar{Y}_i \tag{6.1}$$

In themselves, partial molal quantities are "intensive" properties; they are independent of the total amount of each constituent, but depend on the composition of the system. For the present discussions the systems will be confined to those comprising n_1 moles of solvent and n_2 moles of a single electrolyte.

It is convenient in the analysis of actual measurements to write (6.1) as

$$Y = n_1 \bar{Y}_1^0 + n_2 \phi_Y \tag{6.2}$$

where $\bar{Y}_1^0$ is the partial molal property of the pure solvent (or the value of $\bar{Y}_1$ at infinite dilution), and ϕ_Y is termed the apparent molal property of the solute. With the same terminology,

$$\bar{Y}_2^0 = \phi_Y^0 \tag{6.3}$$

In general, the absolute values of Y, $\bar{Y}_1$ and $\bar{Y}_2$ cannot be determined directly from the experimental data, but it is possible to find their values with respect to chosen reference states, which are conventionally selected as those at infinite dilution, namely Y^0, $\bar{Y}_1^0$, $\bar{Y}_2^0$; then writing

$$Z = Y - Y^0, \bar{Z}_1 = \bar{Y}_1 - \bar{Y}_1^0, \bar{Z}_2 = \bar{Y}_2 - \bar{Y}_2^0, \phi_Z = \phi_Y - \phi_Y^0 \qquad (6.4)$$

from (6.1) and (6.2),

$$Z = n_1\bar{Z}_1 + n_2\bar{Z}_2 = n_2\phi_Z \qquad (6.5)$$

also, since $(\partial Z/\partial n_2)_{n_1, P, T} = \bar{Z}_2$, from (6.5),

$$\bar{Z}_2 = \phi_Z + n_2(\partial\phi_Z/\partial n_2) = \phi_Z + m(\partial\phi_Z/\partial m) = \phi_Z + \sqrt{m}(\partial\phi_Z/2\partial\sqrt{m}) \qquad (6.6)$$

and likewise,

$$\bar{Z}_1 = (n_2/n_1)(\phi_Z - \bar{Z}_2) = -5 \times 10^{-4} M_1 m(\partial\phi_Z/\partial\sqrt{m}) \qquad (6.7)$$

where m is the molality of the solute and M_1 is the molecular weight of the solvent.

Very often, the plot of ϕ_Z against $\sqrt{m}$ is linear or only slightly curved over a wide range of m, so extrapolation of this plot seems a logical way of finding ϕ_Z^0 while its slope will give S_Z where

$$\phi_Z = \phi_Z^0 + S_Z\sqrt{m} \qquad (6.8)$$

In certain instances, ϕ_Z^0 is zero, which is helpful. On the other hand, it must be stressed that it is unusual to find that S_Z obtained in this fashion is in exact agreement with the value predicted by the interionic attraction theory. This is due to the difficulty of obtaining accurate data with sufficiently dilute solutions.

Heats of Dilution and Partial Molal Heats

When $Y = H$ in (6.1), H is termed the total heat content; $\bar{H}_1$ and $\bar{H}_2$ are the partial molal heat contents of the solvent and solute. Similarly, from (6.4), when $Z = L$, $L = H - H^0$, $\bar{L}_1 = \bar{H}_1 - \bar{H}_1^0$, $\bar{L}_2 = \bar{H}_2 - \bar{H}_2^0$, L is termed the relative total heat content, $\bar{L}_1$ and $\bar{L}_2$ are termed the relative partial molal heat contents of the solvent and solute. From (6.5)

$$L = n_1\bar{L}_1 + n_2\bar{L}_2 = n_2\phi_L \qquad (6.9)$$

Another definition is ΔH_d, the molal heat of dilution. This is the heat absorbed (if positive) or evolved (if negative) on diluting a solution to infinite dilution under isothermal conditions. It is related to the above quantities by

$$L/n_2 = \phi_L = \phi_H - \phi_H^0 = -\Delta H_d \qquad (6.10)$$

Heats of dilution are determined by measurements with carefully designed two-section or twin calorimeters whereby minute temperature differences can be determined with a precision of about 2×10^{-7} degree. Lange and his associates [2] have made many important contributions to this branch of calorimetry. Their arrangement consisted of a 2 l. Dewar flask divided into two sections by a 1,000–1,500 junction constantan-iron thermocouple (thermel). Each section of the calorimeter was fitted with matched constantan wire heaters, identical stirrers rotating at the same speeds, and matched metal pipettes which could be opened or closed as desired and containing the solution to be diluted. These parts passed through a brass cover, and the whole was supported in a water bath controlled to $\pm$ 0.0002°C. In order to observe any temperature differences between the bath and the inside of the calorimeter, the cover of the latter was fitted with a small thermel. To obtain the information required for ΔH_d determinations, the minute temperature rise in the section where the dilution was carried out was counterbalanced by activating the heater in the other section. The equilibration was followed by a sensitive galvanometer connected to the large thermel, and the amount of electrical heating gave the required figures.

Gucker *et al.* [3] have also designed and described apparatus for such studies. This comprised a double calorimeter connected by a sixty junction copper-constantan thermel, a special galvanometer and recording potentiometer. Considerable efforts were made to ensure the maximum accuracy and the minimization of sources of errors. For instance, a support free of vibration for the galvanometer involved the building of an 8-ton concrete pier on which rested a heavy iron frame to support a modified Julius suspension by means of a single wire.[4] A set of vanes on the bottom plate of the suspension extended into a tank of oil which formed a damper.

Example 6.1. Gulbransen and Robinson [5] obtained the heats of dilution shown in the third row of Table 6.1 between the molalities shown by the first two rows. The values of $\Delta H_d(0.1 \rightarrow m)$ were obtained by stepwise combination of $\Delta H_d(m_1 \rightarrow m_2)$, e.g., $-29.1 = (-65.1 + 51.1) + (-56.2 + 41.1)$, $-70.2 = -65.1 + (-56.2 + 51.1)$.

According to (6.8) and (6.10), ΔH_d for 0.1 M NaCl could be obtained by extrapolating the plot of $-\Delta H_d(0.1 \rightarrow m)$ against $\sqrt{m}$, especially if the plot is linear. Within experimental error, this is so below $\sqrt{m} = 0.1$, and from the intercept, the value is -86.5 cal./mol. Thus ΔH_d at any value up to 0.1 M is derived from $\Delta H_d = -86.5 - \Delta H_d(0.1 \rightarrow m)$. The limiting slope of the plot is $S_Z = S_{\Delta H_d} = 430$; (see S^0 below, and after (6.38)—also example (6.2)).

In order to take any curvature into account and thus provide a plot extending to higher concentrations, Young and Groenier [6] devised an analytical method which is a modified version of the chord-area method of Young and Vogel.[7] Writing

$$S = -\Delta H_{d(m_1 \rightarrow m_2)}/\sqrt{m} = S^0 + B\sqrt{m} + Cm \tag{6.11}$$

the parameters of the right-hand side are found by the method of least squares [8] as applied to the general equation $y = a + bx + cx^2$. This is an involved process and leads to $S^0 = 476.1$, $B = -1452.2$, $C = 730.4$ up to $\sqrt{m} = 0.4$. Thus

$$-\Delta H_d = \int_0^{\sqrt{m}} S\mathrm{d}\sqrt{m} = 476.1\sqrt{m} - 1452.2m/2 + 730.4m^{\frac{3}{2}}/3 \quad (6.12)$$

for NaCl at 25°C. The figures up to 0.1 M are shown in Table 6.1. The same treatment has been applied to the heats of dilution of several lanthanide salts.[9]

TABLE 6.1

Heats of Dilution and Relative Partial Molal Heats of NaCl at 25°C

$10^3 m_1$		100	100	50	50	25	25	12.5	12.5	6.25	6.25
$10^3 m_2$		2.57	5.07	1.285	2.535	0.642	1.28	0.322	0.634	0.161	0.318
$-\Delta H_d(m_1 \to m_2)$		65.1	55.9	56.2	51.1	46.4	41.1	34.6	30.8	27.6	24.6
$10^3 m$	100	50	25	12.5	6.25	5.07	2.57	1.283	0.638	0.320	
$-\Delta H_d(0.1 \to m)$	0	14.0	29.1	40.9	50.9	55.9	65.1	70.2	75.3	79.1	
$-\Delta H_d$ at m	86.5	72.5	57.4	45.6	35.6	30.6	21.4	16.3	11.2	7.4	
$-\Delta H_d$(6.12)	85.6	72.8	58.1	44.5	33.2	30.3	22.3	16.1	11.6	8.3	
$\bar{L}_2$(6.13)	106	93	76	61	50	46	32.5	24	16.5	11.5	
$\bar{L}_2$(6.15)	99.8	93.9	79.0	62.6	47.6	43.7	32.6	23.6	17.1	12.3	

Wallace and Robinson have simplified the calculations by introducing suitable modifications.[23]

$\bar{L}_1$ and $\bar{L}_2$ can be derived from ΔH_d, and also from e.m.f. measurements. To use the former, (6.6) and (6.9) are combined to produce

$$\bar{L}_2 = \phi_L + \sqrt{m}(\partial\phi_L/2\partial\sqrt{m}) = -\Delta H_d - \sqrt{m}(\partial\Delta H_d/2\partial\sqrt{m}) \quad (6.13)$$

so that $\bar{L}_2$ can be calculated by taking tangents of a plot of ΔH_d against $\sqrt{m}$. This is also of use in the derivation of $\bar{L}_1$ since from (6.7) and (6.13),

$$\bar{L}_1 = -n_2(\Delta H_d + \bar{L}_2)/n_1 = n_2(\sqrt{m}\partial\Delta H_d/2\partial\sqrt{m})/n_1$$
$$= 5 \times 10^{-4} M_1 m^{\frac{3}{2}}\partial\Delta H_d/\partial\sqrt{m} \quad (6.14)$$

In the concentration range where (6.11) is valid, substitution in (6.13) and (6.14) produces expressions for calculating $\bar{L}$:

$$\bar{L}_2 = 3S^0\sqrt{m}/2 + Bm + 5Cm^{\frac{3}{2}}/6 \quad (6.15)$$

$$\bar{L}_1 = -5 \times 10^{-4} M_1(S^0 m^{\frac{3}{2}} + Bm^2 + Cm^{\frac{5}{2}}) \quad (6.16)$$

In order to calculate $\bar{L}$ and other partial molal quantities from e.m.f.s, and to correlate these with the Debye–Hückel treatment, a convenient starting-point is provided by the Gibbs–Helmoltz equation in the form

$$\partial(\Delta\bar{G}_i/T)_{P,m} = -\Delta H_i\partial T/T^2 \quad (6.17)$$

whence, from (1.20),

$$\bar{H}_i = \bar{H}_i^0 - RT^2(\partial \ln a_i/\partial T)_{P,m} \tag{6.18}$$

so that

$$\bar{L}_1 = - RT^2(\partial \ln a_1/\partial T)_{P,m} \tag{6.19}$$

$$\bar{L}_2 = - \nu RT^2(\partial \ln \gamma_\pm/\partial T)_{P,m} \tag{6.20}$$

The equations for e.m.f. cells such as (4.23, 4.29, 4.31) may be generalized into

$$E = E^0 - (\nu RT/Fn) \ln m\gamma_\pm \tag{6.21}$$

Differentiating this, with P and m constant, and equating to (6.20),

$$\mathrm{d} \ln \gamma_\pm/\partial T = - (nF/\nu R)\{\partial(E - E^0)/T\partial T\} = - \bar{L}_2/\nu RT^2 \tag{6.22}$$

whence

$$\bar{L}_2 = - nF(E - E^0) + nFT\partial(E - E^0)/\partial T \tag{6.23}$$

Owing to $\bar{L}_2$ being very sensitive to errors in $\partial(E - E^0)/\partial T$, it is found that the best calorimetric results are superior to any of those derived by potentiometry. To some extent it helps to smooth out slight deviations in $\partial(E - E^0)/\partial T$ by means of empirical equations which relate E and E^0 to T. This has been examined by Harned and Nims;[10] one method is to write $E = a + bT + cT^2$, $E^0 = a' + b'T + c'T^2$ so that

$$(E - E^0)/\partial T = b - b' + 2T(c - c') \qquad (m \text{ constant}) \tag{6.24}$$

This device cannot be applied to the potentials of concentration cells with transference for it would give the difference between L values of different solutions, one being the reference.

Another way, due to Akerlöf and Teare,[11] is to ascertain the constants of

$$\log \gamma_\pm = - A''\sqrt{m}/(1 + \sqrt{2m}) + b'm + c'm^2 + d'm^3 + \ldots \tag{6.25}$$

when by differentiation and insertion into (6.20) there results

$$\bar{L}_2 = - 3 \ln RT\{(1 - b'T)A''\sqrt{m}/(1 + \sqrt{2m})\} - 2 \times 2.303\, RT^2Z \tag{6.26}$$

where

$$Z = (m\partial b'/\partial T) + (m^2\partial c'/\partial T) + (m^3\partial d'/\partial T) + \ldots, \; b'T = - T\partial\epsilon/\partial T \tag{6.27}$$

A similar but more detailed scheme has been tried by Harned *et al.*[12] Alternative and simpler expressions are [13,14]

$$\log \gamma_\pm = A_1 + B_1T + C_1T^2 \qquad (m \text{ constant}) \tag{6.28}$$

$$\log \gamma_\pm = - A_2/T - B_2 \log T + C_2 \qquad (m \text{ constant}) \tag{6.29}$$

From the Gibbs–Duhem equation, (4.43), and from (6.18), recalling that $\bar{H}_1 - \bar{H}_1^0 = \bar{L}_1$,

$$\bar{L}_1 = - 10^{-3} M_1\int_0^m m\mathrm{d}\bar{L}_2 \tag{6.30}$$

so that $\bar{L}_1$ can be computed by graphical integration of a plot of m against $\bar{L}_2$. Akerlöf [11] formed an expression for calculating $\bar{L}_1$ by multiplying the differential of (6.26) by m and substituting the results in (6.30).

In (6.20), $\bar{L}_2$ is equated to $\partial \ln \gamma_\pm/\partial T$, so by taking T as a variable in the Debye–Hückel activity coefficient expression, it should be possible to derive an expression relating $\bar{L}_2$ to m or C and to derivatives such as $\partial\epsilon/\partial T$. Debye and Hückel paid some attention to this,[15] and the subject has been developed and exploited by Harned, Owen, and their associates.[16,17,18] Only 1 : 1 electrolytes will be considered here, and writing (3.42) as

$$-\log f_\pm = A''\sqrt{C}/(1 + B''a\sqrt{C}) - C''C \qquad (C = \text{molarity}) \quad (6.31)$$

where C'' is an empirical parameter, then the final expression is

$$\bar{L}_2 = S_H\sqrt{C}/(1 + B''a\sqrt{C}) + W_H C/(1 + B''a\sqrt{C})^2 + K_H C \quad (6.32)$$

where

$$S_H = -\ 2.303 \times 3\ RT^2A''\{(\partial \ln \epsilon/\partial T) + T^{-1} + \alpha/3\} \quad (6.33)$$

$$W_H = 2.303\ RT^2A''B''a\{(\partial \ln \epsilon/\partial T + T^{-1} + \alpha - 2\partial \ln a/\partial T\} \quad (6.34)$$

$$K_H = 2.303 \times 2\ C''RT^2\{\partial \ln C''/\partial T - \alpha\} \quad (6.35)$$

where α is the coefficient of thermal expansion of the solution; it can be replaced by $\partial \ln C/\partial T$, or [21] $\partial\rho/\rho$, where ρ is the density of the solution.

The limiting form of (6.32) is

$$\bar{L}_2 = S_H\sqrt{C} \qquad (C \text{ very small}) \quad (6.36)$$

Harned and Owen [16] have tabulated values of S_H between 0°C and 85°C, and give S_H as 708 at 25°C. Their figures are based on $\partial\epsilon/\partial T$ values given by Wyman,[19] whereas more recent data [20] are slightly different. Guggenheim and Prue,[22] whose treatment is mentioned later on, by taking the mean of the two sets, find S_H to be 720 at 25°C.

From (6.6), using C instead of m,

$$\bar{L}_2 = \phi_L + C\partial\phi_L/\partial C, \text{ i.e., } \partial(C\phi_L) = L_2\partial C = S_H\sqrt{C}\partial C \quad (6.37)$$

so from (6.10),

$$-\ \phi_L = \Delta H_d = \phi_H^0 - \phi_H = -\ 2S_H\sqrt{C}/3 \quad (6.38)$$

Thus the limiting slope of the plot of ΔH_d against $\sqrt{C}$ at 25°C should be 472 or 480. The experimental values derived earlier on for NaCl are $S_Z = 430$ $S^0 = 476$. The latter is clearly in good agreement with the limiting value(s).

The more complete expression for ΔH_d from theory [16,18] is (ignoring $\partial \ln a/\partial T$ and $\partial \ln C''/\partial T$),

$$\Delta H_d = -\ 2S_H\tau\sqrt{C}/3 - \tfrac{1}{2}W_H\theta C - \tfrac{1}{2}K_H C \quad (6.39)$$

where

$$\theta = 4(\tau - \sigma)/3\kappa a = 1 - \tfrac{8}{5}\kappa a + 2\kappa^2 a^2 + \ldots \;(\kappa a = B'' a\sqrt{C}) \qquad (6.40)$$

$$\tau = 3\{\tfrac{1}{2}\kappa^2 a^2 - \kappa a + \ln(1 + \kappa a)\}/\kappa^2 a^2 = 1 - \tfrac{3}{4}\kappa a + \tfrac{3}{5}\kappa^2 a^2 + \ldots \qquad (6.41)$$

$$\sigma = \partial(\tau\kappa a)/\partial(\kappa a) = 1 - \tfrac{6}{4}\kappa a + \tfrac{9}{5}\kappa^2 a^2 + \ldots \quad \text{(cf., 5.29)} \qquad (6.42)$$

From (6.10),

$$\phi_H - \phi_H^R + \Delta H_d = \phi_H^0 - \phi_H^R, \text{ so } \Delta H_{d(m_R \to m)} + \Delta H_d = (\Delta H_d)_R,$$

where R refers to the reference solution, whence, from (6.39),

$$\Delta H_{d(m_R \to m)} - \tfrac{2}{3} S_H M_H \sqrt{C} = (\Delta H_d)_R + \tfrac{1}{2} K_H C \qquad (6.43)$$

where

$$M_H = \{(\partial \ln \epsilon/\partial T + T^{-1})/(1 + \kappa a) + \sigma\alpha/3\}/\{\partial \ln \epsilon/\partial T + T^{-1} + \alpha/3\} \quad (6.44)$$

As a first approximation in dilute solutions, $M_H = 1/(1 + \kappa a)$, i.e.,

$$\Delta H_{d(m_R \to m)} - \tfrac{2}{3} S_H \{\sqrt{C}/(1 + B'' a\sqrt{C}) - C'' C\} = (\Delta H_d)_R \qquad (6.45)$$

should provide a convenient way of deriving the ΔH_d values of reference solutions and hence ΔH_d values of more dilute solutions. The following example will explain what is meant.

Example 6.2. Taking the experimental figures for NaCl at 25°C given in the third row of Table 6.1, $R = 0.1$ and $m = m_2$, and putting $a = 4.0$ Å and $C'' = 0$ in (6.45) together with the mean theoretical value of $\tfrac{2}{3} S_H = 475$ (see 6.33), extrapolation of the left-hand side of (6.45) against C gives $\Delta H_d = -85.5 \pm 1.5$ for 0.1 M NaCl, and by putting $C'' = 0.41$, ΔH_d averages -86.1 ± 0.8 cal./mol. over the whole range. While it is partly empirical, this method has a sounder theoretical basis and leads to a more reliable result than plotting ΔH_d against $\sqrt{m}$ that was described earlier, and is easier than the method of Young and Groenier.[6]

To calculate a more accurate value of M_H, two more figures are required. For water at 25°C these are [21] $10^6\alpha = 257$ and [19,20] $\partial\epsilon/\epsilon\partial T = -4.58 \times 10^{-3}$.

A rigid application of (6.32) is precluded by a lack of information on $\partial \ln a/\partial T$ and $\partial \ln C''/\partial T$. The former of these cannot be very significant so that W_H at 25°C for NaCl (with $a = 4.0$ Å, i.e., $\kappa a = B'' a\sqrt{C} = 1.313\sqrt{C}$) is -166.

Owen and Brinkley [18] have given the following method for $(\Delta H_d)_R$ and for K_H, namely a plot of the left-hand side of (6.43) against C gives the former quantity at $C = 0$, and the slope gives $\tfrac{1}{2} K_H$ [they express the equation with $\phi_H - \phi_H^R$ for $\Delta H_{d(m_R \to m)}$ and $\phi_H^0 - \phi_H^R$ for $(\Delta H_d)_R$ — see (6.38); in Table 6.1, $R = 0.1$ M]. With $S_H = 471.8$, their plot gives $K_H = -440$.

By taking the $\bar{L}_2$ result from Table 6.1 at 0.1 M, together with the mean theoretical value of $S_H = 715$, and $W_H = -166$, (6.32) gives $K_H = -500$, and using these figures in (6.32) gives the $\bar{L}_2$ values shown in Table 6.2, where ΔH_d derived from (6.45) with $\frac{2}{3}S_H = 475$ and $C'' = 0.41$ is also shown. These may be compared with those in Table 6.1.

TABLE 6.2

Calculated Values of $\bar{L}_2$ and ΔH_d for NaCl at 25°C.

$10^3 m$	100	50	25	12.5	6.25	2.57	1.283	0.320
$\bar{L}_2$	101.5	93.5	78.4	61.9	47.2	32.3	23.6	12.3
$-\Delta H_d$	86.7	72.3	57.3	43.9	32.9	22.1	16.0	8.3

Because $\partial \ln C''/\partial T$ cannot be derived from theory, Guggenheim and Prue [22] have considered how molal heats may be used to derive this term. Writing (6.31) as

$$-\ln \gamma = \alpha\sqrt{m}/(1+\sqrt{m}) + 2\beta' m \tag{6.46}$$

from (6.9), (6.20) and (5.24),

$$H_m = L/n_2 = n_1\bar{L}_1/n_2 + \bar{L}_2 = \nu RT^2\partial\{-\ln\gamma - (1-\phi)\}/\mathrm{d}T \tag{6.47}$$

since at $m = 0$, $\ln\gamma = 0$, $\phi = 1$. Standard functions are defined; (3.42) with $Q'' = 0$, and (5.26), and H_m^{st} defined in terms of (6.33). An expression very similar to (6.43) is derived so that, in effect, $2\partial\beta'/\partial T = \partial C''/C''\partial T$ is obtained. For NaCl at 25°C, this is 0.00104 kg/mole deg, and agrees very well with that derived from e.m.f. data.

Partial Molal Heat Capacities

With $Y = C_p$ (total heat capacity) in (6.1) and (6.2), $\bar{Y}_1 = \bar{C}_1$, $\bar{Y}_2 = \bar{C}_2$ (partial molal heat capacities), $\phi_Y = \phi_c$ (apparent molal heat capacity of the solute),

$$C_p = n_1\bar{C}_1 + n_2\bar{C}_2 = n_1\bar{C}_1^0 + n_2\phi_c \tag{6.48}$$

$$\bar{C}_2 = \phi_c + \tfrac{1}{2}\sqrt{m}(\mathrm{d}\phi_c/\mathrm{d}\sqrt{m}) \tag{6.49}$$

$$\bar{C}_1 = \bar{C}_1^0 + n_2(\phi_c - \bar{C}_2)/n_1 = C_1^0 - m^{\frac{3}{2}}M_1(\mathrm{d}\phi_c/\mathrm{d}\sqrt{m})/2{,}000 \tag{6.50}$$

If the specific heats of the pure solvent, and of the solution are c_1^0 and c_2, and if the solution consists of mM_2 gm of solute in 1,000 gm of solvent, then

$$\phi_c = (C_p - n_1\bar{C}_1^0)/n_2 = \{(1{,}000 + mM_2)c_2 - 1{,}000\, c_1^0\}/m \tag{6.51}$$

Modified forms of the adiabatic twin calorimeter described in the previous section have been used for determining $\bar{C}_1$ and $\bar{C}_2$. Gucker [24] and Hess [25] have given details of these. That of the former consisted of similar calorimeters suspended in a metal air-jacket built into a water thermostat bath.

The lid of each calorimeter held a stirrer and wells for one arm of a twenty-junction thermel, a variable resistance electrical heater and a thermel dipping into the water bath (to ensure that adiabatic control was maintained). One calorimeter, the "working" one, I, was filled with a solution of known molality, while the other or "tare" calorimeter, II, contained a fixed weight of water. The volumes of I and II were kept equal. The heater resistances were adjusted until the temperatures of I and II rose by the same amount (about one degree), and to follow this and to also measure the ratio of the resistances during the "run," the resistances formed two arms of a Wheatstone bridge with a sensitive galvanometer as detector. If x_0 and x_1 are the resistance ratios when I contains first water and then solution, T is the total heat capacity of II, c_1 is that of II when empty, while C_p^w and C_p^s are the heat capacities of I when containing first water and then solution, then $x_0 = T/(C_p^w + c_1)$, $x_1 = T/(C_p^s + c_1)$ whence C_p^s is found since $C_p^w =$ weight of water times c_1^0. To determine c_1, water measurements were made with the aid of a small copper sphere, first empty, then filled. Since c_1 varied slightly with the total volume, this was kept constant.

Ackermann,[52] following the design of Eucken and Eigen,[53] has developed a semi-automatic adiabatic calorimeter for measuring specific heats ($\pm$ 0.1 %). It is fitted with a heating coil and magnetic stirrer, and is surrounded by an adiabatic mantle controlled automatically by a thermel system operating a photocell galvanometer, with a control of $\pm$ 0.005°C. Temperature and time are recorded automatically during the heating period, using a Pt wire thermometer in a bridge combined with a galvanometer relay.

After calculating ϕ_c by (6.51), $\bar{C}_1$ and $\bar{C}_2$ can be ascertained by (6.50) and (6.49) and plots of ϕ_c against $\sqrt{m}$. An error of 0.01 % in c_1 causes an error of 10 cals. at 0.01 M, so the practical limit is around [26] 0.05 M.

If $J = C_p - C_p^0$, $\bar{J}_1 = \bar{C}_1 - \bar{C}_1^0$, $\bar{J}_2 = \bar{C}_2 - \bar{C}_2^0$, values of J, $\bar{J}_1$, $\bar{J}_2$, and ϕ_J are correlated by (6.4) to (6.7).

Since $C_i = \mathrm{d}L/\mathrm{d}T$ and $\bar{C}_i = \mathrm{d}\bar{L}_i/\mathrm{d}T$ where $i = 1$ and 2, experimental values of $\bar{J}_i$ can be obtained from ΔH_d determinations described in the previous section by plotting $\bar{L}_i$ against T and taking tangents. Also, since from (6.19),

$$\bar{J}_i = -R\mathrm{d}\{T^2\partial \ln a_i/\partial T\}/\mathrm{d}T \qquad (6.52)$$

then $\bar{J}_2$ values may be derived by suitable differentiation with respect to T of the e.m.f. and activity coefficient expressions (6.23–29). Numerous examples and references are given by Harned and Owen [16] and others.[11, 24]

Heat capacity expressions can be derived from the Debye–Hückel theory since they involve differentiation of S_H and M_H of (6.33) and (6.34). Guggenheim and Prue [22] have dealt with this in some detail. From (6.3), $\bar{C}_2^0 = \phi_c^0$, and from (6.4), $\phi_J = \phi_c - \phi_c^0$; these, together with the Debye–Hückel theory give

$$\phi_c = \bar{C}_2^0 + \phi_J = \bar{C}_2^0 + \tfrac{1}{2}\nu z^3 R\alpha f\sqrt{m}\tau(z\sqrt{m}) - 2\nu RTm(\partial\beta'/\partial T) - \nu RT^2 m(\partial^2\beta'/\partial T^2) \quad (6.53)$$

where [27]

$$f = 1 + 2\omega x + 5\omega^2 x^2 - 2T\omega x^2 + \tfrac{2}{3}yz + y^2 z^2 - \tfrac{2}{3}T\omega x^2 \quad (6.54)$$

in which $\omega = T/\epsilon$, $x = \mathrm{d}\epsilon/\mathrm{d}T$, $y = T/V$, $z = \mathrm{d}V/\mathrm{d}T$, V is the volume in litres containing m g. mols of solvent, $\alpha = A''$ (see 6.46, which also defines β'), and for 1 : 1 types, $\tau(z\sqrt{m}) = 1.5/(1 + \sqrt{m}) - \tfrac{1}{2}\sigma(\sqrt{m})$, where $\sigma(m)$ is defined by (5.29); τ and σ are also defined, but not so simply, by (6.41, 6.42); Note:[18] $2\tau + \sigma = 3/(1 + \kappa a)$.

Also,

$$\bar{J}_2 = \tfrac{3}{4}\nu z^3 R\alpha f\sqrt{m}/(1 + \sqrt{m}) - 4\nu RTm(\mathrm{d}\beta'/\mathrm{d}T) - 2\nu RT^2 m(\mathrm{d}^2\beta'/\mathrm{d}T^2) \quad (6.55)$$

To obtain f for aqueous solutions at 25°C, the measurements of Wyman, as revised by Wyman and Ingalls,[28] and those of Drake *et al.*[29] are available. These lead to $f = 3.16$ and 1.98 respectively, so $f = 2.57 \pm 0.6$. For a 1 : 1 electrolyte at 25°C,

$$\phi_c - \bar{C}_2^0 = k_c\sqrt{m}\tau(\sqrt{m}) - k_c' - k_c'' \quad (6.56)$$

$$\bar{J}_2 = 1.5\,k_c\sqrt{m}/(1 + \sqrt{m}) - 2k_c' - 2k_c'' \quad (6.57)$$

where $k_c = 6.0 \pm 1.4$, $k_c' = 2.4 \times 10^3\, m(\mathrm{d}\beta'/\mathrm{d}T)$, $k_c'' = 3.6 \times 10^5\, m(\mathrm{d}^2\beta'/\mathrm{d}T^2)$. Also, as derived from (6.47), for NaCl at 25°C, $\mathrm{d}\beta'/\mathrm{d}T = 0.001$, so $\mathrm{d}^2\beta'/\mathrm{d}T^2 = -5 \pm 1 \times 10^{-5}$.

Guggenheim and Prue [22] applied these formulae to the data of Randall and Rossini [30] below 0.1 M, where $\phi_c - \bar{C}_2^0 = (6.0 \pm 1.4)\sqrt{m}\tau(\sqrt{m}) + (15.6 \pm 4)m$. Since $\bar{C}_2^0$ is unknown, arbitrary values of ϕ_c were calculated, and the corresponding curve drawn on a graph of ϕ_c (exptl.) against m. The line best coinciding with ϕ_c (exptl.) was when $\bar{C}_2^0$ was taken as − 22.2 cal/mol. deg, as compared with purely extrapolated values of [30] − 23.8 and − 21.6 obtained by Pitzer [31] by taking $k_c = 8.0$. It is concluded that the Debye–Hückel treatment should be used in helping to derive $\bar{C}_2^0$, but there is an uncertainty of about one cal. which cannot be eliminated.

Harned and Owen [16] suggest that to overcome the difficulty of extrapolating dilute calorimetric data, since $\bar{C}_2^0 = \bar{C}_2 - \bar{J}_2$, calorimetric determinations of $\bar{C}_2$ should be combined with $\bar{J}_2$ values obtained from calorimetric ΔH_d determinations, or from e.m.f. cells.

Partial Molal Volumes

From (6.1) and (6.2), when $Y = V$ (volume of solution), $\bar{Y}_i = \bar{V}_i$ (relative partial molal volumes), $\bar{V}_1^0 = \bar{V}_1$ when $n_2 = 0$, $\phi_Y = \phi_V$ (apparent molal volume of the solute), and

$$V = n_1\bar{V}_1 + n_2\bar{V}_2 = n_1\bar{V}_1^0 + n_2\phi_V \quad (6.58)$$

and by analogy with (6.49) and (6.50),

$$\bar{V}_2 = \phi_V + \tfrac{1}{2}\sqrt{m}(\partial\phi_V/\partial\sqrt{m}) \tag{6.59}$$

$$\bar{V}_1 = \bar{V}_1^0 - 5 \times 10^{-4} M_1 m(\partial\phi_V/\partial\sqrt{m}) \tag{6.60}$$

Considering 1,000 g of solvent, $V = (1{,}000 + mM_2)/d$, $n_1\bar{V}_1^0 = (1{,}000/M_1)(M_1/d_0)$ $n_2 = m$, where d and d_0 are the densities of the solution and solvent. Thus

$$\phi_V = \{1{,}000(d_0 - d) + mM_2d_0\}/d_0md \tag{6.61}$$

If C is the molarity, $V = 1{,}000$, $n_1 = \{1{,}000d - cM_2)/M_1\}(M_1/d_0)$, $n_2 = C$

$$\phi_V = \{1{,}000(d_0 - d) + d_0CM_2\}/d_0C \tag{6.62}$$

The densities are measured by special pycnometers,[32] or by sinkers suspended in a vessel of solution [33] in thermostat baths controlled to $\pm$ 0.001°C. The sinker wire is attached to a balance situated over the bath. If w and w_0 are the vacuum corrected weights required to counterpoise the sinker in solution and solvent, then $d_0 - d = (w - w_0)/v$, where v is the volume of the sinker. With $v = 300$ ml., and w and w_0 determined to 0.1 mg, the sensitivity is about 1 in 10^6. Precision dilatometers [34] also have this order of accuracy. These have been used to derive the density of water; between 0°C and 40°C, this is expressed by

$$1 - d_0 = (t - 3.9863)^2(t + 288.9414)/\{508929.2(t + 68.12963)\}$$

where d_0 is in gm./ml, $t =$ °C. Above this, and up to 85°C, $d_0 = \delta + d_0$ (calc.) where $10^6\delta = 0.2, 4.9, 24.5, 66.2, 127.1$ at 45, 55, 65, 75 and 85°C respectively.

A differential magnetic float method has been developed [35, 36] for finding the densities of very dilute solutions. The floats are made by enclosing soft iron cores in glass pear-shaped vessels which have a Pt point at the base and have the weight adjusted so that they have a slight tendency to rise in solution. This is balanced by the field from a coil fixed under the base of the solution container. A known current is passed and adjusted until the Pt contact just leaves a small Pt dish fixed directly above the coil and inside the container, the movement being observed by a submerged microscope with its axis bent to permit focusing on the dish and point. The current is calibrated by varying the known weight of the sinker in small increments around the buoyancy point. Slight uncertainties in the temperature of the bath are eliminated by simultaneous measurement of solvent and solution in independent systems set up in the same thermostat bath.

Very small density differences (about 10^{-8}) may be measured by [37] comparing the heights of solution and solvent in hydrostatic equilibrium by means of a beat-frequency oscillator and micrometer.

In many instances, the expressions

$$\phi_Y = \phi_Y^0 + S\sqrt{m} = \phi_Y^0 + S'\sqrt{C} \tag{6.63}$$

where S and S' are empirical parameters derived from the data hold over a wide range of concentrations. This is particularly true of ϕ_V values,[16, 38] but in common with other ϕ_Y's, S and S' are usually different to those calculated by formulae derived from the Debye–Hückel theory. This has already been illustrated in the preceding sections. According to Owen and Brinkley,[18]

$$\bar{V}_2 - \bar{V}_2^0 = S_v\sqrt{C}/(1 + B''a\sqrt{C}) + W_vC/(1 + B''a\sqrt{C})^2 + K_vC \quad (6.64)$$

where

$$S_v = -2.303\,\nu RTA''\tfrac{3}{2}\{\partial \ln \epsilon/\partial P - \beta'/3\} \quad (6.65)$$

$$W_v = 2.303\,\nu RT^2A''Ba''\tfrac{1}{2}\{\partial \ln \epsilon/\partial P - \beta - 2\,\partial \ln a/\partial P\} \quad (6.66)$$

where β is the coefficient of compressibility (see 6.71). Also (cf. 6.39, etc.),

$$\phi_v - \tfrac{2}{3}S_v\tau\sqrt{C} - \tfrac{1}{2}W_v\theta C = \phi_v^0 + \tfrac{1}{2}K_vC \quad (6.67)$$

$$\phi_v - \tfrac{2}{3}S_vM_v\sqrt{C} = \phi_v^0 + \tfrac{1}{2}K_vC \quad (6.68)$$

where

$$M_v = \{(\partial \ln \epsilon/\partial P)/(1 + B''a\sqrt{C}) - \sigma\beta/3\}/\{\partial \ln \epsilon/\partial P - \beta/3\} \quad (6.69)$$

(ignoring $\partial \ln a/\partial P$). By plotting the left-hand sides of (6.67) and (6.68) against C, the intercept gives ϕ_v^0 and the slope gives $\tfrac{1}{2}K_v$. At 25°C in water,[39] $\beta = 45.5 \times 10^{-6}$ while for $\partial \ln \epsilon/\partial P$, Mavroides [40] gives $5.3 \pm 1.2 \times 10^{-5}$; from data given by Harned and Owen [16] it is $5.8_5 \times 10^{-5}$.

Owen and Brinkley [18] have applied the above treatments to data for [41] NaCl, KCl and HCl; Wirth and Collier [42] have used them for $HClO_4$ and $NaClO_4$.

Partial Molal Expansibilities and Compressibilities

$\partial\bar{V}_1/\partial T$, $\partial\bar{V}_2/\partial T$ and $\partial\phi_v/\partial T$ at constant pressures are termed partial and apparent molal expansibilities $\bar{E}_1$, $\bar{E}_2$ and ϕ_E respectively. Their mutual relationships can be expressed in similar terms to those used for $\bar{C}_i$ and $\bar{J}$ in an earlier section. Since α, the thermal expansion, is expressed by

$$\alpha = (\partial V/\partial T)_{P,m}/V = -(\partial d/\partial T)_{P,m}/d = -(\partial C/\partial T)_{P,m}/C \quad (6.70)$$

ϕ_E can be derived from α or d data at different temperatures.[43]

Molal compressibilities $\bar{K}_1$, $\bar{K}_2$, ϕ_K are defined by $-(\partial\bar{V}_1/\partial P)_{T,m}$, $-(\partial\bar{V}_2/\partial P)_{T,m}$ and $-d\phi/\partial_vP$. The compressibility coefficient β is defined by

$$\beta = -(\partial V/\partial P)_{T,m}/V = (\partial d/\partial P)_{T,m}/d = (\partial C/\partial P)_{T,m}/C \quad (6.71)$$

Putting K for Y in a combination of (6.1) and (6.2) and substituting from (6.71) and (6.62),

$$\phi_K = (V\beta - n_1\bar{V}_1^0\beta_0)/n_2 = \{1{,}000(\beta - \beta_0) + \beta\phi_v md_0\}/md_0 = \{1{,}000(\beta - \beta_0) + \beta_0 C\phi_v\}/C \quad (6.72)$$

Some of the more accurate information on the compressibilities of electrolyte solutions below 0.5 M has been amassed by Gibson.[44] Up to about 1,000 bars (1 bar $= 10^{-6}$ dynes/cm^2 $= 0.9869$ atm.), volume changes under pressure have been studied by piezometer, which consists of a glass tube with a ground glass stopper and a re-entrant capillary tube at the base drawn out as a tip at its top at about 90° to the main stem. This is filled with solution and inserted into a steel tube at the bottom of which is a pool of mercury. Under pressure, mercury enters the piezometer and the volume of this measures the total volume changes of the solution and the glass. A full account of the experimental and technical details have been given by Adams,[45] while Gibson [44] describes a number of improvements.

The standard pressure to which β, $\bar{K}_2$, etc., are referred is either one atm. or one bar. Owen and Brinkley [46] have given a full account of the expressions for reducing high pressure data to the standard conditions. These are largely based on Tait's equation as modified by Gibson.[44] One of these expressions is

$$\beta_P v_P = x_1 C'/(B' + P + P_e) \tag{6.73}$$

where v_P is the volume of one g of solution containing x_1 g of solvent, β_P is the compressibility coefficient at a pressure, P, C' and B' are parameters characteristic of the solvent, and P_e is a parameter representing a constant effective pressure on the solvent due to the solute. C' and B' are found by measurements on the pure solvent and P_e by obtaining β_P values for one solution at different pressures. A second expression is

$$\bar{K}_2^0 = -[\{(\bar{V}_2^0)_{P_1} - (\bar{V}_2^0)_{P_1}\}/(P_2 + P_1)][(B' + P_2)(B' + P_1)/(B' + 1)^2] \tag{6.74}$$

For water at 25°C, $B' - 2{,}996$ bars, $C' = -434.3 \times 0.3150$ ml per kgm.

The compressibilities of dilute solutions at atmospheric pressure are best measured by ultrasonic apparatus. Various physical effects can be produced by subjecting materials to high frequency pulses from an ultrasonic generator; there are a number of books [47] and articles [48] concerned with the evaluation of molal compressibilities.[49,50] Owen and Simons [50] used an ultrasonic interferometer of the movable reflector type. This consists of quartz crystals driven by a crystal controlled oscillator operating at various fixed frequencies between 0.5 and 10 Mcs. The radiations pass through solution or solvent to a plane metal reflector moving on guides, the height of which can be varied by a precision micrometer. The crystal and reflector are adjusted so that they are exactly parallel as shown by vertical symmetry and maximum depth of voltage dips on a vacuum tube voltmeter produced across the crystal circuit at nodal settings. The micrometer was calibrated by assuming that the velocity of sound along the axis of the liquid is constant throughout the range of 2 to 6 in. above the crystal. The whole apparatus was enclosed and immersed in a bath at 25 $\pm$ 0.003°C.

The velocity u in m/sec of ultrasonic waves in a fluid of density d is related to β_s (in bar^{-1}), the adiabatic compressibility, by

$$u^2 = 100/\beta_s d = 100\ \epsilon_s/d = 100\ c_p/\beta c_v d \tag{6.75}$$

where ϵ_s is the adiabatic elasticity, β is the isothermal compressibility, d is the density in g/cm^3, and c_p/c_v is the ratio of the specific heats at constant P and V. These expressions are given in numerous textbooks.[51] Also,

$$\delta = \beta - \beta_s = 10^6\alpha^2 T/c_p d J_H = 0.02391\ \alpha^2 T/\sigma \tag{6.76}$$

where α is the coefficient of thermal expansion in l./°C, c_p is in cal/g. deg, and J_H is in ergs/cal, while $\sigma = c_p d$.

By analogy with other molal properties ($X = V, C_p, E, K$), if

$$\phi_x = \phi_x^0 - S_x\sqrt{C} \tag{6.77}$$

expresses the variation of ϕ_x with C, then the corresponding directly measurable quantities $y = d, \delta, \sigma, \alpha, \beta$, must follow equations of the form

$$y = y_0 + A_y C + B_y C^{\frac{3}{2}} \tag{6.78}$$

The final equation is

$$\phi_K^0 = 1{,}000\ A_\delta + \delta_0\phi_v^0 + (2\phi_v^0 - M_2/d_0 - 2{,}000\ A_u/u_0)\beta_{os} \tag{6.79}$$

where β_{os} is the compressibility coefficient of the solvent as measured. Owen and Simons [50] have given numerical data which lead to $\phi_K^0 = -46.65 \times 10^{-4}$ for NaCl and -40.72×10^{-4} for KCl.

Expressions derived from the Debye–Hückel treatment relating ϕ_K and ϕ_K^0, $\bar{K}_2$ and $\bar{K}_2^0$, are based on differentiations of the corresponding ϕ_v and $\bar{V}_2$ expressions—(6.64), (6.67) and (6.68). Barnartt [48] has given the resulting equations.

REFERENCES

1. Ref. 10, Ch. 1.
2. Lange, E. and Monheim, J. *Z. phys. Chem.* **149A,** 51 (1930) Lange, E. and Robinson, A. L. *Chem. Rev.* **9,** 89 (1931).
3. Gucker, F. T., Pickard, H. B. and Planck, R. W. *J. Amer. chem. Soc.* **61,** 459 (1939).
4. Johnson, R. P. and Nottingham, W. B. *Rev. sci. Instrum.* **5,** 191 (1934).
5. Gulbransen, E. A. and Robinson, A. L. *J. Amer. chem. Soc.* **56,** 2637 (1934).
6. Young, T. F. and Groenier, W. L. *ibid.* **58,** 187 (1936).
7. Young, T. F. and Vogel, O. G. *ibid.* **54,** 3030 (1932).
8. Plummer, H. C. "Probability and Frequency." MacMillan and Co. Ltd., London (1940); and similar books. Cox, G. J. and Matuschak, M. C. *J. phys. Chem.* **45,** 362 (1941).
9. Spedding, F. H., Naumann, A. W. and Eberts, R. E. *J. Amer. chem. Soc.* **81,** 23 (1959); Eberts, R. E. *Diss. Abs.* **17,** 2834 (1957). KCl in 15 % sucrose and in 5 % urea—Lange, E. and Robinson, A. L. *J. Amer. chem. Soc.* **52,** 4218 (1930). NaCl in ethylene glycol—Wallace, W. E., Mason, L. S. and Robinson, A. L. *ibid.* **66,** 362 (1944).

10. Harned, H. S. and Nims, L. F. *ibid.* **54**, 423 (1932).
11. Akerlöf, G. and Teare, T. W. *ibid.* **59**, 1855 (1937).
12. Harned, H. S., Keston, A. S. and Donelson, J. G. *ibid.* **58**, 989 (1936).
13. Robinson, R. A. and Harned, H. S. *Chem. Rev.* **28**, 419 (1941).
14. Gupta, S. R., Hills, G. J. and Ives, D. J. G. *Disc. Faraday Soc.* **24**, 147 (1957). Grzybowski, A. R. *J. phys. Chem.* **62**, 550 (1958).
15. Debye, P. and Hückel, E. *Phys. Z.* **24**, 185 (1923).
16. Ref. 2, Ch. 1.
17. Harned, H. S. and Ehlers, R. W. *J. Amer. chem. Soc.* **55**, 2179 (1933). Harned, H. S. and Hecker, J. C. *ibid.* **55**, 4838 (1933).
18. Owen, B. B. and Brinkley, S. R. *Ann. N. Y. Acad. Sci.* **51**, 753 (1949). Gucker, F. T. *loc. cit.* 680.
19. Wyman, J. *Phys. Rev.* **35**, 623 (1930).
20. Malmberg, C. G. and Maryott, A. A. *J. Res. nat. Bur. Stand.* **56**, 1 (1956).
21. Tilton, L. W. and Taylor, J. K. *ibid.* **18**, 205 (1937).
22. Guggenheim, E. A. and Prue, J. E. *Trans. Faraday Soc.* **50**, 710 (1954).
23. Wallace, W. E. and Robinson, A. L. *J. Amer. chem. Soc.* 958 (1941). *Chem. Rev.* **30**, 195 (1942).
24. Gucker, F. T., Ayres, F. D. and Robin, T. R. *J. Amer. chem. Soc.* **58**, 2118 (1936).
25. Hess, C. B. and Gramkee, B. E. *J. phys. Chem.* **44**, 483 (1940). Hess, C. B. *ibid.* **45**, 755 (1941).
26. Gucker, F. T. and Schminke, K. H. *J. Amer. chem. Soc.* **54**, 1358 (1932).
27. La Mer, V. K. and Cowperthwaite, L. A. *ibid.* **55**, 1004 (1933).
28. Wyman, J. and Ingalls, E. N. *ibid.* **60**, 1182 (1938).
29. Drake, F. H., Pierce, G. W. and Dow, H. T. **35**, *Phys. Rev.* 613 (1930).
30. Randall, M. and Rossini, F. D. *J. Amer. chem. Soc.* **51**, 323 (1929). Rossini, F. D. *J. Res. nat. Bur. Stand.* **7**, 47 (1931).
31. Pitzer, K. S. *J. Amer. chem. Soc.* **59**, 2365 (1937).
32. Gucker, F. T., Gage, F. W. and Moses, C. E. *ibid.* **60**, 2582 (1938).
33. Wirth, H. E. *ibid.* **59**, 2549 (1937). Baertsch, P. and Thurkamb, M. *Helv. Chim. Acta* **42**, 282 (1959).
34. Owen, B. B., White, J. R. and Smith, J. S. *J. Amer. chem. Soc.* **78**, 3561 (1956).
35. Lamb, A. B. and Lee, R. E. *ibid.* **35**, 1666 (1913). Carton, F. and Anacker, E. W. *J. Chem. Educ.* **37**, 36 (1960).
36. Geffcken, W., Beckman, C. and Kruis, A. *Z. phys. Chem.* **20B**, 398, (1933). Holland, H. G. and Moelwyn-Hughes, E. A. *Trans. Faraday Soc.* **52**, 297 (1956).
37. Wirth, H. E., Thompson, T. G. and Utterbuck, C. L. *J. Amer. chem. Soc.* **57**, 400 (1935).
38. Masson, D. O. *Phil. Mag.* **8**, 218 (1929). Scott, A. F. *J. phys. Chem.* **35**, 2315 (1931).
39. Gibson, R. E. *J. Amer. chem. Soc.* **56**, 4 (1934).
40. Mavroides, J. G. *J. phys. Chem.* **24**, 398 (1956).
41. (a) NaCl: Robinson, A. L. *J. Amer. chem. Soc.* **54**, 1311 (1932). Kruis, A. *Z. phys. Chem.* **34B**, 1 (1936). (b) KCl: Geffcken, W. and Price, D. *ibid.* **26B**, 81 (1934). (c) HCl: Redlich, O. and Bigeleisen, J. *J. Amer. chem. Soc.* **64**, 785 (1942).
42. Wirth, H. and Collier, F. N. *ibid.* **72**, 5292 (1950).
43. Gucker, F. T. *ibid.* **56**, 1017 (1934); **61**, 1558 (1939).
44. Gibson, R. E. *ibid.* **56**, 4 (1934); **57**, 284 (1935); **59**, 1521 (1937); **60**, 511 (1938). *Amer. J. Sci.* **35A**, 49 (1938). *Sci. Monthly* **46**, 103, (1938). Gibson, R. E. and Loeffler, O. H. *Ann. N. Y. Acad. Sci.* **51**, 727, (1948); **63**, 443 (1941).
45. Adams, L. H. *J. Amer. chem. Soc.* **53**, 3769 (1931).

46. Owen, B. B. and Brinkley, S. R. *Chem. Rev.* **29,** 461 (1941).
47. Vigoureux, P. "Ultrasonics." Chapman and Hall Ltd., London; J. Wiley and Sons, New York, (1950). Richardson, E. G. "Ultrasonic Physics." Elsevier Publishing Co., New York (1952). Bergmann, L. "Der Ultraschall und Seine Anwendung in Wissenschaft und Technik." S. Hirzel Verlag, Stuttgart, 6th edn. (1954).
48. Sollner, K. *Chem. Rev.* **34,** 371 (1944). Barnartt, S. *Quart. Rev.* **7,** 84 (1952).
49. Freyer, E. B. *J. Amer. chem. Soc.* **53,** 1313 (1931). Bazhulin, P. A. *J. Phys. Moscow* **1,** 431 (1939). Branca, G. and Carelli, A. *Nuovo Cim.* **7,** 190 (1950). Krishnamurty, B. *J. Sci. Indian Res.* **9B,** 215 (1950); **10B,** 149 (1951). Maffei, F. and Buonsanto, M. *Ann. Chem. appl. Roma* **43,** 804 (1953). Hill, M. D. Univ. Microfilms, U.S.A., Pub. No. 13, 167. C.G. Balachandra, *J. Indian. Inst. Sci.* **38A,** 1 (1956).
50. Owen, B. B. and Simons, H. L. *J. phys. Chem.* **61,** 479 (1957).
51. Partington, J. R. "An Advanced Treatise on Physical Chemistry." Longmans, Green and Co. Ltd., London, I, 820 (1949). Also physics textbooks on sound.
52. Ackermann, Th. *Disc. Faraday Soc.* **24,** 180 (1957); *Z. Elektrochem.* **62,** 411 (1958). Ackermann, Th. and Schreiner, F. *loc. cit.* 1143.
53. Eucken, A. and Eigen, M. *ibid.* **55,** 343 (1951). Eigen, M. and Wicke, E. *loc. cit.* 354.

CHAPTER 7

DIFFUSION

IF A solution is formed which contains regions of differing concentrations and is not stirred, the solute and the solvent particles will migrate or diffuse until uniformity is established throughout the whole system. Representing the concentration at a point in a non-uniform solution by n (amount per unit volume), then if v is the velocity of movement of the solute,

$$nv = J = -D\nabla n \tag{7.1}$$

where J is the flow, or amount migrating in unit time. D is termed the diffusion coefficient, and the concentration gradient ∇n in terms of co-ordinate axes is

$$\nabla n = \partial n/\partial x + \partial n/\partial y + \partial n/\partial z \tag{7.2}$$

In dealing with electrolyte solutions, owing to the volume changes associated with concentration changes, J is defined as being relative to a local frame of reference moving with the solvent.

Equation (7.1) is usually termed Fick's first law of diffusion,[1] and since diffusion is commonly studied by methods which restrict the flow to one direction, which is taken along an x axis (flow in the x direction being taken as positive), (7.1) is written

$$J = -D\partial n/\partial x \tag{7.3}$$

This, and (7.1), only apply under steady state conditions when ∇n or $\partial n/\partial x$ do not change with time. When this is not so, the outcome can be seen by considering a tube of unit area cross-section along which diffusion takes place. It is supposed that two planes cut the tube perpendicular to the flow at distances x and $x + dx$. Then the concentration increase in this segment is the difference between the amount diffusing in and that diffusing out, divided by x. Hence

$$\partial n/\partial t = \{-D(\partial n/\partial x)_x + D(\partial n/\partial x)_{x+\partial x}\}/\partial x$$

Also,

$$(\partial n/\partial x)_{x+\partial x} = (\partial n/\partial x)_x + \{\partial(\partial n/\partial x)/\partial x\}\partial x$$

so that

$$\partial n/\partial t = D(\partial^2 n/\partial x^2) \tag{7.4}$$

This is termed Fick's second law of diffusion.[1] Its three-dimensional form is

$$\partial n/\partial t = \nabla . D\nabla n \tag{7.5}$$

There are two features which have an important influence on the diffusion of electrolytes, namely that cations and anions move in the same direction, and that they move with the same speed. Since moving ions are being considered, and their rates of diffusion are concentration-dependent, at least a part of the theoretical treatment of conductance is applicable. Onsager and Fuoss [2] have taken account of this in their limiting treatment which is described in Chapter 3.

At very low concentrations, if ionic interactions are taken to be negligible, and writing for a single electrolyte composed of i and j ions,

$$J_i = n_i v,\ J_j = n_j v,\ J = nv \tag{7.6}$$

and by analogy with (3.50),

$$v = k_i \omega_i = k_j \omega_j = -\ \omega_i \nabla \bar{G}'_i = -\ \omega_j \nabla \bar{G}'_j \tag{7.7}$$

where ω_i and ω_j are the velocities of the cations and anions under unit force, k_i and k_j are the forces, and it is assumed that the work done by the diffusing ions is equal to the free energy change.[3]

If now

$$k = \nu_i k_i + \nu_j k_j = -\ \nabla \bar{G}'_2 \tag{7.8}$$

then from (7.7) and (7.8),

$$k_j = k_i \omega_i / \omega_j,\ -\ \nabla \bar{G}'_2 = \nu_i k_i + \nu_j k_i \omega_i / \omega_j$$

whence

$$k_i = -\ \omega_j \nabla \bar{G}'_2 / (\nu_i \omega_j + \nu_j \omega_i) \tag{7.9}$$

and

$$v = k_i \omega_i = -\ \omega_i \omega_j \nabla \bar{G}'_2 / (\nu_i \omega_j + \nu_j \omega_i) \tag{7.10}$$

Now for an ideal solution, from (1.19), for one "molecule" of electrolyte,

$$\nabla \bar{G}'_2 = (\nu_i + \nu_j) kT \nabla n / n \tag{7.11}$$

so from (7.1) and (7.10),

$$-\ D \nabla n = J = nv = -\ (\nu_i + \nu_j) \omega_i \omega_j kT \nabla n / (\nu_i \omega_j + \nu_j \omega_i) \tag{7.12}$$

This form of the diffusion law was derived by Nernst,[4] replacing ω_i and ω_j by functions of λ_i^0 and λ_j^0, but it does not allow for interionic forces. On a first consideration these could be allowed for by replacing n by $f'n$ where f' represents activity coefficients for concentrations of n "molecules" per ml. This modification was introduced by Hartley[5] but it is not the complete story for there still remains the computation of the effect of ionic interaction upon the mobility of diffusing ions. Since all the ions in a particular solution migrate with the same velocity, an ionic atmosphere does not need any inducement to follow its central ion and consequently the atmosphere remains symmetrical. The relaxation effect, which is predominant in conductance theory, is absent from the theory of diffusion.

But an electrophoretic effect is still operative since it depends on a volume force attacking in the ionic atmosphere. From (3.50), $k_j = v\rho_j$, and from (3.67), for a simple electrolyte,

$$\Delta v_j = -\frac{2(n_i e_i \rho_i + n_j e_j \rho_j) v e_j}{3\,\eta \epsilon k T \kappa (1 + \kappa a)} + \frac{(n_i e_i^2 \rho_i + n_j e_j^2 \rho_j) v e_j^2 \phi(\kappa a)}{3\,\eta (\epsilon k T)^2} \tag{7.13}$$

where

$$\phi(\kappa a) = \exp\,(2\kappa a)\,.\,\mathrm{Ei}(2\kappa a)/(1 + \kappa a)^2 \tag{7.14}$$

$\mathrm{Ei}(x)$ is defined in (3.66). The factor Δv_j alters the forces k_i and k_j that are needed to make both ions migrate with the same velocity v, i.e.,

$$k_j = \rho_j (v - \Delta v_j) \text{ and } k_i = \rho_i (v - \Delta v_i) \tag{7.15}$$

so that, from (7.8) and (7.13),

$$k = -\nabla \bar{G}_2' = \nu_i \rho_i (v - \Delta v_i) + \nu_j \rho_j (v - \Delta v_i)$$

$$v\left\{\nu_i \rho_i + \nu_j \rho_j + (\rho_i - \rho_j)^2 \frac{\nu_i \nu_j \kappa}{(\nu_i + \nu_j) 6\pi\eta(1 + \kappa a)} - \frac{(\nu_j \rho_i + \nu_i \rho_j)^2 \kappa^4 \phi(\kappa a)}{(\nu_i + \nu_j)^2\, 48\pi^2 n \eta}\right\} \tag{7.16}$$

is the "force" per "molecule" of solute at a concentration of $n = n_i/\nu_i = n_j/\nu_j$ "molecules" per ml. The last two terms of (7.16) are due to deviations from a random arrangement. Neglecting second and higher powers of the correction Δv_j, and replacing ρ by $1/\omega$, (7.10) now becomes, on multiplying by n,

$$nv = J = -\Omega \nabla G_2' = -\{n\omega_i\omega_j/(\nu_i\omega_j + \nu_j\omega_i) + \Delta\Omega\}\nabla \bar{G}_2' \tag{7.17}$$

where

$$\Delta\Omega = -\frac{(\omega_i - \omega_j)^2 \nu_i \nu_j \kappa n}{(\nu_i\omega_j + \nu_j\omega_i)^2 (\nu_i + \nu_j) 6\pi\eta(1 + \kappa a)} + \frac{(\nu_i\omega_i + \nu_j\omega_j)\kappa^4 \phi(\kappa a)}{(\nu_i + \nu_j)(\nu_i\omega_j + \nu_j\omega_i)^2 48\pi^2\eta} \tag{7.18}$$

Since $n = N\bar{n}$, where $\bar{n}$ is the concentration in g. molecules per ml, and using $\bar{G}_2 = N\bar{G}_2'$, where $\bar{G}_2$ is the free energy in ergs per g molecule, then

$$\bar{J} = \bar{n}v = -(\Omega/N^2)\nabla\bar{G}_2 = \bar{\Omega}\nabla\bar{G}_2 \tag{7.19}$$

where $\bar{J}$ is the flow in g molecules per ml., and $\bar{\Omega} = \Omega/N^2$. Also, from (3.73),

$$\omega_j = c^* \bar{u}_j^0/10^8 e_j = \lambda_j^0 c^*/10^8 F e_j = \lambda_j^0 c^*/10^8\,|\,z_j\,|\,e \tag{7.20}$$

whence, from the last four equations,

$$\bar{\Omega} = (\lambda_i^0 \lambda_j^0 \bar{n})/(10^7 F^2 \nu_i\,|\,z_j\,|\,\Lambda^0) + \Delta\bar{\Omega}' + \Delta\bar{\Omega}'' \tag{7.21}$$

where

$$\Delta\bar{\Omega}' = -\frac{(\,|\,z_j\,|\,\lambda_i^0 - |\,z_i\,|\,\lambda_j^0)^2 \kappa \bar{n}}{(\Lambda^0)^2\,|\,z_i z_j\,|\,(\nu_i + \nu_j) N 6\pi\eta(1 + \kappa a)} \tag{7.22}$$

and

$$\Delta\bar{\Omega}'' = \frac{(z_j^2 \lambda_j^0 + z_j^2 \lambda_1^0)^2 N^2 e^4 \bar{n}^2 \phi(\kappa a)}{(\Lambda^0)^2 3\eta(\epsilon R T)^2} \tag{7.23}$$

In these expressions, κ is defined by (3.12) and since $1{,}000\,\bar{n} = C$, the molarity,

$$\kappa^2 = N^2 e^2 4\pi \mid z_i z_j \mid (\nu_i + \nu_j) C/1{,}000\,\epsilon RT \tag{7.24}$$

and taking the appropriate values of the constants (Appendix, p. 293), with R in ergs/deg. mole,

$$\kappa = 3.5563 \times 10^9 \sqrt{\Gamma}/(\epsilon T)^{\frac{1}{2}} \tag{7.25}$$

$$\bar{\Omega} = 1.0740 \times 10^{-20}(\lambda_i^0 \lambda_j^0 C/\nu_i \mid z_i \mid \Lambda^0) + \Delta\bar{\Omega}' + \Delta\bar{\Omega}'' \tag{7.26}$$

$$\Delta\bar{\Omega}' = \frac{(\mid z_j \mid \lambda_i^0 - \mid z_i \mid \lambda_j^0)^2\, 3.1321 \times 10^{-19} C\sqrt{\Gamma}}{(\Lambda^0)^2 \mid z_i z_j \mid (\nu_i + \nu_j)\eta(1 + \kappa a)(\epsilon T)^{\frac{1}{2}}} \tag{7.27}$$

$$\bar{\Omega}'' = \frac{(z_i^2 \lambda_j^0 + z_j^2 \lambda_i^0)^2\, 9.3051 \times 10^{-13} C^2 \phi(\kappa a)}{(\Lambda^0)^2 \eta (\epsilon T)^2} \tag{7.28}$$

Recalling the definition of activity, from (1.19), (1.24) and (1.27), differentiation leads to

$$\partial\bar{G}_2 = (\nu_i + \nu_j) RT\{(\partial C/C) + \partial \ln y_\pm\} \tag{7.29}$$

so that

$$C\partial\bar{G}_2/\partial C = (\nu_i + \nu_j) RT\{1 + C\partial \ln y_\pm/\partial C\} \tag{7.30}$$

It is now possible to equate the free energy function of (7.17) to the diffusion coefficient by means of (7.1), for the two equations are equal in that

$$-D\nabla n = -\bar{\Omega}\nabla\bar{G}_2', \text{ i.e., } D\partial C = 1{,}000\bar{\Omega}\partial\bar{G}_2 \tag{7.31}$$

so from (7.30),

$$D = (\nu_i + \nu_j) 1{,}000\, RT\bar{\Omega}\{1 + \mathrm{d} \ln y_\pm/\mathrm{d} \ln C\}/C \tag{7.32}$$

At infinite dilution, $\Delta\bar{\Omega}'$ and $\Delta\bar{\Omega}''$ are both zero, and denoting this value of D by D_0, from (7.28) and (7.32), since $\ln y_\pm = 0$,

$$D_0 = \lambda_i^0 \lambda_j^0 (\nu_i + \nu_j) RT/10^7 F^2 \nu_i \Lambda_0 \mid z_i \mid$$

$$= 8.9298 \times 10^{-10}\, \lambda_i^0 \lambda_j^0 (\nu_i + \nu_j) T/\nu_i \Lambda_0 \mid z_i \mid \tag{7.33}$$

$$= 8.9298 \times 10^{-10}\, T_i^0 T_j^0 \Lambda_0 (\mid z_i + z_j \mid)/\mid z_i z_j \mid \tag{7.34}$$

since

$$T_{i/j}^0 = \lambda_{i/j}^0 \Lambda^0 \text{ and } \mid \nu_i z_i \mid = \mid \nu_j z_j \mid .$$

Diffusion Coefficients in Dilute Solutions

Accurate diffusion coefficients can be determined below 0.01 M by a method devised by Harned and French.[6] This is based upon conductance measurements across the bottom and top portions of a cell in which upward diffusion takes place. Following improvements,[7] the diffusion coefficients of

KCl at 25°C from 0.0013 to 0.33 M were obtained, and since then, a number of similar investigations have been reported by Harned and his associates.[8] The experimental arrangement, which is described in full in the original articles [1,2] consists of a special cell made of lucite fitted with upper and lower pairs of electrodes (marked as *E* in Fig. 7.1a) made of sheet Pt soldered to a Cu block. The centres of the electrodes are such that the bottom pair are set at $a/6$ and the top pair set at $5a/6$ from the cell base where a is the cell height. The reason for these dimensions is explained later on.

To fill this cell it is inverted, the plates *F* and *C* removed and the cell and cup *B* filled with conductivity water. The cup has the exact area of the cell base, and on shearing plate *C* into place, excess water is sheared off; this passes out through an exit hole in *B* without introducing air bubbles. The next operation is to place plate *F* in position and to introduce salt solution into cup *A* via the filling hole *G* which is lined up with *A*. *F* is then shifted to one side as shown in Fig. (7.1a).

The filled cell is mounted on a heavy brass block enclosed in a lucite box which is placed on a platform in a large thermostat controlled to $\pm$ 0.01°C. The platform is suspended from the ceiling by a three-point spring suspension. These arrangements ensure good temperature control and eliminate turbulent flow, and also mechanical disturbance over the period required, which may be as long as a week.

After the cell has been in position for 24 h, the sliding plate *C* is moved by a wire and capstan system so that the salt solution in *A* is sited at the base of the cell. When a sufficient amount has diffused in, as shown by the conductance across the lower electrodes, plate *C* is moved so that cup *A* regains the position shown in Fig. 7.1a. After another 36 h, measurements are taken across each pair of electrodes at 2h intervals for about 5 days. Finally, the solution is stirred to uniformity by placing a heating lamp near the cell. This concentration, *C*, is derived by interpolation of its conductance from a plot with known data.

A simpler type of cell, made of 8–mm diameter soft glass has been tested by Harned and Blander [47] and found satisfactory with KCl. This is shown in Fig. 7.1b. The electrodes *H* are Pt tubes, 2 cm long, 3 mm diameter; their ends are closed and made flat by spinning and are platinized after fixing in position (positions as for the cell described above). A ground-glass plate *J*, lightly greased, clamps off the top of the cell and shears off excess water. *K* is a special stopcock; the hole *L* serves to introduce salt solution. Firstly, the cell is inverted and the cup *L* is turned to L_1 (Fig. 7.1c), filled and air bubbles removed by a sharp rod. Next, *L* is turned to L_2, the cell leads attached, the cell turned upright and placed in its lucite box (mounted on a brass block as described above). After 48 h, *L* is turned to L_3, left for 1 h, then turned to L_4. Measurements were taken at 2h intervals for 5 days.

Considering unidirectional flow in the x direction, from (7.4), since $J = 0$ for $x = 0$ and $x = a$, and the boundary conditions are $\partial C/\partial x = 0$ at these two values of x, the solution of (7.4) can be arranged in a series form,

$$C = \sum_{n=1}^{\infty} A_n \exp(-n^2\pi^2 Dt/a^2) \cos(n\pi x/a) + C_0 \tag{7.35}$$

where A_n represents the Fourier coefficients which satisfy the boundary conditions, and C is the concentration at a height x. The difference between the concentrations at the bottom electrodes (ξ) and the top electrodes ($a - \xi$), is given by

$$C(\xi) - C(a - \xi) = 2A_1 \exp(-\pi^2 Dt/a^2) \cos(\pi\xi/a) + 2A_3 \exp(9\pi^2 Dt/a^2)$$
$$\cos(3\pi\xi/a) + 2A_5 \exp(25\pi^2 Dt/a^2) \cos(5\pi\xi/a) + \ldots \tag{7.36}$$

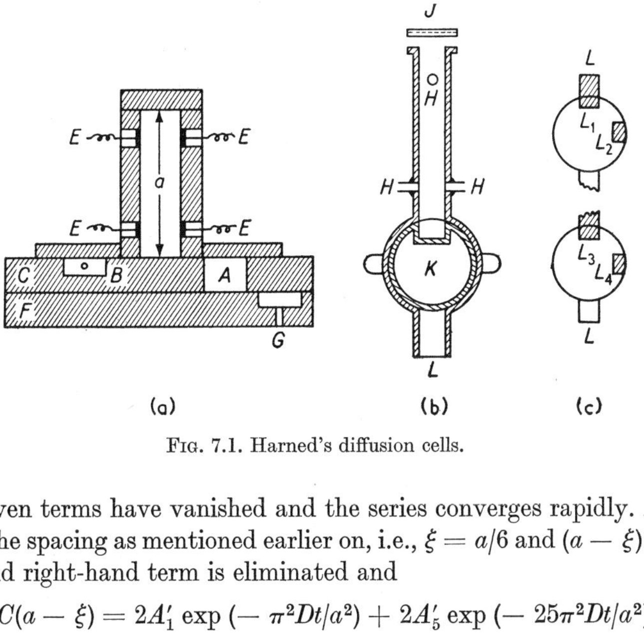

FIG. 7.1. Harned's diffusion cells.

All the even terms have vanished and the series converges rapidly. Also, by making the spacing as mentioned earlier on, i.e., $\xi = a/6$ and $(a - \xi) = 5a/6$, the second right-hand term is eliminated and

$$C(\xi) - C(a - \xi) = 2A_1' \exp(-\pi^2 Dt/a^2) + 2A_5' \exp(-25\pi^2 Dt/a^2) \tag{7.37}$$

where A_1' and A_5' include the now constant trigonometric terms. In practice, after a short time has elapsed, only the first term is of significance. Then if $1/\tau = \pi^2 D/a^2$,

$$\ln\{C(\xi) - C(a - \xi)\} = -t/\tau + \text{constant} \tag{7.38}$$

Now if K_B and K_T are the reciprocals of the resistances across the bottom and the top sets of electrodes, it can be assumed that $K_B - K_T$ is proportional to $C(\xi) - C(a - \xi)$, (for KCl, the error is [2] $\not> 0.02\ \%$), so (7.38) becomes

$$\ln(K_B - K_T) = -t/\tau + \text{constant} \tag{7.39}$$

and the slope of the line obtained on plotting ln $(K_B - K_T)$ against t is equal to $1/\tau$, so D is obtained from

$$D = a^2/\tau\pi^2 \tag{7.40}$$

and only conductance measurements at a set of suitable time intervals are required. Owing to mechanical imperfections, corrections are required to allow for differences in the cell constants of the two sets of electrodes. If K_B^* and K_T^* are the measured reciprocals,

$$\ln\{(K_B^* - K_T^*) - (K_B^\infty - K_T^\infty)\}/t = 1/\tau \tag{7.41}$$

where K_B^∞ and K_T^∞ refer to $t = \infty$, i.e., when the concentration is uniform.

Example 7.1. With $C = 0.00585$, $a = 5.697$ cm, $K_B^\infty - K_T^\infty = 0.1584 \times 10^{-4}$. At $t = 259200$ sec, $K_T^* = 15.1939 \times 10^{-4}$, $K_B^* = 26.0967 \times 10^{-4}$, whence from (7.41), $1/\tau = 5.8719 \times 10^{-6}$, so from (7.40), $D = 1.9309 \times 10^{-5}$ cm^2/sec.

Example 7.2. Harned and Nutall[7] obtained the diffusion coefficients of KCl in aqueous solution at 25°C. Their figures are shown in Table 7.1. In order to test (7.32), d ln $y_\pm$/d ln C must be evaluated; since only dilute solutions are involved, differentiation of a suitable form of the D.H. equation such as (3.40) together with the conversion d ln $C = 2\,\mathrm{d}\sqrt{C}/\sqrt{C}$ gives a suitable expression, namely

$$\mathrm{d}\ln f_\pm/\mathrm{d}\ln C = -\,0.5 \times 2.303 \times 0.5091\,\sqrt{C}/(1 + 1.15\,\sqrt{C})^2 + 0.15\,C \tag{7.42}$$

(from example (3.2), $B'\mathring{a} = 1.15$, $\mathring{a} = 3.50$ Å, $Q = 0.065$, so $2.303\,Q = 0.15$).

TABLE 7.1

Observed and Calculated Diffusion Coefficients of Dilute KCl Solutions at 25°C (cm^2/sec.)

10^3C	1.25	1.94	3.25	5.85	7.04	9.80	12.61
$\phi(\kappa a)$	2.20	1.82	1.56	1.35	1.26	1.13	1.02
$-\,10^3(\mathrm{d}\ln f/\mathrm{d}\ln C)$	18.95	23.10	29.63	37.01	39.86	45.31	49.74
$-\,10^{24}\Delta\bar{\Omega}'$	—	—	0.01	0.02	0.03	0.05	0.07
$10^{24}\Delta\bar{\Omega}''$	0.65	1.30	3.1	8.8	11.9	20.6	30.8
$10^{22}\bar{\Omega}$	5.0324	7.8131	13.098	23.609	28.424	39.609	51.009
10^5D(calc.)	1.9583	1.9507	1.9390	1.9269	1.9220	1.9131	1.906
10^5D(obs.)	1.9612	1.9545	1.9433	1.9308	1.9246	1.9155	1.908
$10^8(D'' - D_0)$ calc.	2.6	3.3	4.8	7.4	8.3	10.4	12.1
$10^8(D'' - D_0)$, obs.	5.6	7.2	9.1	11.5	11.2	12.9	14.4

Taking $\lambda_K^0 = 73.52$, $\lambda_{Cl}^0 = 76.35$ int. ohm^{-1} cm^2, from (7.26), after conversion to absolute[9] ohms, $\bar{\Omega} = 4.0207 \times 10^{-19}\,C + \Delta\bar{\Omega}' + \Delta\bar{\Omega}''$. Also, from (7.23, 7.28 and 7.32), $\Delta\bar{\Omega}' = 5.766 \times 10^{-23}\,C^{\frac{3}{2}}/(1 + 1.15\,\sqrt{C})$, $\Delta\bar{\Omega}'' = 1.8961 \times 10^{-19}\,C^2\phi(\kappa a)$, $D = 4.9580 \times 10^{13}\,\bar{\Omega}(1 + \mathrm{d}\ln f_\pm/\mathrm{d}\ln C)/C$. Values of $\phi(\kappa a)$ were calculated by means of (7.14) and (3.66); alternatively, they may be interpolated from Tables.[10]

On the average, the observed diffusion coefficients are 0.17 % higher than the calculated values. From (7.32), $D_0 = 1.9935 \times 10^{-5}$; this will be used to apply Guggenheim's test,[9] namely to examine (7.32) in the form

$$(D'' - D_0)/D_0 = -(\lambda_i^0 + \lambda_j^0)^2(b_i + b_j)\kappa a/\Lambda_0^2 2a(1 + \kappa a) + e^2(b_i + b_j)(\kappa a)^2\phi(\kappa a)/4\epsilon kTa^2 \quad (7.43)$$

where $b_{i/j} = 10^7 F^2/6\pi N\lambda_{i/j}^0$, and $D'' = D/(1 + \mathrm{d}\ln f/\mathrm{d}\ln C)$ for 1 : 1 electrolytes. Taking the numerical values for KCl,

$$D'' - D_0 = D_0\{-1.43 \times 10^{-4}\sqrt{C}/(1 + 1.15\sqrt{C}) + 0.4720\, C\phi(\kappa a)\}.$$

Figures calculated from this are shown in Table 6.1 together with $(D'' - D_0)$, obs. obtained by means of D (obs.) and the activity coefficient function values given in the third row of the Table. Guggenheim has made similar calculations [9] for NaCl, and considers that in dilute solutions, values of D can be calculated from D_0 and Onsager's formula more accurately than it has yet been possible to measure them.

It has been shown by Harned [11,12] that the process of deriving diffusion coefficients with the aid of known activity coefficients can be reversed. By re-arrangement of (7.32), for 1 : 1 electrolytes,

$$DC/(\nu_i + \nu_j)1{,}000\, RT\bar{\Omega} - 1 = D' = (\mathrm{d}\ln f_\pm)\sqrt{C}/2\,\mathrm{d}\sqrt{C} \quad (7.44)$$

so that

$$2.303 \log f_\pm = 2\int_0^C (D'/\sqrt{C})\mathrm{d}\sqrt{C} \quad (7.45)$$

Now from the limiting D.H. equation, $\log f_\pm = -A'\sqrt{C}$, so that $\mathrm{d}\log f_\pm/\mathrm{d}\sqrt{C} = -A' = 2D'/2.303\sqrt{C}$. Hence the limiting value of $D'/\sqrt{C}$ as $C \to 0$ is $2.303\,\mathrm{d}\log f/2\,\mathrm{d}\sqrt{C} = -2.303\,A'/2$, and by graphical integration of $D'/\sqrt{C}$ against $\sqrt{C}$, activity coefficients are derived.

Example 7.3. Taking the KCl data of example (7.2), and since $D' = DC/4.9580 \times 10^{13}\,\bar{\Omega} - 1$, and $A' = -0.5091$, the limiting value of

$$D'/\sqrt{C} = -2.303 \times 0.5091/2 = -0.5861,$$

and the following figures are obtained:

10^3C	1.25	1.94	3.25	5.85	7.04	9.80	12.61
$-10^3D'$	17.48	21.16	27.48	35.00	38.59	44.10	48.65

A smooth curve was drawn through the plot of $D'/\sqrt{C}$ against $\sqrt{C}$, and by tabular integration (see Example 5.3), the following values of $f_\pm$ were obtained:

10^3C	1.00	2.00	5.00	10.0
$f_\pm$	0.966	0.953	0.930	0.906

These are about 0.3 % higher than the corresponding figures in Table 3.1.

Diffusion Coefficients in Concentrated Solutions

A number of methods have been developed for measuring the diffusion coefficients of electrolytes in concentrated solutions. They have been described and reviewed by Harned,[13] and by Robinson and Stokes.[14] The brief account given here is confined to the two that have been used the most extensively, namely the Goüy interference method, and the use of sintered glass diaphragm cells.

1. The Goüy Interference Method

If light from a horizontal slit marked as *S* in Fig. 7.2a passes through a lens system inside of which is a cell *C* containing solution, initially, up to *D*, where there is a sharp boundary with the solvent, then as diffusion occurs, an interference pattern is generated and focused in the plane *P*. This pattern consists of horizontal bands of varying intensity. The intensity of these bands depends on the refractive index gradients and hence on the concentration gradients. This is of greatest change at the initial boundary *D*; on either side of this the refractive index gradients will be equal at equal distances from *D*, but they will be smaller as the distance increases.

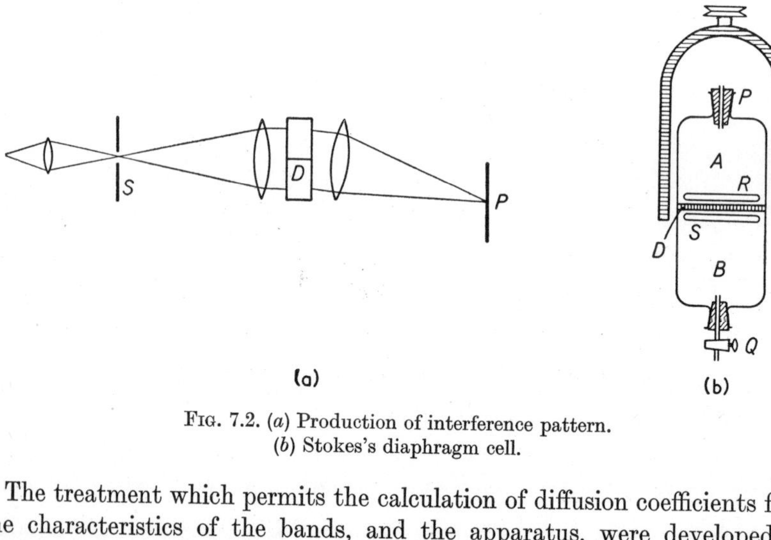

Fig. 7.2. (*a*) Production of interference pattern.
(*b*) Stokes's diaphragm cell.

The treatment which permits the calculation of diffusion coefficients from the characteristics of the bands, and the apparatus, were developed by Kegeles and Gosting,[15] Coulson *et al.*,[16] and Gosting and Onsager.[17] Gosting [18] was the first to obtain diffusion coefficients of electrolytes by this method; he investigated aqueous KCl between 0.1 N and 3.9 N and these may be compared with the results obtained by Harned and Nutall [7] up to 0.5 N by the conductance method. The results are in good agreement except at 0.1 N.

The very elaborate optical system used by Gosting was a modification of that devised by Longsworth [19] for the study of electrophoresis. Rayleigh interferometers may also be used; Creeth [20] has reported a detailed comparison, and Robinson and Stokes [3] mention and discuss a number of other methods. These authors have also given an excellent account of the theory and practical aspects of the Goüy method. Hall *et al.* [21] have used a Goüy

TABLE 7.2

Diffusion Coefficients of KCl at 25°C

$10^2 C$	10.00	22.50	33.20	50.00	1000	1500
$10^5 D$(Goüy)	1.8512	1.8382	1.8400	1.8497	1.8923	1.9427
$10^5 D$(Conductance)	1.845	1.939	1.842	1.850	—	—
$10^5 D$(calc.)	1.8469	1.8396	1.8443	1.8588	1.9206	2.0024

diffusiometer (and described its construction in detail) for KCl, NH_4Cl and $CaCl_2$ solutions. Their results for $CaCl_2$ are in close agreement with those of Lyons and Riley [22] who have also studied CsCl. They have observed that the method gives results which are too high at concentrations below and up to 0.1 N (see the first column of Table 7.2). NaCl, $BaCl_2$ and other electrolytes [23] have also been examined.

In order to calculate d ln $y_{\pm}$/d ln C, it follows from (3.44) that a further term is required to express d ln $y_{\pm}$/d ln C; this is

$$-\,C[\partial d/\partial C + 0.001(2M_1 - M_2)]/[d + 0.001(2M_1 - M_2)C].$$

Gosting [18] has applied a more elaborate expression but from about 0.3 N onwards, the calculated diffusion coefficients increase more rapidly than do the experimental values. Stokes [21] suggested that the reason for this is that allowance must be made for the fact that hydrated ions and water molecules diffuse in opposite directions. As a consequence of this, Wishaw and Stokes [23] have replaced (7.32) by

$$D = (D_0 + \Delta_1 + \Delta_2)(1 + \mathrm{d}\ln\gamma_{\pm}/\mathrm{d}\ln m)(1 + 0.018\,\nu m(D^*/D - n)\eta^0/\eta \qquad (7.46)$$

where $D_2 + \Delta_1 + \Delta_2$ is equivalent to 1,000 $\nu RT\bar{\Omega}$ of (7.32), m is the molality, n represents the number of moles of water "bound up" in ion hydration, D^* is the self-diffusion coefficient of water, while η^0 and η are the viscosities of the solvent and solution. The expression is adapted from an equation due to Agar [24] which in turn is based on one which Hartley and Crank [25] derived to express the mutual diffusion coefficient in binary non-electrolytes. A full discussion is given by Robinson and Stokes.[3] The values of $1 + (\mathrm{d}\ln\gamma_{\pm}/\mathrm{d}\ln m)$ are obtained by taking tangents from plots of experimental $\log\gamma_{\pm}$ values against $\log m$, or since $1 + d \ln \gamma_{\pm}/\partial \ln m = \phi + \partial\phi/\sqrt{m}/2\partial\sqrt{m}$ (see 5.23), by taking tangents from a plot of $\phi/\sqrt{m}$ against $\sqrt{m}$. The average value of D^* at 25°C from measurements with water containing H^2, H^3 or O^{18} is [3] 2.44×10^{-5} cm^2/sec, and since it is assumed that D is unknown, D^*/D is replaced by D^*/D_0. There are two adjustable parameters, a and n. The former enters into the calculation of Δ_1 and Δ_2 (or $\Delta\bar{\Omega}'$ and $\Delta\bar{\Omega}''$), but these are relatively minor terms and an adequate assessment of a can be taken from other sources such as activity coefficient expressions. The other parameter, n, must be adjusted until the calculated values of D agree with those determined by measurement. The test of (7.46) is therefore whether or no a single value of n can be found which is satisfactory. This has been found to be so for a number of 1 : 1 electrolytes,[3] the limiting concentration being about 1.0 M; it fails when it is applied to $CaCl_2$, the only strong 2 : 1 electrolyte that has been studied at high concentrations.

Harned and Owen [26] have made the criticisms that D^* varies with the salt concentration and that the $\Delta\bar{\Omega}'$ and $\Delta\bar{\Omega}''$ corrections are not applicable to concentrated media. Another comment is that made by Stokes,[27] namely that in the derivation of the electrophoretic terms (7.27, 7.28), Onsager and Fuoss [2] only took the first and second terms of the expansion of an exponential series (see 3.67). This is satisfactory for symmetrical electrolytes since the series converges rapidly, but is unsatisfactory for unsymmetrical types, for which the first order term only is justifiably acceptable, i.e., only the Δ_1 or $\Delta\bar{\Omega}'$ correction should be retained. Stokes demonstrated this by calculating diffusion coefficients for dilute solutions of $CaCl_2$ and obtained values in better agreement with the measured ones [8] than are given by (7.32). Harned and Parker [8] have since then repeated the experimental work and these fresh results agree more closely with those calculated by means of (7.32) than do those derived by Stokes, although the limiting concentration is about 0.01 M.

2. The Diaphragm Cell Method

This method is based on the rate of change of concentration which takes place when solution diffuses through a porous barrier separating two different concentrations. Northrop and Anson [28] who are generally credited with having initiated this method, used a vertical cell with a horizontal membrane made of powdered glass, the more concentrated solution being placed in the upper compartment. The various improvements made up to 1950 have been reviewed by Gordon,[29] Harned [30] and Stokes.[31] From a consideration of the results then to hand, Stokes [31] concluded that the effects of stirring, bulk flow as distinct from diffusion, the choice of a suitable standard for cell calibration and the possible effects of the large surface area of the diaphragm pores all needed further investigation. The outcome of his work is the diaphragm cell shown in Fig. 7.2b. Each compartment, A and B, has a volume of about 50 ml., the diaphragm D is made of sintered glass 40 mm diameter 2 to 3 mm thick and No. 4 porosity. R and S are stirrers formed of glass tubing inside of which lengths of iron wire are sealed. These stirrers are weighted in such a way that they rest against the diaphragm when the cell is filled. M is a rotating magnet which causes R and S to wipe both sides of D. The rubber stoppers P and Q are pushed in to reference marks so that the volumes are reproduced to within 0.02 ml. In order to avoid disturbance of liquid in the diaphragm, Q is fitted with a capillary tube and stopcock while P contains an open capillary which extends above the level of the liquid in the thermostat bath.

The cell is filled with solution which has been freed from gas by pumping, and suction is also applied at P to remove bubbles from D. Since gas may be left in B, A is refilled, P replaced by a stopper, the cell inverted and B refilled. The cell is then left in the bath for 10 min, solution A drawn off, the compartment rinsed several times with gas-free distilled water and then filled with this, taking care not to disturb the solution in D. After allowing diffusion to proceed for several hours, A is again filled with fresh gas-free water. Diffusion is allowed to proceed for one to four days, then samples withdrawn from A and B for analysis.

Stokes [31] found that a minimum stirring rate of 25 r.p.m. was required, and that even at the fastest stirring rate of 80 r.p.m. liquid was not forced through D. Operating the cell with solution in A and water in B, and no stirring, the results were 6 to 10 % too high. It was also found that it is important to keep the diaphragm as horizontal as possible to avoid bulk flow—within two degrees to keep the error below 0.2 %. In addition, on comparison with Harned's conductimetric method for KCl, the diaphragm cell does not give reliable results below 0.05 N; below this, the results are on the high side. This is attributed to transport effects on the surface of the sintered glass.

The theoretical treatment [3, 29-32] is developed on the assumption that the diaphragm consists of parallel pores of length l and effective cross-sectional area A. Then if a quantity $\mathrm{d}q$ of solute diffuse through in time $\mathrm{d}t$,

$$\mathrm{d}q = D(A/l)(C' = C'')\mathrm{d}t \tag{7.47}$$

where C' and C'' are the concentrations in B and A respectively, while if V_1 and V_2 are the volumes of B and A, and $\mathrm{d}C'$ and $\mathrm{d}C''$ are the respective concentration changes in the two compartments,

$$V_1\mathrm{d}C' + (D''A/l)(C' + C'')\mathrm{d}t = 0 \tag{7.48}$$

and

$$V_2\mathrm{d}C'' + (D''A/l)(C'' - C')\mathrm{d}t = 0 \tag{7.49}$$

where D'' is an "effective" diffusion coefficient. It follows that

$$d(\varDelta C)/\varDelta C + \beta D'' dt = 0 \tag{7.50}$$

where

$$\beta = (A/l)(V_1^{-1} + V_2^{-2}) \tag{7.51}$$

If C_1 and C_3 are the concentrations C' at $t = 0$ and $t = t$, while C_2 and C_4 are those of C'' at $t = 0$ and $t = t$, then by integration of (7.50),

$$\ln \{(C_1 - C_2)/(C_3 - C_4)\} = \int_0^t \bar{D}(t) dt \tag{7.52}$$

where $\bar{D}(t)$ is a "concentration average" diffusion coefficient at time t. Also, if

$$\bar{D} = t^{-1} \int_0^t \bar{D}(t) dt \tag{7.53}$$

where $\bar{D}$ is termed the "diaphragm cell" diffusion coefficient, then

$$\bar{D} = (\beta t)^{-1} \ln \{(C_1 - C_2)/(C_3 - C_4)\} \tag{7.54}$$

It has been shown by Gordon [29] and confirmed by Stokes [31, 32] that there is not more than 0.02 % error in taking

$$\bar{D} = \{(1/(C'_m - C''_m)\} \int_{C''_m}^{C'_m} D dC \tag{7.55}$$

where $C'_m = (C_1 + C_3)/2$, $C''_m = (C_2 + C_4)/2$, and D is the true diffusion coefficient.

This expression is used to establish the cell constant, β. The conductimetric values of D for KCl obtained by Harned and Nutall are plotted against C and the area found under the curve between $C''_m = 0.0125$ and $C'_m = 0.0875$. Then from (7.55), $\bar{D}$ is found to be 1.866×10^{-5} cm²/sec. By measuring t, V_1, V_2, C_3, C_4 and calculating C_1 from the last four quantities, since C_2 is zero (A is initially filled with water), substitution in (7.54) gives β. A second method is described below.

The next stage is to derive D. If [32] $C_1 = C$, $C_2 = 0$ and $\bar{D}^0$ is the integral diffusion coefficient which would be obtained in a run of vanishingly small duration, then

$$\bar{D}^0 = C^{-1} \int_0^C D dC \tag{7.56}$$

Before developing this for the derivation of D, its use for obtaining β will be analyzed. To do this, let $\bar{D}_a^0$ and $\bar{D}_b^0$ be the corresponding values of $\bar{D}^0$ in hypothetical experiments where $C = C'_m$ and C''_m respectively. Then as Fig. 7.3 shows, on considering the areas under the curve,

$$\int_0^{C'_m} D dC = \int_0^{C''_m} D dC + \int_{C'_m}^{C''_m} D dC \tag{7.57}$$

whence, from (7.55) and (7.56),

$$C'_m \bar{D}^0_a = C''_m \bar{D}^0_b + \bar{D}(C'_m - C''_m) \tag{7.58}$$

so that

$$\bar{D}^0_a = \bar{D} - (C''_m/C'_m)(\bar{D} - \bar{D}^0_b) \tag{7.59}$$

or

$$\bar{D} = \{\bar{D}^0_a - \bar{D}^0_b(C''_m/C'_m)\}/\{1 - (C''_m/C'_m)\} \tag{7.60}$$

$\bar{D}^0_a$ and $\bar{D}^0_b$ can be obtained for KCl by graphical integration, obtaining D_0 via (7.34); this follows from (7.56). Thus for $C''_m = 0.0125$, $\bar{D}^0_b = 1.938 \times 10^{-5}$ and for $C'_m = 0.0875$, $\bar{D}^0_a = 1.878 \times 10^{-5}$, whence from (7.60), $\bar{D}$ for 0.1 N KCl $= 1.867 \times 10^{-5}$. (cf. 1.866×10^{-5} derived earlier on from (7.55)); this is placed in (7.54) to establish β.

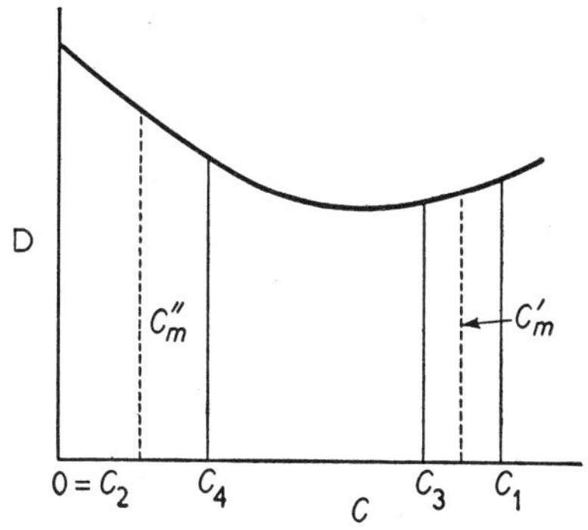

FIG. 7.3.

By combining the values of D for KCl obtained by Goüy method [18] between 0.1 N and 3.9 N with the conductimetric values [7,6] which extend up to 0.5 N, Stokes [32] has calculated $\bar{D}^0$ values at a series of round concentrations between 0 and 3.9 N. These are more convenient for cell calibration than the above data since the concentration and the duration of the run can be varied considerably.

In order to derive the diffusion coefficients D of unknowns, the first step is to calculate $\bar{D}^0$ by means of (7.59). To do this, values of $\bar{D}$ obtained by means of (7.54) are plotted against $\sqrt{C_1}$ and extrapolated to D_0 (given by 7.34). Then from this curve, first values of $\bar{D}^0_b$ are read off and substituted into (7.59). The resulting values of $\bar{D}^0_a$ are plotted against $\sqrt{C'_m}$ and the curve used to get second values of $\bar{D}^0_b$ and hence of $\bar{D}^0_a$. This procedure is made possible because the plot of $\bar{D}^0_a$ against $\sqrt{C'_m}$ lies within 1 % of that of $\bar{D}$ against $\sqrt{C_1}$. The final step is to insert the figures into the differential of (7.56):

$$D = \bar{D}^0 + C(\mathrm{d}\bar{D}^0/\mathrm{d}C) \text{ or } D = \bar{D}^0 + \tfrac{1}{2}\sqrt{C}(\mathrm{d}\bar{D}^0/\mathrm{d}\sqrt{C}) \tag{7.61}$$

The second terms on the right-hand side of these equations are obtained by graphical or tabular methods.

Example 7.4. Stokes [32,33] obtained the diffusion coefficient of a number of 1 : 1 electrolytes by the diaphragm cell. The figures given in Table 7.3 are some of those for HCl solutions at 25°C. Taking $\lambda^0_{\mathrm{H}} = 349.8$, $\lambda^0_{\mathrm{Cl}} = 76.35$, from (7.34), $D_0 = 3.337 \times 10^{-5}$. From this and the column of $\bar{D}$ values, first values of $\bar{D}^0_b$ were derived; these are given in the 5th column. Stokes has not given the values of C''_m; those given in the Table assume that the two parts of the cell have the same volume. The figures of the last column were derived by means of (7.59). As an example of a calculation of D, at $C = 0.05$, from the plot of $\bar{D}^0$ against $\sqrt{C}$, $\partial\bar{D}^0/\partial\sqrt{C} = -0.35 \times 10^{-5}$, and $\bar{D}^0 = 3.113 \times 10^{-5}$, so from (7.61), $D = 3.074 \times 10^{-5}$.

TABLE 7.3

Integral Diffusion Coefficients for HCl Solutions

C_1	$10^5\bar{D}$	C'_m	C''_m	$10^5\bar{D}^0_b$	$10^5\bar{D}_a$
0.01013	3.198	0.00864	0.00149	3.29	3.212
0.02079	3.152	0.01765	0.00319	3.25	3.170
0.07926	3.095	0.06871	0.01055	3.20	3.111
0.09722	3.066	0.08386	0.01366	3.185	3.085

Self Diffusion and Tracer Diffusion

The availability of radio-isotopes has, in recent years, stimulated the study of self diffusion, which is the diffusion of particles in their own uniform solution, and tracer diffusion, which is the diffusion of a particle in a relatively large concentration of another solute. The study of these processes in electrolyte solutions has been founded on two methods, namely the capillary tube method and the use of diaphragm cells.

1. The Capillary Tube Method

This was introduced by Anderson and Saddington [34] and consists of allowing a salt solution containing a tracer, which is usually radioactive, to diffuse from a capillary tube open at its top into salt solution of the same concentration. Numerous investigations have been reported by Wang,[35] Mills [36] and others, but many of these are suspect since various sources of errors have gradually been revealed although by now it is believed that most of these have been minimized or eliminated.

The diffusion tubes consist of even bore capillary tube usually 0.3 to 0.5 mm internal bore and 2 to 5 cm long. The ends are ground flat and smooth, a glass seal is made at one end and the internal length measured by e.g., micrometer calipers and a steel plunger. The tubes are calibrated by filling them with de-gassed solution by means of fine pipettes drawn from glass tubing, taking precautions to avoid contaminating the outside of the capillary, then centrifuged into a centrifuge tube. Several washings are needed to extract all the activity from the tube. The contents of the centrifuge tube can either be made up to a known volume and a portion taken, or the entire contents taken for counting.

It has been found necessary to stir the electrolyte solution into which diffusion takes place because this disperses local concentrations near the exit of the capillary. This stirring must be non-turbulent and within specified limits; this was demonstrated by Mills [37] who has designed a cell of 6 to 7

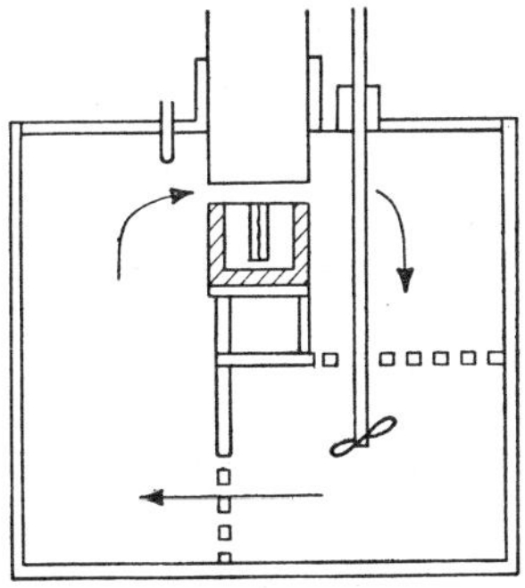

FIG. 7.4. Capillary tube diffusion apparatus of Mills and Godbole.[38]

litres capacity for this purpose. The essentials are shown in Fig. 7.4. The streamlined flow over the top of the capillary should be between 1 and 3 mm per sec. Without stirring, the results are 4 % too low and at 6 to 8 mm per sec they are 1 % to 3 % too high; the chosen rate is that which gives results for the self diffusion of the Na^+ ion in NaCl solutions that agrees with those obtained with the stirred diaphragm cell described in the preceding section.

To ensure that the tube and its contents are equilibrated before starting the diffusion, and to prevent mechanical disturbance, the capillary was fixed in a holding device moving on a vertical thread. This was lowered into the solution until only the top was exposed, then immersed to a few mm below surface after a suitable time. Mills and Godbole [38] have improved the precision by fixing the capillary in the centre of a plastic scintillator shielded by lead and locating a photomultiplier tube within a few mm of the top of the capillary. In this way the rate of decrease of activity could be continuously monitored and the accumulated data enable D to be found to within $\pm$ 0.2 % compared with $\pm$ 1 % by the older procedures.

There are several minor corrections. For instance, the background count was found to rise rapidly at first and then remained constant. Also, to correct for any fluctuations in the counting equipment, a standard source was fixed in the brass cover of the light-tight tank which surrounded the diffusion vessel. Furthermore, since the gradient of the activity is not symmetrical about the mid-point of the capillary, a correction is needed. Mathematical analysis showed this to be about 0.3 %, but in a revised procedure [39] for calculating D from the data, they showed that there is < 0.1 % error by taking the measured counting rate. Furthermore, they have shown that the initial count C_0 need not be known. The details are given later on.

In general, D is found by the appropriate solution of (7.4), where C is the concentration of tracer at time t. The concentration of tracer at the upper end of the capillary is effectively zero, and if l is the length of the capillary and C_0 the initial concentration of tracer in the tube, the limiting conditions are

$$\text{(i) } x = 0,\ \partial C/\partial x = 0;\ \text{(ii) } x = l,\ C = 0;\ \text{(iii) } t = 0,\quad C = C_0 \qquad (7.62)$$

Assuming that [40]

$$C = \exp(\alpha x + \beta t) \qquad (7.63)$$

since $\partial^2 C/\partial x^2 = \alpha^2 \exp(\alpha x + \beta t)$ and $\partial C/\partial t = \beta \exp(\alpha x + \beta t)$, then from (7.4), $\beta = D\alpha^2$. Under these conditions, (7.63) is a solution of (7.4). Taking $\alpha = i\mu$, where $i = \sqrt{-1}$, then

$$C = \exp(i\mu x + \beta t) = \exp(i\mu x - D\mu^2 t)$$
$$= \tfrac{1}{2}\exp(-D\mu^2 t)(\exp i\mu x \pm \exp - i\mu x) \qquad (7.64)$$

are both solutions of (7.4). Then from Euler's sine and cosine series,

$$C = a \exp - D\mu^2 t \cos \mu x = b \exp - D\mu^2 t \sin \mu x$$
$$= \exp - D\mu^2 t(a \cos \mu x + b \sin \mu x) \qquad (7.65)$$

are also solutions of (7.4), where a and b are constants. From this, taking (i) of (7.62), since $x = 0$, then $\partial C/\partial x = \exp - D\mu^2 t(- a\mu \sin \mu x + b\mu \cos \mu x) = \exp - D\mu^2 t(- a \sin \mu x)$. Taking (ii), in order to satisfy the last part of (7.65), $\cos 1 = 0$, or in general, since $\cos \tfrac{1}{2}(2n - 1)\pi$ is also zero when $\pi = 180°$, $\mu = (2n + 1)\pi/2l$, i.e.,

$$C = a_1 \exp\{(-\pi/2l)^2 Dt\} \cos(\pi x/2l) + a_2 \exp\{-(3\pi/2l)^2 Dt\} \cos(3\pi x/2l) + \ldots \qquad (7.66)$$

and taking (iii), from (7.66), since $\exp(0) = 1$,

$$C_0 = a_1 \cos(\pi x/2l) + a_2 \cos(3\pi x/2l) + \ldots \qquad (7.67)$$

By means of a Fourier series it can be shown that in general, for $n = 0$ to $n = \infty$,

$$a_n = (2C_0/l)\int_0^l \{(-1)^n \cos(2n+1)\pi x/2l\}dx = (-1)^n 4C_0/(2n+1)\pi \quad (7.68)$$

so that

$$C/C_0 = \sum_{n=0}^{\infty} \{4(-1)^n/\pi(2n+1)\} \exp\{-(2n+1)^2\pi^2 Dt/4l^2\} \cos(2n+1)\pi x/2l \quad (7.69)$$

Now C refers to the concentration at time t at the bottom of the tube, where $x = 0$, whereas the mean concentration, C_m, is measured. Since $C_m = l^{-1}\int_0^l C dx$

$$C_m/C_0 = \sum_{n=0}^{\infty} \{8/(\pi^2(2n+1)^2\} \exp - \pi^2(2n+1)^2 Dt/4l^2 \quad (7.70)$$

$$= (8/\pi^2)\{\exp(-\theta) + \tfrac{1}{9}\exp(-9\theta) + \tfrac{1}{25}\exp(-25\theta) + \ldots\}$$
$$\text{where } \theta = \pi^2 Dt/4l^2 \quad (7.71)$$

Providing t and l are sufficiently large, the series converges rapidly and few terms are needed. The practical limitation is that if the tube is too long, the time required for C_m to be of the order of $\frac{1}{2}C_0$ is excessive. When Dt/l^2 is > 0.24, only $n = 0$ need be taken for an accuracy of $> 0.1\%$ i.e.,

$$\pi^2 Dt/4l^2 = \ln(8C_0/C_m\pi^2) \quad (7.71)$$

This can be arranged as

$$(\pi^2 D/4l^2)t + \ln C_m = \ln(8C_0/\pi^2) \quad (7.72)$$

and with the continual monitoring method of Mills and Godbole,[38,39] a plot of $\ln C_m$ against t has a slope of $\pi^2 D/4l^2$. Others have left the capillary to diffuse for several days, and have measured C_m for a single specified time (as well as measuring C_0).

Wang and Kennedy [41] have devised a diffusion cell where diffusion takes place in tubes of "infinite" length. The cell contained eight capillaries and was made in three sections which could be sheared so that the capillary segments are lined up. Each capillary was first filled with inactive solution, the cell turned so than an end segment was isolated, the inactive solution in this replaced by an active one, the cell re-aligned and diffusion allowed to proceed into the other two segments.

2. The Diaphragm Cell Method

A number of studies have been made with the stirred diaphragm cell described earlier on. Mills [36] used this for a careful study of the self diffusion of Na^+ ions in NaCl solutions (via Na [22]). The data, which are given in Table

7.4 are useful references for assessing the precision of the capillary tube method under various conditions of stirring, etc.

TABLE 7.4

Self-Diffusion Coefficients of Na^+ in NaCl Solutions at 25°C, in cm^2/sec.

C	0.230	0.286	0.400	0.500	0.638	0.810	1.00	1.69	2.25	2.72	3.00	4.00	4.373
$10^5 D$	1.294	1.290	1.285	1.279	1.270	1.247	1.234	1.161	1.110	1.061	1.033	0.934	0.895

The average deviation is $\pm$ 0.5 %. Two important practical points are that the solutions should be de-fibred by filtration through glass sinters before filling the cell, and that a suitable stirring rate is 60 to 110 r.p.m. (cf. a minimum stirring rate of 25 r.p.m. mentioned by Stokes [31]).

The self diffusion of the Cl^- ion in LiCl, NaCl and KCl solutions and the tracer diffusion of the Na^+ ion in KCl solutions [43] have been followed by this method, and likewise Stokes, Woolf and Mills [44] have measured the tracer diffusion of the I^- ion in alkali metal chloride solutions by both chemical and radiotracer means. For the former, the iodide was introduced as inactive KI at concentrations between 0.01 and 0.1 M, and the results extrapolated to zero concentration of KI.

Onsager [45] derived a limiting equation for the diffusion coefficient D_j of minute concentration of an ion j whose transference number is small. The equation has the form [46]

$$D_j = \omega_j\{kT - (\kappa e_j^2/3\epsilon)(1 - \sqrt{\mathrm{d}\omega_j})\} \tag{7.73}$$

where

$$\mathrm{d}\omega_j = (\bar{\lambda}/\bar{\Gamma})\sum_i T_i/(\omega_i + \omega_j) \tag{7.74}$$

and $\bar{\lambda} = \sum_i n_i e_i^2 \omega_i$, $T_i = n_i e_i^2 \omega_i/\bar{\lambda}$, $\kappa^2 = 4\pi\bar{\Gamma}/\epsilon kT$, $\bar{\Gamma} = \sum_i n_i e_i^2$, $\omega_i = 10^{-8}$ $c^*\lambda_i^0/\mid z_i \mid Fe$ (see 3.73). Using the ω_i term, $n_i = NC_i/1{,}000$ (see 3.29) and $Ne/c^* = F/10$,

$$D_j = 10^{-7}\lambda_j^0 RT/\mid z_j \mid F^2 - \{10^{-9}\lambda_j^0 \mid z_j \mid c^{*2}e\sqrt{4\pi}/3\epsilon(1{,}000\,\epsilon RT)^{\frac{1}{2}}\}$$
$$(1 - \sqrt{\mathrm{d}\omega_j})\sqrt{\Gamma} \tag{7.75}$$

where $\Gamma = \sum_i C_i z_i^2$, and R is in ergs/deg. mole. Substituting the numerical values of the constants,

$$D_j = 8.930 \times 10^{-10}\lambda_j^0 T/\mid z_j \mid - 1.769 \times 10^{-3}\lambda_j^0 \mid z_j \mid (1 - \sqrt{\mathrm{d}\omega_j})\sqrt{\Gamma}/\epsilon^{\frac{3}{2}}\sqrt{T}) \tag{7.76}$$

$$\mathrm{d}\omega_j = (1/\Gamma)\sum C_i \mid z_i \mid \lambda_i^0/(\lambda_j^0 \mid z_j \mid + \lambda_i^0 \mid z_i \mid) \tag{7.77}$$

For the tracer diffusion of minute concentrations of an ion j of concentration C_1 and valence z_1 in a single electrolyte consisting of non-common ions C_2, z_2 and C_3, z_3, since $C_2 \mid z_2 \mid = C_3 \mid z_3 \mid$,

$$\mathrm{d}\omega_j = \{ \mid z_1 \mid /(\mid z_2 \mid + z_3 \mid)\}\{ \mid z_2 \mid \lambda_2^0/(\mid z_1 \mid \lambda_2^0) + \mid z_2 \mid \lambda_1^0) + \mid z_3 \mid \lambda_3^0/(\mid z_1 \mid \lambda_3^0 + \mid z_3 \mid \lambda_1^0)\} \quad (7.78)$$

and for the self diffusion of 1 : 1 electrolytes, if ions 1 and 2 are common,

$$\mathrm{d}\omega_j = (\lambda_2^0 + 3\lambda_3^0)/4(\lambda_2^0 + \lambda_3^0) \quad (7.79)$$

Example 7.5. Wang [46] obtained the tracer diffusion of Na^+ ions in aqueous KCl at 25°C by the capillary tube method. From (7.78), and the λ^0 data, $\mathrm{d}\omega_1 = 0.5943$, and from (7.76), $10^5 D_j = 1.334 - 0.239\sqrt{C}$. Some of the measured and calculated results are:

C_{KCl}	0	0.00494	0.0125	0.0375	0.0550
$10^5 D_{\mathrm{obs.}}$	—	1.31 ± 0.035	1.32 ± 0.035	1.30 ± 0.02	1.29 ± 0.02
$10^5 D_{\mathrm{calc.}}$	1.334	1.317	1.307	1.288	1.278

The limited range of validity of the equations, and the large uncertainties in the experimental results rule out any critical assessment, but the results are of the order predicted. This is true of other results obtained by this method and up to the present, no results have been published on tracer and self diffusion in dilute solutions of sufficient precision to supply a rigid test of the theory.

Turning to concentrated solutions, Mills and Kennedy [36] have found that by plotting $D\eta/\eta_0$ against $\sqrt{C}$, straight lines are obtained for the tracer diffusion of the I^- ion in concentrated alkali metal iodides. On replacing $\sqrt{C}$ in (7.76) by $\sqrt{C}/(1 + B''a\sqrt{C})$, the figures fall on a smooth curve with not more than 2 % scatter. This is similar to the attempt to extend the limiting Onsager conductance equation (see 3.81). However, it has been shown by Robinson and Stokes [3] that the viscosity correction is of a different form to that used by Mills and Kennedy, and is characteristic of ion sizes.

REFERENCES

1. Fick, A. *Ann. Phys., Lpz.* **94,** 59 (1855).
2. Onsager, L. and Fuoss, R. M. *J. phys. Chem.* **36,** 2689 (1932).
3. Ref. 11, Ch. 1.
4. Nernst, W. *Z. phys. Chem.* **2,** 613 (1888).
5. Hartley, G. S. *Phil. Mag.* **12,** 473 (1931).
6. Harned, H. S. and French, D. M. *Ann. N.Y. Acad. Sci.* **44,** 267 (1945).
7. Harned, H. S. and Nutall, R. L. *J. Amer. chem. Soc.* **69,** 736 (1947).
8. Harned, H. S. *J. Amer. chem. Soc.* (a) LiCl and NaCl: with Hildreth, C. L. **73,** 650 (1951). (b) RbCl: with Blander, M. **75,** 2853 (1953). (c) CsCl: with Blander, M. and Hildreth, C. L. **76,** 4219 (1954). (d) KCl: Ref. 7 and with Nutall, R. L. **71,** 1460,

(1949): with Blake, C. A. **72,** 2265, (1950). (e) $LiClO_4$ and $KClO_4$: with Parker, H. W. and Blander, M. **77,** 2071 (1955). (f) KNO_3: with Hudson, R. M. **73,** 652 (1951). (g) $AgNO_3$: with Hildreth, C. L. **73,** 3292 (1951). (h) Li_2SO_4 and Na_2SO_4: with Blake, C. A. **73,** 2448 (1951). (i) Cs_2SO_4: as (h), 5882. (j) $MgCl_2$ and $BaCl_2$: with Polestra, F. M. **76,** 2064, (1954). (k) $CaCl_2$: with Levy, A. L. **71,** 2781 (1949); with Parker, H. S. **77,** 265 (1955). (l) $SrCl_2$: with Polestra, F. M. **75,** 4168 (1953). (m) $LaCl_3$: with Blake, C. A. **73,** 4255 (1951). (n) KCl in 0.25 M sucrose: with Shropshire, J. A. **80,** 5652 (1958).

9. Guggenheim, E. A. *Trans. Faraday Soc.* **50,** 1048 (1954).
10. Stokes, R. H. *J. Amer. chem. Soc.* **75,** 4563 (1953). Ref. 11, Ch. 1.
11. Harned, H. S. *Proc. nat. Acad. Sci.* **40,** 551 (1954).
12. Harned, H. S. *Disc. Faraday Soc.* **24,** 7 (1957).
13. Harned, H. S. *Chem. Rev.* **40,** 461 (1947).
14. Ref. 11, Ch. 1.
15. Kegeles, G. and Gosting, L. J. *J. Amer. chem. Soc.* **69,** 2516 (1947).
16. Coulson, C. A., Cox, J. T., Ogston, A. G. and Philpot, J. St. L. *Proc. roy. Soc.* **192A,** 382 (1948).
17. Gosting, L. J. and Onsager, L. *J. Amer. chem. Soc.* **74,** 6066 (1952).
18. Gosting, L. J. *ibid.* **72,** 4418 (1950).
19. Longsworth, L. G. *Ind. Eng. Chem. Anal.* **18,** 219 (1946). *J. Amer. chem. Soc.* **69,** 2510 (1947).
20. Creeth, J. M. *ibid.* **77,** 6428 (1955).
21. Hall, J. R., Wishaw, B. F. and Stokes, R. H. *ibid.* **75,** 1556 (1953).
22. Lyons, P. A., and Riley, J. F. *ibid.* **76,** 5216 (1954).
23. Vitagliano, V. and Lyons, P. A. *ibid.* **78,** 1549 (1956). Wishaw, B. F. and Stokes, R. H. *ibid.* **76,** 2065 (1954).
24. Agar, J. N. private communication to R. H. Stokes.
25. Hartley, G. S. and Crank, J. *Trans. Faraday Soc.* **45,** 801 (1949).
26. Ref. 2, Ch. 1.
27. Stokes, R. H. *J. Amer. chem. Soc.* **75,** 4563 (1953).
28. Northrop, J. H. and Anson, M. L. *J. gen. Physiol,* **12,** 543 (1929).
29. Gordon, A. R. *Ann. N.Y. Acad. Sci.* **46,** 285 (1945).
30. Harned, H. S. *Chem. Rev.* **40,** 461 (1947).
31. Stokes, R. H. *J. Amer. chem. Soc.* **72,** 763 (1950).
32. Stokes, R. H. *ibid.* **72,** 2243 (1950); **73,** 3527 (1951).
33. Dunlop, P. J. and Stokes, R. H. *ibid.* **73,** 5456 (1951).
34. Anderson, J. S. and Saddington, K. *J. chem. Soc. Suppl.* 381 (1949).
35. Wang, J. H. *J. Amer. chem. Soc.* **73,** 510 (1951), **74,** 1182; Wang, J. H. and Miller, S. *ibid.* **74,** 1611 (1952); Wang, J. H. *ibid.* **75,** 1769 (1953); Wang, J. H. and Polestra, F. M. *ibid.* **76,** 1584 (1954).
36. Mills, R. and Kennedy, J. W. *ibid.* **75,** 5696 (1953); Mills, R. and Adamson, A. W. *ibid.* **77,** 3454 (1955); Friedman, A. F. and Kennedy, J. W. *ibid.* **77,** 4499 (1955); Matuura, R. and Shimozawa, R. *Mem. Fac. Sci. Kyūshū Univ.* Series C, **2,** 53 (1955).
37. Mills, R. *J. Amer. chem. Soc.* **77,** 6116 (1955).
38. Mills, R. and Godbole, E. W. *Aust. J. Chem.* **11,** 1 (1958).
39. Mills, R. and Godbole, E. W. *ibid.* **112,** 102 (1959).
40. Mellor, J. W. "Higher Mathematics for Students of Chemistry and Physics." Longmans, Green and Co., London, 1916, Ch. VIII.
41. Wang, J. H. and Kennedy, J. W. *J. Amer. chem. Soc.* **72,** 2080 (1950).

42. Adamson, A. W. *J. chem. Phys.* **15,** 762 (1947); Adamson, A. W., Cobble, J. W. and Nielson, J. M. *ibid.* **17,** 740 (1949); Nielson, J. M., Adamson, A. W. and Cobble, J. W. *J. Amer. chem. Soc.* **74,** 446 (1952).
43. Mills, R. *J. phys. Chem.* **61,** 1258 (1957).
44. Stokes, R. H., Woolf, L. A. and Mills, R. *J. phys. Chem.* **61,** 1634 (1957).
45. Onsager, L. *N.Y. Acad. Sci.* **46,** 241 (1945).
46. Gosting, L. J. and Harned, H. S. *J. Amer. chem. Soc.* **73,** 159 (1951); Wang, J. H. *ibid.* **74,** 1182 (1952).
47. Harned, H. S. and Blander, M. *J. phys. Chem.* **63,** 2078 (1959).

CHAPTER 8

INCOMPLETE DISSOCIATION, PART I: CONDUCTANCE METHODS

A. Application of the Limiting Onsager Equation

1. Salts. The Method of Davies

Although the limiting Onsager–Fuoss conductance equation is only valid at very low concentrations (see Example 3.4), precise measurements can be obtained in these regions, and when the experimental conductances are less than those calculated by (3.79), it may be assumed that this is due to incomplete dissociation. This can be discussed in a quantitative manner. Taking the simplest examples, the 1 : 1 valent electrolytes, Davies [1] has used the following procedure. The limiting equation (cf. 3.82), is written as

$$\Lambda_x = \Lambda_0 - (\alpha\Lambda_0 + \beta)\sqrt{c_i} = \Lambda_0 - (\alpha\Lambda_0 + \beta)(\Lambda c/\Lambda_x)^{\frac{1}{2}} \tag{8.1}$$

where Λ_x is the sum of the conductances of the cation and anion in a completely dissociated electrolyte solution of equivalent concentration c_i, and c is the actual equivalent concentration. To solve for Λ_x and hence for c_i where

$$c_i = \Lambda c/\Lambda_x = \alpha_c c \tag{8.2}$$

where α_c is the degree of dissociation, an approximate value of Λ_x such as Λ_0 is inserted on the right-hand side of (8.1) and the resulting calculation for Λ_x is used for a second approximation. This procedure of successive approximations is continued until Λ_x is constant, and the dissociation constant K then calculated by means of (8.2) and

$$K = c_i^2 f_{\pm}^2/c_u \tag{8.3}$$

where $c_u = c - c_i$. The mean activity coefficients are calculated by using the limiting form of the D.H. equation (3.39).

Example 8.1. Dunsmore and Speakman [2] have measured the conductances of dilute solutions of KCl in 40 % dioxan-water at 25°C. For such solutions, ϵ is [3] 44.54 and η_0 is [4] 0.01740, while from a plot of Λ against $\sqrt{c}$, $\Lambda_0 = 73.9$. So from (3.79), $\alpha\Lambda_0 + \beta = 80.7$, and [5] a plot of $\Lambda + 80.7\sqrt{c}$ against c gives a more exact value of $\Lambda_0 = 73.85$, whence $\Lambda_x = 73.85 - 80.72(\Lambda c/\Lambda_x)^{\frac{1}{2}}$. Also, from (3.36) and (3.39),— $\log f_{\pm} = 1.192\sqrt{c_i}$. The final calculations are given in Table 8.1.

According to these calculations, KCl does not dissociate completely in this solvent although the fraction existing as ion-pairs is very small. Similar considerations [6] indicate that a number of 1 : 1 salts, particularly those of the oxyanions do not fully ionize even in aqueous solution. Since the derived values of K are of the order of unity, very precise measurements are called for, together with accurate estimates of Λ_0. Sometimes the method of deriving this which has been described above [5] gives rise to a curved plot. A useful alternative has been found by Owen [7] providing $K > 0.005$ and $c < 0.005$. This consists of plotting $\{\Lambda + (\alpha\Lambda_0 + \beta)\sqrt{c} - \Lambda_0\}/c$ against log c (on the basis

TABLE 8.1

Dissociation Constant of KCl in 40% Dioxan at 25°C, using the Limiting Onsager Equation and the Method of Davies

10^4c	3.00	4.00	5.00	7.00	10.00
Λ(exptl.)[2]	72.39	72.13	71.92	71.54	71.08
Λ_x	72.45	72.24	72.04_5	71.74	71.30
K	0.27	0.29	0.24	0.22	0.28

that the extension terms to (3.82) are $Ac \log c + Bc$, where A and B are empirical constants) and adjusting Λ_0 until the plot is linear. An alternative way is described shortly.

When dissociation is almost complete, the undissociated fraction may not be significant until the valid concentration range of the Onsager equation is exceeded. To deal with this, Davies [8] has used the empirical form

$$(\Lambda_0 - \Lambda\sqrt{\eta}/\sqrt{\eta_0})/(\alpha\Lambda_0 + \beta) = f(c) \tag{8.4}$$

to calculate Λ_x; (η and η_0 are the viscosities of solution and solvent). This expression is justified on the grounds that it can reproduce the conductances of HCl, LiCl, NaCl and KCl in water up to $c = 0.5$, and these electrolytes are regarded as ones which dissociate completely. The values of $f(c)$ at 25°C are:

c	0.005	0.01	0.02	0.05	0.10	0.20	0.50
$f(c)$	0.070	0.094	0.125	0.175	0.224	0.279	0.355

The dissociation constants of a number of 1 : 1 electrolytes in water [1, 8] have been estimated by this method.

Symmetrical salts composed of higher valent ions show much greater departures from the limiting equation. Typical examples are [9] $MgSO_4$ and [10]$LaFe(CN)_6$.

Example 8.2. Dunsmore and James [9] obtained the conductances of dilute aqueous solutions of $MgSO_4$ at 25°C. The relevant equations are (a) $\Lambda_x = 133.06 - 342.7(\Lambda c/\Lambda_x)^{\frac{1}{2}}$, (b) $-\log f_\pm = 2.0364\sqrt{2c_i}$, (c) $K = c_i^2 f_\pm^2/2c_u$, where c, c_i and c_u are in equiv./l.

There is a systematic drift in the answers, and by extrapolating a plot of K against c, $K = 0.0057 \pm 0.0002$ at $c = 0$. The drift could be eliminated by a slight adjustment of Λ_0, which in this case is taken as the sum of the limiting conductances of the two ions as estimated from strong electrolytes via $\Lambda_0(MgSO_4) = \Lambda_0(MgCl_2) - \Lambda_0(NaCl) + \Lambda_0(Na_2SO_4)$ (or the K^+ salts). This slight adjustment of Λ_0 could be justified on the grounds that the conductances of these strong salts are slightly in error,[11] or that the $MgSO_4$ conductances

TABLE 8.2

Dissociation Constant of $MgSO_4$ in Water at 25°C using the Limiting Onsager Equation and the method of Davies

10^4c	1.7021	3.9871	6.1798	8.5401	11.194	14.393	17.692
Λ(exptl.)[9]	127.11	123.13	120.33	117.89	115.50	113.14	111.02
Λ_x	128.62	126.30	124.69	123.27	121.90	120.46	119.15
10^3K	5.96	5.81	5.95	6.13	6.21	6.41	6.57

are based on a slightly incorrect cell constant. Alternatively it might improve matters to use an extended form of the D.H. equation (e.g., 3.40) with suitable parameters, but since the conductance equation is of the limiting form, there is no sound reason for doing this. The discussion on this aspect is resumed later on.

The adjustment of Λ_0 so that K is constant over a range of concentrations has proved useful when the exact value of Λ_0 is unknown or in doubt. Thus in the study of [10] $LaFe(CN)_6$, K was found to be constant at 25°C ($K = 1.82 \times 10^{-4}$) with $\Lambda_0 = 168.9$. This is in good agreement with a subsequent independent value of [12] 168.7 derived from the conductances of $LaCl_3$ and $K_3Fe(CN)_6$. Yet another way of finding Λ_0 under such circumstances is to plot $f_{\pm}^2 \Lambda_x c$ against $1/\Lambda_x$ and to extrapolate to $c - 0$.

A number of multivalent symmetrical electrolytes, both in water and in mixed solvents have been examined by the above method,[8, 14, 15] but in order to apply it to electrolytes which form charged ion-pairs, e.g., $NaSO_4^-$, $LaSO_4^+$, allowance must be made for the contribution of such ion-pairs to the total conductance. The mobilities of these cannot be determined so it has to be assumed that their conductance is a certain fraction of that of the ion of like charge, in proportion to the charge ratio, e.g., $\lambda^0_{LaSO_4^+} = \frac{1}{3}\lambda^0_{La^{3+}}$. There is a certain amount of information to support such assumptions, for instance Jeffery and Vogel,[16] by comparing some dibasic organic acid anions with the corresponding amine monobasic acid anions found the ratio to be 0.53. The ratio $HCO_3^- : CO_3^{2-}$ is [17] 0.64, that of $HSO_4^- : SO_4^{2-}$ is [18] 0.64 also, while for $Co(NH_3)_4(NO_2)_2^+ : CO(NH_3)_5(NO_2)^{2+}$ it is [19] 0.54. The ratio of the limiting conductances of some ions with a charge ratio of 2 : 3 is [19,20] $Co(NH_3)_5(NO_2)^{2+}$: $Co(NH_3)_6^{3+} = 0.68$, and $Fe(CN)_5(NO)^{2-} : Fe(CN)_6^{3-} = 0.70$, while for the 3 : 4 ratio it is [12] $Fe(CN)_6^{3-} : Fe(CN)_6^{4-} = 0.88$, $P_3O_9^{3-} : P_4O_{12}^{4-} = 0.89$. On the whole, the measured ratios are higher than expected but calculation shows the mobilities of the ion-pairs can be varied considerably without altering the derived value of K very much.

Righellato and Davies [8] examined the conductances of a number of 1 : 2 and 2 : 1 valent electrolytes and concluded that many of them do not dissociate completely in water, the dissociation constants ranging from 0.5 for $CdNO_3^+$ down to 0.01 for $CdCl^+$. Similar investigations of 2 : 3, 3 : 2, 1 : 3 and other valence types [8, 21, 22] indicate that none of these ionize completely.

The conductance treatment can be generalized into the following expressions:

$$\Lambda_{\text{obs.}}\, c = 1{,}000\, L^*_{\text{obs.}} = \sum_i z_i m_i \lambda_i; \quad c = \nu_+ z_+ m = |\nu_- z_- m| \tag{8.5}$$

where z refers to ion valence, m_i to ion molarity and m to electrolyte molarity. For the conductance of an electrolyte solution containing M^{z+} and L^{z-} ions and $ML^{z+ - z-}$ ion-pairs of concentration x,

$$\begin{aligned} 1{,}000\, L^*_{\text{obs.}} &= z_+\lambda_M(\nu_+ m - x) + |z_-|\lambda_L(\nu_- m - x) + x\lambda_{ML}(z_+ - |z_-|) \\ &= c(\lambda_M + \lambda_L) - x\{z_+\lambda_M + |z_-|\lambda_L + (z_+ - |z_-|)\lambda_{ML}\} \end{aligned} \tag{8.6}$$

Alternatively, if α_c is the fraction of, say, L^{z-} ions which associate with M^{z+} ions, then $x = \nu m\alpha_c$, and from (8.5) and (8.6),

$$|z_-|\Lambda_{\text{obs.}} = |z_-|(\lambda_M + \lambda_L)(1 - \alpha_c) - \alpha_c(z_+ - |z_-|)(\lambda_M + \lambda_{ML}) \tag{8.7}$$

Example 8.3. An aqueous solution of lanthanum sulphate of $c = 0.0003943$ has a conductance [21] of 103.81 at 25°C. $\lambda^0_{La} = 69.6$, $\lambda^0_{SO_4} = 80.0$, and $\lambda^0_{LaSO_4}$ will be taken as 69.6/3. From (3.78), $\lambda_{La} = 69.6 - 128.6\sqrt{\Gamma}$, $\lambda_{SO_4} = 80.0 - 117.0\sqrt{\Gamma}$, and $\lambda_{LaSO_4} = 23.2 - 29.3\sqrt{\Gamma}$, calculating q^* from (3.76) with $z_i = 1$, $z_j = 2$. Also,

$$\Gamma = 5c - 12x, \text{ and } \Lambda_{\text{obs.}} c = c(\lambda_{La} + \lambda_{SO_4}) - x(3\lambda_{La} + 2\lambda_{SO_4} + \lambda_{LaSO_4}).$$

This is solved by approximations, taking $x = 0$ in the first instance. The final value of x is 0.0000380, and K is calculated by means of $K = (\frac{1}{3}c - x)(\frac{1}{2}c - x)f_{La}f_{SO_4}/xf_{LaSO_4}$, with the aid of the D.H. limiting equation, whence $K = 2.52 \times 10^{-4}$. Some other figures are given in Table 8.3. If (8.7) is applied, the resulting equation is

$$2\Lambda_{\text{obs.}} = 2(\lambda_{La} + \lambda_{SO_4}) - \alpha_c(\lambda_{La} + \lambda_{LaSO_4}).$$

As with $MgSO_4$ in Example 8.2, there is a slight drift of K with c and extrapolation to $c = 0$ gives $K = 2.45 \times 10^{-4}$. On the other hand, if Λ_0 is adjusted till K is constant, $K = 2.75 \times 10^{-4}$.

There are times when it is not practicable to prepare a sample of the salt whose conductance is required. A useful alternative is to measure mixtures of two available salts that provide the two ions. For example, the conductances of mixtures of MCl_2 and $Na_3P_3O_9$ enable the dissociation constants

TABLE 8.3

Dissociation Constant of the Ion-Pair $LaSO_4^+$ in Water at 25°C

10^4c	3.943	5.909	7.056	10.75	12.55	15.24	18.20
$\Lambda_{obs.}$	103.81	95.42	91.77	82.79	80.13	76.22	72.81
10^5x	3.80	7.00	8.96	15.90	19.33	24.82	31.10
10^4K	2.52	2.59	2.62	2.64	2.71	2.73	2.75

of the $MP_3O_9^-$ ion-pairs to be found.[23] In a solution containing c_1 equiv./l. of $Na_3P_3O_9$ and c_2 of MCl_2, if a g ions of $NaP_3O_9^{2-}$ and b g ions of $MP_3O_9^-$ are present per l., then from (8.5),

$$1{,}000\, L^*_{obs.} = \lambda_{Na}(c_1 - a) + 3\lambda_{P_3O_9}(\tfrac{1}{3}c_1 - a - b) + 2\lambda_{NaP_3O_9}(a) + + 2\lambda_M(\tfrac{1}{2}c_2 - b) + \lambda_{Cl}(c_2) + \lambda_{MP_3O_9}(b)$$

$$= c_1\Lambda_{Na_3P_3O_9} + c_2\Lambda_{MCl_2} - 2\tfrac{2}{3}b\lambda_{P_3O_9} - 2b\lambda_M \quad (8.8)$$

if it is assumed that $\lambda_{MP_3O_9} = \frac{1}{3}\lambda_{P_3O_9}$. The ionic conductances are calculated via (3.78), taking $\Gamma = 4c_1 + 3c_2 - 6a - 12b$, ignoring a and b in the first approximation. If no $MP_3O_9^-$ formed,

$$1{,}000\, L^*_{calc.} = c_1\Lambda_{Na_3P_3O_9} + c_2\Lambda_{MCl_2} \quad (8.9)$$

Experimental Λ values of each of these two salts are plotted against $\sqrt{c}$, and interpolations made at Γ. Subtraction of (8.8) from (8.9) gives b. This implies that the values of a is the same for both, which is not quite true. The actual values can be calculated by means of the known value[23] of $K_{NaP_3O_3}$ This is also needed in calculating the true values of Γ and of $K_{MP_3O_3}$. A number of condensed phosphates and ferricyanides have been studied by this mixture method.[23]

2. The Methods of Fuoss and Kraus and of Shedlovsky

Fuoss and Kraus[24] adapted the limiting Onsager–Fuoss equation by combining (8.1) and (8.2) into

$$\alpha_c = \Lambda/\Lambda_0\{1 - [\alpha\Lambda_0 + \beta](\alpha_c c)^{\frac{1}{2}}/\Lambda_0\} \quad (8.10)$$

The terms in the { } brackets are replaced by

$$F(z) = 1 - z\{1 - z\{1 - z\{\ldots\}^{-\frac{1}{2}}\}^{-\frac{1}{2}}\}^{-\frac{1}{2}} = \tfrac{4}{3}\cos^2 \tfrac{1}{3}\cos^{-1}[(-3\sqrt{3}z)/2] \quad (8.11)$$

where

$$z = [\alpha\Lambda_0 + \beta](\Lambda c)^{\frac{1}{2}}/\Lambda_0^{\frac{3}{2}} \tag{8.12}$$

so that

$$\alpha_c = \Lambda/\Lambda_0 F(z)$$

Also,

$$K_F = \alpha_c^2 f_\pm^2 c/(1 - \alpha_c) \tag{8.13}$$

Fuoss [25] has given a Table of $F(z)$ values ranging from $z = 0$ to $z = 0.209$ at 0.001 intervals and has estimated that (8.13) holds up to $c = 3.2 \times 10^{-7}\epsilon^3$ for 1: 1 electrolytes at 25°C. To obtain Λ_0 a first value is found by the usual plot of Λ against $\sqrt{c}$, or some other convenient way; z is then calculated, $F(z)$ found and α_c found at each concentration so that each $f_\pm$ can be derived by means of the D.H. limiting equation. Now combination of the expressions for α_c and K_F leads to

$$F(z)/\Lambda = 1/\Lambda_0 + c\Lambda f_\pm^2/F(z)K_F\Lambda_0^2 \tag{8.14}$$

so that a plot of $F(z)/\Lambda$ as ordinate against $c\Lambda f_\pm^2/F(z)$ yields a better value of $1/\Lambda_0$ at the intercept on the y axis. New calculations are now made and the slope of the final plot gives $1/K_F\Lambda_0^2$. If K_F is extremely small, the plot is rather insensitive; this is discussed later on.

Equations (8.1) and (8.2) can also be combined to form

$$1/\Lambda_0 = \alpha_c/\Lambda - \alpha_c(\alpha\Lambda_0 + \beta)(\alpha_c c)^{\frac{1}{2}}/\Lambda\Lambda_0 \tag{8.15}$$

This is Shedlovsky's treatment [27] after replacing $\Lambda\Lambda_0$ by Λ_0^2 on the right-hand side and putting

$$\alpha_c = S(Z)\Lambda/\Lambda_0 \tag{8.16}$$

where

$$S(Z) = \{\tfrac{1}{2}Z + (1 + \tfrac{1}{4}Z^2)^{\frac{1}{2}}\}^2 = 1 + Z + \tfrac{1}{2}Z^2 + \ldots \tag{8.17}$$

and

$$Z = (\alpha\Lambda_0 + \beta)(c\Lambda)^{\frac{1}{2}}/\Lambda_0^{\frac{3}{2}} \tag{8.18}$$

Accordingly, (8.15) can be written as

$$1/\Lambda S(Z) = 1/\Lambda_0 + c\Lambda S(Z)f_\pm^2/K_s\Lambda_0^2 \tag{8.19}$$

where K_s is the dissociation constant. A plot of $1/\Lambda S(Z)$ as ordinate against $c\Lambda S(Z)f_\pm^2$ has a slope of $1/K_s\Lambda_0^2$ and intercept of $1/\Lambda_0$. Values of $S(Z)$ have been recorded [28] for various values of Z.

K_s and K_F are related by [29]

$$1/K_s = 1/K_F + (\alpha\Lambda_0 + \beta)^2/\Lambda_0^2 \tag{8.20}$$

In the range $10^{-3} \leqslant K \leqslant 1$, the Shedlovsky function is recommended; when $K < 10^{-3}$ the two methods are of equal standing. When the electrolyte is very weak ($K \sim 10^{-6}$), the plots do not establish Λ_0 and K with much precision. To deal with this [27] Λ_0 might be found by applying Kohlrausch's rule, e.g.,

$\Lambda_0^{\text{HA}} = \Lambda_0^{\text{HCl}} - \Lambda_0^{\text{NaCl}} + \Lambda_0^{\text{NaA}}$ but the success of this depends on the HCl, NaCl and NaA being at least largely dissociated in the solvent being used. If they are weak, the rule does not help. The problem is considered in greater detail in the next section (on weak acids).

Even when all possible precautions have been taken to ensure that the solvent is pure and that CO_2 is excluded, the solvent conductance is often a relatively large fraction of the whole. Shedlovsky and Kay [27] have developed a procedure which avoids the measurement of the solvent conductance and also allows for any variation of it as the concentration of electrolyte increases. Writing

$$\Lambda = (L_s - L_o)/c \tag{8.21}$$

where $L_s/1{,}000$ is the specific conductance of a solution (as measured) and $L_0/1{,}000$ is the (unknown) specific conductance of the solvent, combining (8.21) with

$$\sqrt{K_s} = (\alpha_c f_\pm \sqrt{c})/(1 - \alpha_c)^{\frac{1}{2}} \tag{8.22}$$

and replacing α_c by (8.16),

$$L_s - L_o = (\Lambda_0 \sqrt{K_s c}/S(Z) f_\pm)(1 - S(Z)\Lambda^*/\Lambda_0)^{\frac{1}{2}} \tag{8.23}$$

where $\Lambda^* = L_s/c$. A plot of L_s against the right-hand function with Λ_0 and $\sqrt{K_s}$ omitted has a slope of $\Lambda_0\sqrt{K_s}$ (and intercept of L_0).

Fuoss and Kraus have amassed the dissociation constants of a large number of 1 : 1 electrolytes, particularly those of substituted ammonium salts in a wide variety of organic solvents by application of their method. Much of the work has been summarized by Fuoss [24] and others.[30] A series designed to examine the effects of structure and constitution upon ionic interaction in various solvents has been systematically explored by Kraus and his school.[31] Evers and Knox [32] have found that in methanol, for NaCl, $K_s = 0.083$, $K_F = 0.22$ and for KCl, $K_s = 0.068$, $K_F = 0.121$, while in ethanol Wardlaw and others [33] conclude that $K_{\text{HCl}} = 0.0082$, $K_{\text{CsCl}} = 0.0066$, $K_{\text{MgCl}_2} = 1.9 \times 10^{-4}$ (approx. see also Ref. 67) and $LaCl_3$ is very weak. Griffiths [94] and his co-workers have studied $AgNO_3$ and $AgClO_3$ in a wide variety of solvents (cf. similar work [68] with NaI). Ordinary strong electrolytes prove to be very weak in acetic acid solutions. The K values of the alkali metal bromides, formates and acetates are [35] from 0.65 to 6×10^{-7}. Other figures are $NaClO_4 = 0.0046$ (91 % acetic), 0.00053 (95 % acetic), $HClO_4 = 9.1 \times 10^{-7}$ (100 % acetic). In ethanolamine [63] ($\epsilon = 37.72$) the values of some 1 : 1 salts range from 0.5 to 0.03.

3. Weak Acids

The conductance method has been applied on a large scale to study the dissociation of acids. A useful review of this topic up to about 1939 has been written by Dippy [37] who has made further contributions since then [38] which are aimed at finding relationships between dissociation and the ionizing properties of organic groups and constituents of organic acids. Harned and Owen,[11] and Robinson and Stokes [11] have given extensive Tables in their books which include most of the reliable information obtained up to the present time; this is also true of the compilation of Bjerrum *et al.*[14]

Many of the common monocarboxylic acids have been subjected to very careful investigation. The first of these, that of MacInnes and Shedlovsky [39] on acetic acid in water, will be used as a basis for discussing a number of points of interest and importance. Firstly, it will be recalled that it has been mentioned that when the electrolyte is very weak, it is difficult to derive Λ_0 with sufficient accuracy, and a common alternative is to derive it from the conductances of electrolytes which are at least extensively dissociated in the solvent being used, e.g., $\Lambda_0^{HA} = \Lambda_0^{HCl} - \Lambda_0^{NaCl} + \Lambda_0^{NaA}$. In order to suppress hydrolysis, sufficient HA is added to NaA solutions to make the pH around 7.

On the other hand, it is claimed by Ives [40] and Ives and Sames [41] that if $K \nless 2 \times 10^{-5}$ Λ_0 may be derived directly from the conductances of the weak acid. At the same time K can be derived. The procedure is to add Λ_0 to both sides of

$$\Lambda - \Lambda_x = - \Lambda^2 f_\pm^2 \, c/\Lambda_x K \tag{8.24}$$

which is derived from (8.1, 8.2, 8.3). Accordingly,

$$\Lambda + (\alpha\Lambda_0 + \beta)(\alpha_c c)^{\frac{1}{2}} = \Lambda_0 - \Lambda^2 f_\pm^2 c/\Lambda_x K \tag{8.25}$$

and writing this as

$$y = \Lambda_0 - m'x \tag{8.26}$$

a plot of y against x has a gradient of $m' = 1/K$ and an intercept on the ordinate of Λ_0. A reasonable value of Λ_0 is taken for the first plot, α, β, $\alpha_c c$, and $f_\pm$ calculated (using the limiting D.H. equation), then y and x are calculated for each measurement and a new value of Λ_0 obtained by the method of least squares:

$$\Lambda_0 = \{\sum(x)\sum(xy) - \sum(x^2)\sum(y)\}/\{[\sum(x)]^2 - n\sum(x^2)\} \tag{8.27}$$

where n is the number of experimental points. The process is repeated until Λ_0 is constant (about 4 series may be needed). The upper concentration limit is 0.003 to 0.005 N. K is also calculated by the method of least squares:

$$-1/K = \{\sum(x)\sum(y) - n\sum(xy)\}/\{[\sum(x)]^2 - n\sum(x^2)\} \tag{8.28}$$

Belcher [42] has expressed the view that such calculations are reliable if $K \nless 10^{-3}$ but that when $K < 10^{-5}$ considerable errors are possible; Kilpatrick [43] has placed the limit at about $K \sim 10^{-6}$ which is a little lower than that given by Ives and Sames.[41]

MacInnes and Shedlovsky [39] calculated Λ_x by an extended form of (8.1), namely

$$\Lambda_x = \Lambda_{HCl} - \Lambda_{NaCl} + \Lambda_{NaAc} = 390.59 - 148.61\sqrt{c_i} + 165.5\,c_i(1 - 0.2274\sqrt{c_i}) \quad (8.29)$$

This is derived from the limiting Onsager equation and an empirical extension term, i.e., from $\Lambda = \Lambda_0 - (\alpha\Lambda_0 + \beta)\sqrt{c_i} + Bc_i\,(1 - \alpha\sqrt{c_i})$, with B adjusted to fit the conductances of each of the strong electrolytes. The true values of Λ_x were found by approximations and K calculated via (8.3) and the limiting D.H. equation. The results are shown in Table 8.4. The average value of K from the six lowest concentrations is 1.752×10^{-5}, but as c increases from thereon, there is a systematic decrease in K. Harned and Owen [11] derived the true answer by extrapolating a plot of log K against c; this is linear and extrapolates to $K = 1.750 \times 10^{-5}$.

This decrease in K as c increases cannot be attributed to the use of limiting conductance and activity coefficient expressions. MacInnes and Shedlovsky [27] suggested that the cause is the formation of a solvent of decreasing dielectric constant while Davies [44] found that the trend in K is diminished if Λ is replaced by $\Lambda\eta/\eta_0$ (see 3.117); these [45] "corrected" conductances together with the corresponding values of K^* are shown in Table 8.4. They are constant, within experimental error, for the first eleven concentrations, which average $1.751_5 \times 10^{-5}$, but there is still a decrease at the higher concentrations.

TABLE 8.4

Dissociation and Dimerisation Constants of Acetic Acid in Water at 25°C from Conductances

$10^4 c$	0.2801	1.1135	1.5321	2.1844	10.283	13.634	24.140	34.407
Λ	210.32	127.71	112.02	96.466	48.133	42.215	32.208	27.191
Λ_x	390.02	389.08	389.01	389.49	388.94	388.81	388.52	388.32
$10^5 K$	1.752	1.754	1.750	1.751	1.751	1.753	1.750	1.750
$\Lambda\eta/\eta_0$	210.32	127.71	112.02	96.468	48.138	42.222	32.216	27.202
$10^5 K^*$	1.752	1.754	1.750	1.751	1.751	1.753	1.751	1.751

$10^4 c$	59.115	98.421	128.29	200.0	500.0	1000	1194.5	2000	2307.9
Λ	20.956	16.367	14.371	11.563	7.356	5.200	4.759	3.650	3.391
Λ_x	387.99	387.61	387.41	387.05	386.19	385.29	385.07	384.41	384.15
$10^5 K$	1.749	1.747	1.743	1.738	1.721	1.695	1.689	1.645	1.633
$\Lambda\eta/\eta_0$	20.970	16.385	14.391	11.589	7.397	5.257	4.822	3.731	3.478
$10^5 K^*$	1.752	1.751	1.750	1.747	1.740	1.731	1.732	1.716	1.715
K_d	—	—	—	—	15.5	16.5	21.7	19.2	21.5

Katchalsky *et al.* [46] pointed out that a reasonable explanation is that dimerised molecules may be present. He and his partners calculated that the dimerisation constant, $K_d = [HA]^2/[H_2A_2]$ is 6.2, in agreement with 5.4 which

MacDougall and Blumer [47] derived from vapour pressures. However, this figure is based on erroneous information,[48] and by recalculation is 20 ± 3. This is in reasonable agreement with 28 derived from freezing-point and distribution studies.[48]

An analysis of the conductance data in which both viscosity and dimerisation are taken into account has been made.[49] Writing $[HAc] = (\alpha_c c)^2 f_\pm^2 / K_a$, then $[H_2Ac_2] = \frac{1}{2}(c - \alpha_c c - [HAc])$, $K_d = 2[HAc]^2/(c - \alpha_c c - [HAc])$. Taking $K_a = 1.751 \times 10^{-5}$, and the viscosity corrected conductances to calculate α_c, the resulting values of K_d are shown in the last row of Table 8.4. Only the last five concentrations provide sufficient concentrations of H_2Ac_2. Formic, propionic and butyric acids have also been discussed by this treatment.[49] The average answers are (at 25°C):

	Formic	Acetic	Propionic	*n*-Butyric
$10^5 K_a$	18.24	1.751	1.347	1.519_5
K_d	120 ± 12	19 ± 2	20 ± 3	11 ± 0.6

The K_d figures for acetic and *n*-butyric acids are in good accord with those calculated from freezing-points and distributions.[48, 50] The viscosity correction is supported by the fact that Λ_0^{HCl} is almost constant in water and in dilute aqueous dioxan and acetone when multiplied by the viscosity ratios.[49]

TABLE 8.5

Dissociation Constant of Acetic Acid in 80.03% (w/w) of Methanol at 25°C by the Conductance Methods of Davies, Fuoss, and Shedlovsky

$10^4 c$	4.743	10.406	19.25	30.81	44.56	87.86	163.81
$1000L$	0.937	1.385	1.880	2.373	2.855	4.015	5.479
Λ_x	95.20	95.12	95.03	94.97	94.90	94.77	94.63
$f_\pm$	0.9908	0.9889	0.9870	0.9854	0.9840	0.9811	0.9779
$1000z$	4.14	5.05	5.90	6.66	7.31	8.66	10.13
$F(z)$	0.9959	0.9949	0.9941	0.9933	0.9927	0.9913	0.9898
$10^3 c\Lambda f_\pm^2/F(z)$	0.903	1.341	1.822	2.300	2.746	3.878	5.272
$10S(Z)$	10.041_5	10.050	10.059	10.067	10.073	10.087	10.102
$10^3 c\Lambda f_\pm^2 S(Z)$	0.903	1.341	1.822	2.299	2.745	3.877	5.272
$10^7 K_D$	1.957	1.960	1.957	1.950	1.950	1.955	1.949

Example 8.4. Shedlovsky and Kay [27] obtained the conductances of dilute solutions of HCl and of acetic acid in water-methanol mixtures at 25°C. They derived Λ_0^{HA} by a slight modification of the Kohlrausch rule, namely by using

$$\Lambda_0^{HAc} = \Lambda_0^{HCl} - (\Lambda_c^{NaCl} - \Lambda_c^{NaAc})(1 + \alpha\sqrt{c})$$

where $c = 0.05$; (cf. 8.29). Λ_0^{HCl} was obtained by means of (8.19). Some data and calculations where the solvent is 80.03 % by weight of methanol are given in Table 8.5. From (3.78), $\alpha = 0.5733$ and $\beta = 72.8$; other required figures are:[27] $\Lambda_0^{HA} = 95.6$, $\epsilon = 42.60$, $\eta = 0.01006$, and from (3.36, 3.39), $-\log f_{\pm} = 1.275\sqrt{\alpha_c c}$. Also, from (8.23), $L_0 = 0.021 \times 10^{-3}$.

A comparison has been made between the methods of Davies, to which Λ_x and K_D refer, and those of Fuoss and Shedlovsky. The average value of $K_D = 1.95_4 \times 10^{-7}$, while from the data it is found that $K_F = K_s = 1.95 \times 10^{-7}$.

B. Application of the Extended Form of the Limiting Onsager-Fuoss Equation

By incorporating a parameter which takes ion sizes into account, the limiting Onsager equation can reproduce the conductances of 1 : 1 electrolytes up to relatively high concentrations providing dissociation is complete, but the treatment is not very convincing when applied to higher valent types. This is illustrated by Table 3.2. For the present purpose, (8.1) can be suitably modified by writing

$$\Lambda_x = \Lambda_0 - (\alpha\Lambda_0 + \beta)\sqrt{c_i}/(1 + B'a\sqrt{c_i}) \qquad (8.30)$$

where B' is given by (3.37). If this is used to estimate degrees of dissociation, the extended form of the D.H. activity coefficient expression should also be used when deriving dissociation constants. Robinson and Stokes,[11] by these expressions, obtained $K_{KNO_3} = 1.73$, $K_{AgNO_3} = 1.97$, $K_{TlCl} = 0.322$, taking the parameter a as being 3.57 Å, which is the critical Bjerrum distance (see Chapter 14). For the first two salts, the concentration range examined was from 0.005 to 0.1 N, and K increased from 1.42 to 2.14 for KNO_3 and decreased from 2.38 to 1.76 for $AgNO_3$. Now to obtain these figures, the Λ_0 values obtained by Shedlovsky were taken (after adjusting the conductances to the Jones and Bradshaw standard—see Chapter 2). To derive Λ_0, Shedlovsky [51] plotted $\Lambda_0' = (\Lambda + \beta\sqrt{c})/(1 - \alpha\sqrt{c}) = \Lambda_0 + Bc$ against c, but in a number of cases, especially when dissociation is incomplete, the plot develops a curvature near $c = 0$ which makes a precise extrapolation difficult. An alternative device is to modify Onsager's method (see Example 8.1) and to plot $\Lambda + (\alpha\Lambda_0 + \beta)\sqrt{c}/(1 + B'a\sqrt{c})$ against c. This leads to $\Lambda_0^{KNO_3} = 144.90 \pm 0.05$ and $\Lambda_0^{AgNO_3} = 133.40 \pm 0.05$, the uncertainties being due to the scatter in the experimental information.[51] These estimates are 0.06 lower and 0.04 units higher respectively than the earlier estimates.[11] A very sensitive way of selecting Λ_0, proposed by Davies,[1] is to select that value of Λ_0 which produces constant values of K. This has been adopted in applying (8.30) to the data of Shedlovsky,[51] and the selected values are $\Lambda_0^{KNO_3} = 144.86$, $\Lambda_0^{AgNO_3} = 133.43$, which with $a = 3.57$ Å give $K_{KNO_3} = 1.90 \pm 0.18$ and $K_{AgNO_3} = 1.72 \pm 0.03$ between $c = 0.005$ and 0.1 N. Robinson and Stokes also used (8.30) to estimate dissociation constants of $MgSO_4$, $CdSO_4$, $ZnSO_4$, and $CuSO_4$ in water, taking $a = 14.28$ Å, the critical Bjerrum distance. The value of $\Lambda_0^{MgSO_4}$ is well established from the conductances of $MgCl_2$, Na_2SO_4 and K_2SO_4 as 133.06. An examination of the figures [52] for $MgCl_2$ by the modified Onsager method described above would give $\lambda_0^{Mg} = 129.35 - 76.35 = 53.00$, which is 0.06 lower than the published figure.[11] The value of $\lambda_0^{SO_4}$ has been established with reasonable certainty by the present methods as 80.00 ± 0.02, i.e., $\Lambda_0^{MgSO_4} = 133.00$. On the other hand, Owen's method [7] (see above Table 8.1), when applied to the figures of Table 8.2, indicates that $\Lambda_0 = 133.10 \pm 0.1$. Under the circumstances therefore, there is no real cause to shift Λ_0 from 133.06 which was used for this Table.

It is much more difficult to assess the size of the parameter a; if an ion-pair consists of a cation and an anion in actual contact, 5.0 Å is a reasonable estimate, but since ions need not be in actual contact to influence one another beyond that expected from interionic forces, the true value of the parameter will be some sort of average between a certain minimum distance and the dimension corresponding to ions in contact. In Table 8.6 three sets of values based on different ion-pair dimensions are given; these are calculated for the measurements and concentrations of Table 8.2.

TABLE 8.6

Dissociation Constants of $MgSO_4$ using Table 8.2, Equations (3.81), (3.31) and (8.30) . $K \times 10^4$

$a = 5.0$ Å	56.1	53.4	54.6	55.9	56.5	58.0	59.1
$a = 10.0$ Å	51.7	50.1	51.0	52.1	52.6	53.9	54.9
$a = 14.3$ Å	49.0	47.5	48.5	49.6	50.1	51.4	52.4

There is a slight drift with concentration, and assuming that the lowest concentration is slightly in error, the extrapolated results are 0.0053, 0.0049 and 0.0047 respectively. It was mentioned in connection with Table 8.2 that where ion-size is ignored, extrapolation leads to $K = 0.0057$, so in general, K becomes smaller as a is increased, and for an uncertainty of 5 Å, there is a corresponding uncertainty of 8 % in K. Judging by the figures for [11] $ZnSO_4$ and $CuSO_4$, the uncertainty decreases when K is smaller. Further discussion on ion-pair sizes is contained in Chapters 9 and 14.

C. Application of the Fuoss-Onsager Extended Conductance Equation

When $\kappa a \nless 0.2$, Fuoss [53] has given (3.110) as expressing the conductances of unassociated 1 : 1 electrolytes. Replacing Λ_x of (8.2) by Λ_i and K of (8.3) by

$1/K_A$ where K_A is the "association" constant to conform to the nomenclature of Fuoss, (8.2) and (8.3) can be converted to

$$\Lambda = \Lambda_i - c_i \Lambda f^2 K_A$$

so if Λ and c of (3.110) are replaced by Λ_i and c_i,

$$\Lambda_0 - S\sqrt{c_i} + Ec_i \log c_i + J_1 c_s(1 - \alpha\sqrt{c_i})(1 - J_2/c_i J_1) = \Lambda + c_i \Lambda f^2 K_A \tag{8.31}$$

or

$$(\Lambda + S\sqrt{c_i} - Ec_i \log c_i - \Lambda_0)/c_i(1 - \alpha\sqrt{c_i}) = J_1 - J_2\sqrt{c_i} - \Lambda f^2 K_A/(1 - \alpha\sqrt{c_i}) \tag{8.32}$$

This is written as

$$y = \Delta\Lambda/c_i(1 - \alpha\sqrt{c_i}) = J_1 - J_2\sqrt{c_i} - K_A x \tag{8.33}$$

The slope of the plot of y against x gives K_A and the intercept on the y axis at $x = \Lambda_0$ gives $J_1 - \Lambda_0 K_A$ on ignoring the small $J_2\sqrt{c_i}$ term, so J_1 is obtained. Interpolation of this on a plot of calculated J_1 values against a (using 3.106) gives the correct value of a (unless K_A is small—below about 1.5).

A first value of Λ_0 obtained by any of the methods described in previous sections is used in (8.32–3), but since the plot of y against x should be linear, a final adjustment to Λ_0 may be needed. A first value of $\gamma c = c_i$ is found by means of the limiting Onsager–Fuoss equation (3.82), and after J_1 and a are found, J_2 is calculated via (3.110). Next, fresh values of c_i are calculated by ignoring the small $\alpha\sqrt{c}$ term in (3.110):

$$c_i'' = c\Lambda/(\Lambda_0 - S\sqrt{c_i} + Ec_i \log c_i + J_1 c_i - J_2 c_i^{\frac{3}{2}}) \tag{8.34}$$

New values of y are x are then obtained for plotting $y' = y + J_2\sqrt{c_i}$ against x, the slope giving the final value of K_A.

When K_A is small, $\gamma = 1$ is an adequate approximation, and when K_A is near or less than unity, the plot of y against x becomes almost horizontal so that it is necessary to assume that a is the same as in solvents of lower dielectric constant, where K_A is larger, in order to calculate J_1 and J_2. In order to obtain these small values of K_A, Λ''' is defined by

$$\Lambda''' = (\Lambda + \beta\sqrt{c} - Ec \log c)/(1 - \alpha\sqrt{c}) - J_1 c + J_2 c^{\frac{3}{2}} \tag{8.35}$$

so from (8.31), with $c = c_i$,

$$\Lambda''' = \Lambda_0 - K_A c \Lambda f^2/(1 - \alpha\sqrt{c}) = \Lambda_0 - K_A z \tag{8.36}$$

A plot of the right-hand side of (8.35) against z as defined by (8.36) has a slope of K_A and intercept of Λ_0.

When the ions are large, e.g., those of the quaternary ammonium or picrate types, a viscosity correction is needed. If the viscosities of the solutions are unknown, Einstein's formula [54] is invoked, i.e., $\eta = \eta_0(1 + 5\phi/2)$ where

$\phi = 4\pi NcR_i^3/3{,}000 = \delta c$ and R_i is the radius of the large ion. Approximating $1/(1 + 5\phi/2)$ by $1 - 5\phi/2$, and ignoring the small $J_2\sqrt{c}$ term leads to

$$y = J_1 - 5\Lambda_0\delta/2 - K_A x;\ y(0) = J_1 - 5\Lambda_0\delta/2 - K_A\Lambda_0 \qquad (8.37)$$

so J_1, and hence δ can be found once a is known. Fuoss and Kraus [55] derived this for $NR_4^+X^-$ salts ($X^- = Br^-, I^-, NO_3^-$) by plotting first values of log K_A against $1/\epsilon$. Since such plots are linear it is suggested that the slope represents $e^2/2.303\,a\epsilon kT$, thus giving a. When viscosity data are available, $\eta = \eta_0(1 + F'c)$ can be used (see 3.112 to 3.115).

Fuoss [56] has simplified the Fuoss–Onsager treatment into the form shown in (3.112). As a result, (8.34) is replaced by

$$\Lambda = \gamma(\Lambda_0 - S\sqrt{c_i} + Ec_i \log c_i + Jc_i)/(1 + F'c) \qquad (8.38)$$

where the viscosity term F' is given by (3.116). Considering simple inorganic electrolytes where viscosity changes in dilute solutions are negligible, (8.32) becomes

$$(\Lambda + S\sqrt{c_i} - Ec_i \log c_i - \Lambda_0)/c_i = J - K_A\Lambda f^2 \qquad (8.39)$$

which is written as

$$y = \Delta\Lambda/c_i = J - K_A x \qquad (8.40)$$

A plot of y against x evaluates J and K_A. By neglecting J and E, (8.39) reduces to the limiting Onsager–Fuoss equation and any of the ways described in the first section can be used to calculate preliminary values of Λ_0 and c_i. For a second approximation, a new form of (8.34) is used,

$$c_i'' = c\Lambda/(\Lambda_0 - S\sqrt{c_i} + Ec_i \log c_i + Jc_i) \qquad (8.41)$$

Small values of K_A are derived by the method described in connection with (8.35) and (8.36), replacing J_1 by J (as defined by 3.113) and omitting J_2.

Accascina, Fuoss and their associates [57] have used the J equations to study the association of tetraethylammonium picrate in methanol-water ($K_A = 0.8$ in water, $K_A = 18.9$ in methanol), and of tetrabutylammonium tetraphenylboride in acetonitrile-carbon tetrachloride ($K_A = 14.3$ in MeCN, $K = 3.0 \times 10^6$ in 95 % CCl_4).

Example 8.5. Martell and Kraus [58] obtained the conductances of $NaBrO_3$ in dioxan-water mixtures at 25°C. Fuoss and Kraus [55] have derived K_A values from these by means of (8.30) to (8.37). Figures relating to 50 % dioxan are shown in Table 8.7. The values of the various terms are: $\epsilon = 35.85$, $\eta = 0.01915$, $\Lambda_0 = 50.74$; from (3.83), $\alpha = 0.7424$; from (3.84), $\beta = 41.66$, so $S = 79.33$; from example (3.6), $\alpha' = 2.408$, $\beta' = 19.81$, $\kappa/\sqrt{c} = 4.865 \times 10^7$, $ab = 1.563 \times 10^7$. From (3.31), (3.36) and (3.37), $-\log f_i = 1.651\sqrt{c_i}/(1 + 0.486a\sqrt{c_i})$.

For the first approximation, $0.486a$ was taken as 1.5, i.e., $a = 3.0$ Å. From the plot of y against x as defined by (8.33), the slope corresponded to $K_A =$

7.5, at $x = \Lambda_0$, $y(0) = J_1 - K_A\Lambda_0 = 220$, so $J_1 = 660$. Proceeding as in example (3.6), with $a = 3.0$ Å, $\theta_1 = 8.211$, $\theta_2 = 36.35$, so $J_1 = 453$. With $a = 5.0$ Å, $\theta_1 = 12.23$, $\theta_2 = 52.2$ and $J_1 = 672$. Then by interpolation $J_1 = 600$ corresponds to $a = 4.3$ Å. This, as in example (3.6), leads to $\theta_1 = -3.2$, $J_2 = -0.8$. Using (8.34) gives the values of c_i'' shown in the Table and by

TABLE 8.7

Dissociation Constant of $NaBrO_3$ in 50% Dioxan-Water at 25°C using the Extended Conductance Treatment of Fuoss and Onsager

10^4c	5.8954	11.805	23.595	35.401	47.211	94.357
Λ	48.555	47.566	46.228	45.259	44.486	42.304
10^4c_i	5.859	11.68	23.17	34.62	46.02	91.35
$\Delta\Lambda$	0.181	0.343	0.747	1.195	1.662	3.536
y	315	301	334	361	380	417
x	41.4	39.0	35.2	34.4	30.6	25.6
$10^4c_i''$	5.876	11.72	23.27	34.72	46.08	90.20
$\Delta\Lambda_i'$	0.186	0.352	0.760	1.206	1.668	3.536
y'	322	308	339	363	381	422
$f_\pm$	0.9139	0.8855	0.8467	0.8193	0.7979	0.7399
x'	41.5	38.3	34.4	31.8	29.8	24.9

means of these and $a = 4.3$ Å, the second set of $\Delta\Lambda'$, y', $f_\pm$ and $x(= x')$ was obtained. The final value of K_A is 7.0 ± 1.0; Fuoss and Kraus [55] give $K_A = 6.87$, with $a = 4.03$ Å.

TABLE 8.8

Dissociation Constant of $NaBrO_3$ in Water at 25°C. Extend Treatment of Fuoss and Onsager.[58] E from (3.111) and example (3.6) = 35.52; $\Lambda_0 = 105.755$; $a = 4.0$ Å.

10^4c	2.60004	5.20539	10.4396	20.8887	41.8148	83.6224
Λ	104.465	103.840	103.071	101.989	100.508	98.566
Λ'''	105.801	105.728	105.714	105.657	105.552	105.392
z	0.0264	0.0516	0.1008	0.2026	0.3430	0.6270

Example 8.6. In water [55, 58] for $NaBrO_3$, $\Lambda_0 = 105.755$, $\epsilon = 78.54$. From example (3.6), $\alpha' = 0.2291$, $\beta' = 8.828$, $b = 1.784$ (taking $a = 4.0$ Å), from (3.83), $\alpha = 0.2290$, from (3.84), $\beta = 60.24$. From (3.106) to (3.109), $\theta_1 = 1.199$, $\theta_2 = 64.36$, $\theta_3 = 0.1525$, $J_1 = 193.6$, $J_2 = 109.0$. The values of Λ''' and z as defined by (8.35) and (8.36) are given in Table 8.8. From the plot of the right-hand side of (8.35) against z, omitting the most dilute point since this is the only one which does not fall on a straight line, the slope corresponds to $K_A = 0.52$.

Comparisons

A comparative analysis [59] of the conductance data of $NaBrO_3$, NaCl, KCl and HCl in dioxan-water using the various methods described in the present chapter reveals certain general features. Firstly, the dissociation constants of $NaBrO_3$, when calculated by the method of Davies or of Fuoss and Kraus with the aid of the limiting Onsager equation (section A) are almost identical, but are lower than those calculated by the extended Fuoss–Onsager treatment (section C). Thus, in 50 % dioxan, $K = 0.097$ by the method of Davies, 0.098 by the method of Fuoss and Kraus whereas the value obtained in Example (8.5) is 0.145.

When K is about 0.0001, as it is for NaCl in 82 % dioxan, it is difficult to assess Λ_0 by the Fuoss treatment used in Example (8.5). For instance, the plots of y' against x' remain linear at 25°C when Λ_0 is varied between 36.5 and 39.0. Consequently K and a cannot be ascertained with precision. Fuoss and Kraus [55] have made similar observations with regard to the J_1, J_2 and a values of NBu_4I in 80 % dioxan. Since other methods provide more sensitive ways of finding Λ_0, as mentioned in the earlier parts of this chapter, it might be of advantage to apply these.

Methods from sections A, B and C have been tested with conductances of NaCl and HCl in 70 % and 82 % dioxan. At 25°C the results are:[59]

	A	B	C
NaCl, 70 % dioxan	0.0063	0.0059	0.0083
HCl, 70 % dioxan	0.0074	0.0070	0.010
NaCl, 82 % dioxan	0.000145	0.000164	0.000154
HCl, 82 % dioxan	0.000203	0.000219	0.000220

In 70 % dioxan, method B, which is based on (8.30), gives the lowest answers, while in 82 % dioxan, method A, based on the limiting Onsager equation, gives the lowest. The explanation is that the ionic parameter term in the activity coefficient expression becomes an important factor at low dielectric constants. If the extended D.H. equation were used instead of the limiting form when applying the methods of section A, the answers would be closer to those given in the other columns.

Spiro [60] has evaluated the dissociation constants of $KClO_4$, KIO_3, LiOH and sulphamic acid in water at 25°C; the dissociation constants of these electrolytes are large, and different methods are found to given different answers. For instance, for $KClO_4$, $K = 1.07$, 1.44 and 1.8. The first of these is derived by an "external" method which utilizes Kohlrausch' rule for deriving the conductance equation of a strong electrolyte; this is illustrated by (8.29). The second is calculated by Shedlovsky's method (see 8.15 to 8.19) of applying the limiting Onsager equation, and the third answer is calculated by means of the Fuoss method of applying the extended Fuoss–Onsager treatment. A much higher answer of 3.0 ± 0.7 has been estimated by applying the method of Davies.[6] The corresponding figures for KIO_3 are 1.8, 3.9 and 0.55, which may be compared with 1.7 derived by the Davies method and the limiting Onsager equation.[5] One of Spiro's comments,[60] that the measurements are not accurate enough may preclude attaching much significance to the wide range of these answers.

High Field Conductances

Patterson and his associates,[61] particularly D. Berg, have measured the conductance increases of a number of polyvalent and weak electrolytes on applying fields up to 200 kV/cm. The increments of the polyvalent electrolytes could only be reconciled with the theoretical treatments of Onsager, Wilson and Kim (see Harned and Owen [11]) if allowance is made for ion-association,[62] but the required dissociation constant of $LaFe(CN)_6$ is 3.7×10^{-4} compared with 1.8×10^{-4} obtained by Davies and James [10] by the method given in section A of this Chapter, although the answers obtained by this method for $MgSO_4$, $ZnSO_4$ and $CuSO_4$ are able to satisfy the high field conductance requirements.

Diffusion

Harned and his co-workers have applied the conductance cell method (Chapter 7) of determining diffusion coefficients to [64] $ZnSO_4$ and $MgSO_4$ solutions at 25°C. In using (7.32) to calculate theoretical values, they obtained $d \ln y_{\pm}/d \ln C$ from the experimental values of $\gamma_{\pm}$ for $ZnSO_4$ calculated from the potentials of cell [65] (4.31). The outcome is that the calculated values of D are considerably lower than those obtained by experiment, and to correlate the differences with the dissociation constants of these salts, as obtained by conductance,[1, 66] it was deduced that

$$D_{\text{obs.}}/D_{\text{calc.}} = 1 + (1 - \alpha_c)[(\lambda_m^0(1/\lambda_1^0 + 1/\lambda_2^0) - 1]$$

in which λ_m^0, the mobility of the ion-pair emerges as 44 for $ZnSO_4$ and 46.5 for $MgSO_4$.

The experimental values of the activity coefficients are lower than those calculated by means of the D.H. formulae, and using (9.9) with $Q'' = 0.3$ and $a = 3.64$ Å, from (7.42),

$$d \ln f_{\pm}/d \ln C = -1.317\sqrt{C}/(1 + 2.392\sqrt{C})^2 + 2.76\,C$$

and with this in (7.32), as the following figures for $ZnSO_4$ show, there is little difference between the observed and calculated diffusion coefficients. The reason for this is that $D_{obs.}$ is based upon conductance measurements which assume that the salts are fully dissociated.

10^3C (molarity)	1.08	1.39	1.92	2.56	3.08	4.39	4.71
$10^6D_{obs.}$	7.47	7.39	7.33	7.28	7.21	7.14	7.07
$10^6D_{calc.}$	7.55	7.45	7.32	7.20	7.12	6.96	6.93

REFERENCES

1. Davies, C. W. *Trans. Faraday Soc.* **23,** 351 (1927).
2. Dunsmore, H. S. and Speakman, J. C. *ibid.* **50,** 236 (1954).
3. Critchfield, F. E., Gibson, J. A. and Hall, J. L. *J. Amer. chem. Soc.* **75,** 1991 (1953).
4. Geddes, J. A. *ibid.* **55,** 4832 (1933).
5. Onsager, L. *Phys. Z.* **28,** 277 (1927).
6. Monk, C. B. *J. Amer. chem. Soc.* **70,** 3281 (1948).
7. Owen, B. B. *ibid.* **61,** 1393 (1939).
8. Righellato, E. C. and Davies, C. W. *Trans. Faraday Soc.* **26,** 592 (1930). Banks, W. H. Righellato, E. C. and Davies, C. W. *ibid.* **27,** 621 (1931). Robinson, R. A. and Davies, C. W. *J. chem. Soc.* 574 (1937).
9. Money, R. W. and Davies, C. W. *Trans. Faraday Soc.* **28,** 609 (1932). Dunsmore, H. S. and James, J. C. *J. chem. Soc.* 2925 (1951).
10. Davies, C. W. and James, J. C. *Proc. roy. Soc.* **195A,** 116 (1948).
11. Refs. 2 and 11, Ch. 1.
12. James, J. C. and Monk, C. B. *Trans. Faraday Soc.* **46,** 141 (1950).
13. Davies, C. W. *J. chem. Soc.* 645 (1933).
14. Bjerrum, J., Schwarzenbach, G. and Sillen, L. G. " Stability Constants of Metal Ion Complexes, with Solubility Products of Inorganic Substances." Spec. Pub. Nos. 6 and 7, The Chemical Society, London, (1957-8).
15. Davies, C. W. *J. chem. Soc.* 271, 2093 (1938). James, J. C. *J. chem. Soc.* 1094 (1950); 153 (1951). Topp, N. E. and Davies, C. W. *ibid.* 87 (1940). Bevan, J. R. and Monk, C. B. *ibid.* 1392 (1956). Davies, C. W. and Thomas, G. O. *ibid.* 3660 (1958).
16. Jeffery, G. H. and Vogel, A. I. *ibid.* 21 (1935).
17. Monk, C. B. *ibid.* 429 (1949).
18. Kerker, M. *J. Amer. chem. Soc.* **79,** 3664 (1957).
19. Extrapolated from the data of Harkins, W. D., Hall, R. E. and Roberts, W. A. *ibid.* **38,** 2643 (1916).
20. Jenkins, I. L. and Monk, C. B. *J. chem. Soc.* 68 (1951).
21. Davies, C. W. *J. Amer. chem. Soc.* **59,** 1760 (1937). Wise, W. C. A. and Davies, C. W. *J. chem. Soc.* 273 (1938). Davies, C. W. *ibid.* 460 (1945). Davies, C. W. and Monk, C. B. *ibid.* 413 (1949). Spedding, F. H. and Jaffe, S. *J. Amer. chem. Soc.* **76,** 882 (1954).
22. Jenkins, I. L. and Monk, C. B. *ibid.* **72,** 2695 (1950).
23. Jones, H. W., Monk, C. B. and Davies, C. W. *J. chem. Soc.* 2693 (1949). Jones, H. W. and Monk, C. B. *ibid.* 3475 (1950). Gibby, C. W. and Monk, C. B. *Trans. Faraday Soc.* **48,** 632 (1952).
24. Fuoss, R. M. *Chem. Rev.* **17,** 27 (1935). Mead, D. J., Fuoss, R. M. and Kraus, C. A. *Trans. Faraday Soc.* **32,** 594 (1936).
25. Fuoss, R. M. *J. Amer. chem. Soc.* **57,** 488 (1935).

26. Fuoss, R. M. *loc. cit.* 2605.
27. Shedlovsky, T. *J. Franklin Inst.* **225,** 739 (1938). Shedlovsky, T. and Kay, R. L. *J. phys. Chem.* **60,** 151 (1956).
28. Daggett, H. M. Jr., *J. Amer. chem. Soc.* **73,** 4977 (1951).
29. Fuoss, R. M. and Shedlovsky, T. *ibid.* **71,** 1497 (1949).
30. Refs. 2 and 11, Ch. 1.
31. Kraus, C. A. and others, *J. Amer. chem. Soc.* **69,** 451, 454, 814, 1016, 1731, 2472, 2481 (1947); **70,** 706, 1707, 1709 (1948); **71,** 1565, 2405, 2694, 2983, 3049, 3288, 3803 (1949); **72,** 166, 3676 (1950); **73,** 799, 2170, 2459, 3293, 4732 (1951). See also Hartley, H. and others, *Proc. roy. Soc.* **109A,** 351 (1925); **127A,** 228 (1930); **132A,** 427 (1932). *J. chem. Soc.* 2488 (1930); 1207 (1933).
32. Evers, E. C. and Knox, A. G. *J. Amer. chem. Soc.* **73,** 1739 (1951).
33. El-Aggan, A. M., Bradley, D. C. and Wardlaw, W. *J. chem. Soc.* 2092 (1958).
34. Griffiths, V. S. and others, *J. chem. Soc.* 686 (1954); 1208, 2797 (1955); 473 (1956); 3243 (1957); 309, 3998 (1958).
35. Jones, M. M. and Griswold, E. *J. Amer. chem. Soc.* **76,** 3247 (1954).
36. Wiberg, K. B. and Evans, R. J. *ibid.* **80,** 3019 (1958).
37. Dippy, F. J. F. *Chem. Rev.* **25,** 151 (1939).
38. Dippy, F. J. F., Barry, L. G., Hughes, S. R. C. and others, *J. chem. Soc.* 550 (1941); 411 (1944); 1470, 4102 (1954); 2995 (1956); 265, 2405 (1957); 1441, 1717 (1959).
39. MacInnes, D. A. and Shedlovsky, T. *J. Amer. chem. Soc.* **54,** 1429 (1932).
40. Ives, D. J. G. *J. chem. Soc.* 731 (1933).
41. Ives, D. J. G. and Sames, K. *ibid.* 511 (1943).
42. Belcher, D. *J. Amer. chem. Soc.* **60,** 2744 (1938).
43. Kilpatrick, M. L. *J. chem. Phys.* **8,** 306 (1940).
44. Davies, C. W. *J. Amer. chem. Soc.* **54,** 3776 (1932).
45. Viscosities interpolated from "International Critical Tables", McGraw-Hill Book Co., New York, Vol. 5 (1928).
46. Katchalsky, A., Eisenberg, H. and Lifson, S. *J. Amer. chem. Soc.* **73,** 5889 (1951).
47. MacDougall, F. H. and Blumer, D. R. *ibid.* **55,** 2236 (1933).
48. Davies, M. and Griffiths, D. M. L. *Z. phys. Chem.* Neue Folge, **2,** 353 (1954).
49. Cartwright, D. R. and Monk, C. B. *J. chem. Soc.* 2500 (1955).
50. Davies, M. and Griffiths, D. M. L. *Z. phys. Chem.* Neue Folge, **6,** 143 (1956).
51. Shedlovsky, T. *J. Amer. chem. Soc.* **54,** 1405 (1932).
52. Shedlovsky, T. and Brown, A. S. *ibid.* **56,** 1066 (1934).
53. Fuoss, R. M. *ibid.* **79,** 3301 (1957).
54. Einstein, A. *Ann. Phys.* **19,** 289 (1906); **34,** 591 (1911).
55. Fuoss, R. M. and Kraus, C. A. *J. Amer. chem. Soc.* **79,** 3304 (1957).
56. Fuoss, R. M. *ibid.* **81,** 2659 (1959).
57. Accascina, F., D'Aprano, A. and Fuoss, R. M. *ibid.* **81,** 1058 (1959); Accascina, F., Petrucci, S. and Fuoss, R. M. *ibid.* **81,** 1301 (1959).
58. Martell, R. W. and Kraus, C. A. *Proc. nat. Acad. Sci. Wash.* **41,** 9 (1955).
59. Nash, G. R. and Monk, C. B. *Trans. Faraday Soc.* **54,** 1650 (1958).
60. Spiro, M. *ibid.* **55,** 1946 (1959).
61. Patterson, A. Jr., *J. Amer. chem. Soc.* (a) $MgSO_4$, $ZnSO_4$: with Bailey, F. E. Jr., **74,** 4426, 4428, (1952); (b) $CuSO_4$: with Berg, D, **74,** 4704 (1952); (c) $CdCl_2$: with Bailey, F. E. Jr., **75,** 1471 (1953); (d) $LaFe(CN)_6$: with Berg, D, **75,** 1484 (1953); (e) Glycine: with Berg, D, **75,** 1483 (1953); (f) aminocaproic acid: with Berg, D, **75** 4834 (1953); (g)NH_4OH: with Berg, D, **75,** 5731 (1953); carbonic acid: with Berg, D, **75,** 5197 (1953). With Wissbrun, K. F. and French, D. M., *J. phys. Chem.* **58,** 693 (1954).

62. Bailey, F. E. Jr. and Patterson, A. Jr. *J. Amer. chem. Soc.* **74,** 4428 (1952).
63. Brewster, P. W., Schmidt, F. C. and Schaap, W. B. *ibid.* **81,** 5532 (1959).
64. Harned, H. S. and Hudson, R. M. *ibid.* **73,** 3781, 5880 (1951).
65. Cowperthwaite, I. A. and La Mer, V. K. *ibid.* **53,** 4333 (1931).
66. Owen, R. B. and Gurry, R. W. *ibid.* **60,** 3078 (1938).
67. Golben, M. and Dawson, L. R. *J. phys. Chem.* **64,** 37 (1960). (K_{diss} of $MgCl_2$ and $CdCl_2$ methanol from 20°C down to − 40°C and − 70°C respectively.)
68. Sukhotin, A. M. and Timofeeva, Z. N. *J. phys. Chem. Moscou* (English trans., Chem. Soc., London), No. 7, 72 (1959).

CHAPTER 9

INCOMPLETE DISSOCIATION, PART II: E.M.F. AND pH METHODS. SUCCESSIVE COMPLEXITY CONSTANTS AT CONSTANT IONIC STRENGTHS

E.M.F. Methods

1. Acids

The "buffered" cell

$$\mathrm{Pt} \mid \mathrm{H_2} \mid \mathrm{HA},\ m_1;\ \mathrm{MA},\ m_2;\ \mathrm{MX},\ m_3 \mid \mathrm{AgX} \mid \mathrm{Ag} \tag{9.1}$$

where A is the anion of an incompletely dissociated acid HA, M is a cation (usually Na^+ or K^+) and X^- is usually Cl^-, has been used extensively by Harned and his associates [1] for determining the dissociation constants of weak acids, and is one of the most exact methods available. It was first used by Harned and Ehlers [2] for acetic acid in aqueous solution. The required information is attained by combining

$$E = E^0_{\mathrm{AgCl}} - k' \log m_{\mathrm{H}} m_{\mathrm{Cl}} \gamma_{\mathrm{H}} \gamma_{\mathrm{Cl}} \tag{9.2}$$

and

$$K = m_{\mathrm{H}} m_{\mathrm{A}} \gamma_{\mathrm{H}} \gamma_{\mathrm{A}} / m_{\mathrm{HA}} \gamma_{\mathrm{HA}} \tag{9.3}$$

to form

$$(E - E^0)/k' + \log (m_{\mathrm{HA}} m_{\mathrm{Cl}}/m_{\mathrm{A}}) = -\log K - \log (\gamma_{\mathrm{Cl}} \gamma_{\mathrm{HA}}/\gamma_{\mathrm{A}}) = -\log K' \tag{9.4}$$

where $m_{\mathrm{HA}} = m_1 - m_{\mathrm{H}}$, $m_{\mathrm{Cl}} = m_3$, $m_{\mathrm{A}} = m_2 + m_{\mathrm{H}}$. One convenient way of finding m_{H} for these is by means of (9.2), and a general form of the D.H. equation, e.g. (3.42), with $I = m_2 + m_3$; if the acid is very weak, m_{H} may be small enough to ignore.

In dilute solutions, $\gamma_{\mathrm{HA}} = 1$, and $\gamma_{\mathrm{Cl}}/\gamma_{\mathrm{A}}$ should also be close to unity; e.g., in [3] 0.1 M aqueous solution at 25°C, $\gamma_{\mathrm{NaCl}}/\gamma_{\mathrm{NaAc}} = 0.778/0.791$. Consequently, K' should be practically constant up to $I = 0.1$. In practice this is often not so, and the true value of K is found by extrapolation, e.g., of $\log K'$ against I.

Example 9.1. Harned and Ehlers [2] obtained data for acetic acid from 0° to 60°C. The molalities used at 25°C and the obtained potentials (in abs. mV) are shown in Table 9.1. $E^0 = 222.46$ mV, $k' = 59.156$ mV.

Leaving out the second result since it is lower than its neighbours would suggest, a plot of K' against I or any of the m values extrapolates to $K = 1.750 \times 10^{-5}$, which is in excellent agreement with $K = 1.751 \times 10^{-5}$ derived from conductances (Chapter 8).

Three possible reasons why K' is not constant are [4] (a) E^0 depends on the composition of the solvent, particularly on m_{HA}, (b) a fraction of the HA molecules are present as dimers, H_2A_2 (this has been considered in Chapter 8), and (c) the dielectric constant of the medium decreases as m_{HA} increases. The effect of m_{HA} upon E^0 has been examined by Harned and Owen [3] by means of the data obtained by Harned and Robinson [5] with the cell

$$\text{Pt} \mid H_2 \mid \text{HA}, m_1; \text{NaCl}, m_3 \mid \text{AgCl} \mid \text{Ag} \tag{9.5}$$

This was originally studied for the purpose of deriving K_{HAc}, but it suffers from the disadvantage that both m_H and m_A depend on the potential so that

TABLE 9.1

K' Values for Aqueous Acetic Acid at 25°C

$10^3 m_1$	4.779	12.035	21.006	49.22	81.01	90.56
$10^3 m_2$	4.599	11.582	20.216	47.37	77.96	87.16
$10^3 m_3$	4.896	12.426	21.516	50.42	82.97	92.76
E(mV)	639.80	616.03	601.74	579.96	567.31	564.42
$10^5 m_H$	2.2	2.4	2.6	3.0	3.2	3.2
$10^5 K'$	1.749	(1.731)	1.739	1.730	1.719	1.720

any error in the potential is squared, and furthermore, m_A is very small if K is small whereas in cell (9.1), it is controlled by m_2. Harned and Owen [3] found that the potentials of cell (9.5) point to E^0 in acetic acid solutions at 25°C changing by over 1 mV for every 1 % increment of acid.

A better way of obtaining E^0 is by means of the cell [4]

$$\text{Pt} \mid H_2 \mid \text{HA}, m_1; \text{HCl}, m_4 \mid \text{AgCl} \mid \text{Ag} \tag{9.6}$$

since the dissociation of HA is suppressed, and m_H can be varied considerably. Results have been obtained [4] for acetic, propionic and butyric acids using the method of calculation described in connection with (4.23), replacing $2k' \log m$ by $k' \log (m_4 + m') + k' \log m_4$, where m' is the $[H^+]$ due to the dissociation of HA, and replacing 0.50 in the generalized D.H. equation (3.43) by the appropriate factor as given by (3.36).

The dimerisation constant K_d can be calculated from measurements with cell (9.1) providing m_1 is large enough to form useful concentrations of H_2A_2. The relevant expressions are:

$$\log m_{HA} = \log m_H m_A - \log K + 2 \log \gamma_\pm \tag{9.7}$$

$$m_{H_2A_2} = \tfrac{1}{2}(m_1 - m_{HA} - m_H); \; K_d = m_{HA}^2 / m_{H_2A_2} \tag{9.8}$$

where m_H is obtained by means of (9.2).

Example 9.2. With [4] $m_1 = 0.1000$, $m_2 = 0.007213$, $m_4 = 0.008262$, E of cell (9.1) for acetic acid at 25°C was found to be 560.02 mV. Since [4] $E^0 = 221.45$

mV, $K = 1.750 \times 10^{-5}$ (example 9.1) and $k' = 59.156$, from (9.2) after two approximations, $m_H = 2.957 \times 10^{-4}$. From (9.7), $m_{HA} = 0.09890$, so $m_{H_2A_2} = 0.00040$ and $K_d = 24$.

The average values of K_d obtained in this way are: acetic acid = 18, propionic = 10.2, butyric = 6.7. These are of the same order but somewhat lower than figures derived from conductance and other methods (Chapter 8).

Considering point (c), there are a few figures showing that the dielectric constant is lowered as m_{HA} increases, e.g.,[6] $\epsilon = 6.2$ for acetic acid, and 2.9 for butyric acid. Tentative calculations,[4] assuming that ϵ of 0.6 m_1 is 3 units lower than that of water and that K_{HAc} is lowered from 1.750×10^{-5} to 1.68×10^{-5} at 25°C raises K_d from 18 to 28, and likewise if K_{HBu} is lowered from 1.515×10^{-5} to 1.37×10^{-5}, K_d is raised from 6.7 to 11.

A large number of dissociation constants of monobasic acids have been found by the Harned method. Most of these are recorded in the books of Harned and Owen,[3] Robinson and Stokes,[7] and Bjerrum, Schwarzenbach and Sillen.[8] In solvents of low dielectric constant even HCl is not completely dissociated, and this has to be taken into account when deriving E^0 before K_{HA} can be calculated. This is illustrated by the calculations of Danyluk and others [9] concerning formic, acetic and propionic acids in 82 % dioxan (they used the data of Harned [3]). Thus at 25°C, the recalculated values are 7.231×10^{-10}, 3.10×10^{-11} and 1.77×10^{-11} respectively.

The dissociation constants of polybasic acids may also be found by means of cell (9.1). Either sodium or potassium salts are used since these are strong electrolytes and the salt: weak acid ratio has to be selected in order to eliminate or minimize certain of the acid anions. For instance, by taking a 2 : 1 molal ratio of $Na_2Ox : H_2Ox$, very little of H_2Ox is present and the determination of $K_2 = a_H a_A / a_{HA}$ is straightforward, but with many weak dibasic and tribasic acids the ratios K_1/K_2 and K_2/K_3 where $K_1 = a_H a_{HA}/a_{H_2A}$ and $K_3 = a_H a_A / a_{HA}$, ($A = A^{3-}$, $HA = HA^{2-}$), are such that there is a considerable overlap of the successive stages of dissociation in spite of carefully selecting the ratio of acid to salt. The exact determination of each dissociation constant then requires considerable calculations by approximation methods; Bates [10] has formulated and discussed appropriate treatments.

Examples of the determination of the dissociation constants of polybasic acids are the K_1 and K_2 values of oxalic,[11] carbonic,[12] sulphuric,[13] glycerol 2-phosphoric,[14] glucose-, aminoethanol- and glycerol-1-phosphoric acids [14] (Datta and his associates have described their careful experimental work and methods of calculation in great detail), phosphoric [15] and citric [16] (also K_3). With regard to carbonic acid, it cannot be assumed that all the CO_2 is present as H_2CO_3, HCO_3^- and CO_3^{2-}. Investigation shows that only a small fraction, e.g., about [17] 0.4 % of the gas is present in the hydrated form. Carini and Martell [18] have determined K_3 and K_4 of ethylene diamine tetra-acetic acid

(H_4V) by means of K_2H_2V, K_3HV and KCl mixtures in the ratios 1 : 1 : 3.20 for K_3 and 1 : 4 : 25 for K_4.

Hamer [13] used mixtures of Na_2SO_4 and HCl for determining K_2 of sulphuric acid with cell (9.1). Such mixtures can contain $NaSO_4^-$ ion-pairs, and if this is taken into account, it has some influence upon the answer.[19] This unwanted feature can be eliminated by using mixtures of H_2SO_4 and HCl, and the data obtained in the first investigation [19] are in good agreement with those obtained by spectrophotometry [20] (Chapter 10) between 5° and 50°C. Hamer's data at 25°C, after allowing for $NaSO_4^-$ formation, and also those from conductance [19, 21] are in agreement with the HCl, H_2SO_4 potential [19, 22] and spectrophotometric values. On the other hand, Hamer's figures at other temperatures, even when allowance is made for $NaSO_4^-$, are in disagreement with the other results. The second set of e.m.f. measurements with H_2SO_4, HCl mixtures [22] should help to clarify the position, but these answers are distinctly higher than the others. A major reason is, as these authors have shown,[22] that the values are critically dependent on the choice of the constants and parameters used in the D.H. equation. Taking (3.41), i.e.,

$$-\log \gamma_i = A''z_i^2\{\sqrt{I}/(1 + B''a\sqrt{I}) - Q''I\} \tag{9.9}$$

Nair and Nancollas took the theoretical values of A'' (3.36), $B''a = 1.0$ and $Q'' = 0.2$, whereas slightly different values of A'' were used by Davies *et al.*[19] Now it would be desirable to test (9.9) with mixtures of strong electrolytes under conditions as close as possible to those existing in H_2SO_4, HCl mixtures, e.g., to derive $\gamma_\pm$ data for HCl from cells containing HCl, MCl_2 solutions. Harned and Paxton [23] have studied one such system, namely HCl(0.01 M), $SrCl_2$ in cell (9.1) and calculation shows that by using the theoretical values of A'', $B''a = 1.0$, $Q'' = 0.3$ is required in (9.9) to reproduce the experimental values of $\gamma_\pm$ of HCl up to $I = 0.1$, although $Q'' = 0.2$ is satisfactory when the cell contains only HCl (up to $I = 0.1$). Alternatively, $B''a$ could be varied although the upper limit is about 1.2 for agreement between experimental and calculated values. Further support for taking $Q'' = 0.3$ rather than 0.2 comes from a consideration of the e.m.f. data of Nair and Nancollas for the HCl, H_2SO_4 cell at 25°C; they studied 16 solutions in all, 10 at $I = 0.0045$ to 0.01, and 6 between $I = 0.016$ to 0.026. With $Q'' = 0.2$, the 1st set average $K_{HSO_4} = 0.0108_5$, the 2nd set average 0.0114, and the overall average is 0.0110 although there is a definite change with ionic strength. Taking $Q'' = 0.3$, the averages of the two sets are 0.0105_5 and 0.0108, so the drift with ionic strength is much less, and at $I = 0$, $K = 0.0105$. The calculations for other temperatures together with results from other sources are given in Table 9.2.

The dissociation constants of amino-acids, peptides and related substances may be determined by Harned's method with cell (9.1). These zwitterions, which will be represented by $Z^{\pm}$ (e.g., glycine is $NH_3^+ . CH_2 . COO^-$) undergoes acidic or basic dissociation according to the conditions. The equilibrium in acidic conditions is $ZH^+ \rightleftharpoons Z^{\pm} + H^+$, the constant of which is represented by

$$K_A \text{ or } K_{1a} = m_Z m_H \gamma_Z \gamma_H / m_{ZH} \gamma_{ZH} \tag{9.10}$$

TABLE 9.2

Dissociation Constants of the HSO_4^- Ion in Water (× 100). (a) Ref. 19, recalc., (b) Ref. 22, recalc., (c) from spectrophotometry,[20] (d) from conductances and transference numbers,[19, 21] (e) selected averages, (f) calculated from $\log K = -1.645 - 0.0134t$ (t = °C).

°C	0	5	10	15	20	25	30	35	40	45	50	55
(a)	—	1.90	1.70	1.51	1.27	1.05	0.91	0.78	0.68	—	0.52	—
(b)	2.55	2.17	—	1.51	—	1.05	—	0.80	—	0.585	—	—
(c)	—	1.85	—	1.39	—	1.04	—	0.77	—	0.565	—	0.413
(d)	2.2	—	—	—	—	1.02	—	—	—	—	0.53	—
(e)	—	1.9	—	1.47	—	1.05	—	0.78	—	0.575	0.52	—
(f)	2.27	1.94	1.66	1.43	1.22	1.05	0.90	0.77	0.66	0.56	0.48	0.42

and under basic conditions the equilibrium is $Z^{\pm} \rightleftharpoons Z^- + H^+$, or $ZOH^- \rightleftharpoons Z^{\pm} + OH^-$, for which

$$K_{2a} = m_H m_{Z^-} \gamma_H \gamma_{Z^-} / m_Z \gamma_Z \tag{9.11}$$

$$K_B = m_Z m_{OH} \gamma_Z \gamma_{OH} / m_{ZOH} \gamma_{ZOH} \tag{9.12}$$

To obtain K_A, the following cell is used:

$$\text{Pt} \mid H_2 \mid Z^{\pm}, m_1; \text{HCl}, m_2 \mid \text{AgCl} \mid \text{Ag} \tag{9.13}$$

and for K_{2a} or K_B,

$$\text{Pt} \mid H_2 \mid Z^{\pm}, m_1; \text{NaOH}, m_2; \text{NaCl}, m_3 \mid \text{AgCl} \mid \text{Ag} \tag{9.14}$$

Glycine has been studied by Owen,[24] King,[25] and by Datta and Grzybowski[26] (K_B only) who have corrected the Cl^- and Z^- concentrations since, like other amino compounds, soluble Ag^+ complexes form. Datta [26] also devised an extrapolation method which has a linear plot; this consists of plotting $-\log K'_{2a}$ against $y + BI + CI^{\frac{3}{2}}$ where $y = (E - E^0)/k' + \log (m_{Cl} m_Z / m_{Z^-})$, while B and C are found from a plot of y against I by the method of least squares. A simpler way which has been applied to triglycine,[27] is to plot $\log K_{2a}$ against m_Z.

Numerous results have been obtained by these e.m.f. methods.[28] A useful cell [27] when only small quantities are available is shown in Fig. 9.1.

Example 9.5. With [26] 0.010376 M glycine hydrochloride and 0.015617 M KOH in cell (9.14) at 25°C, $E - E^0 = 696.54$ abs. mV. Thus $m_1 = m_3 =$

0·010376, m_2 0.005241. From (9.4) and (9.11), $(E - E^0)/k' + \log (m_Z m_{Cl}/m_{Z^-}) = -\log K'_{2a}$. Since $[Z^-] + [Cl^-] + [OH^-] = [K^+] + [H^+]$, $m_{Z^-} + m_{OH} = m_2$, $m_Z = m_1 - m_{Z^-}$. The value of $m_{OH} = K_w/m_H\gamma_{\pm}^2$ (see the next section), so $\log m_{OH} = \log K_w m_{Cl} + (E - E^0)/k'$. Since $k' = 59.156$ mV, and $K_w = 1.008 \times 10^{-14}$, $m_{OH} = 0.000062$, $m_{Z^-} = 0.005179$, $m_Z = 0.005197$, $m_{Cl} = 0.010376$, and $-\log K'_B = 9.792$. Corrections for the formation of silver glycinates alters this [26] to 9.780. Plotting all the results [26] against m_Z gives, on extrapolation, $K_B = 9.779 \pm 0.002$.

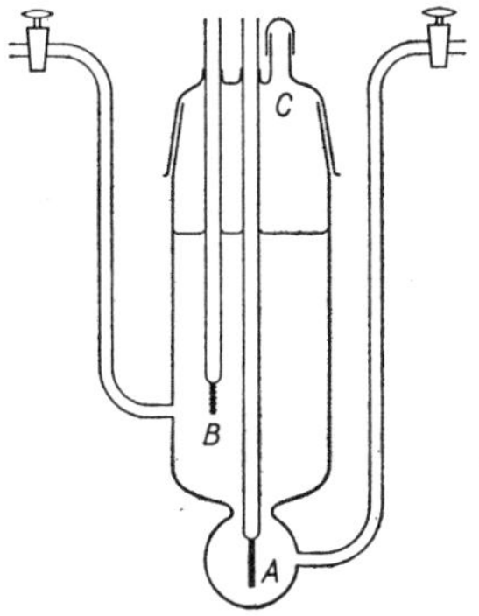

FIG. 9.1. A = AgCl Ag electrode; B = Pt electrode; C = entrance for stock solutions.

Similar types of cell to (9.1) which have been used to obtain the dissociation constants of acids are:

$$\text{Pt} \mid H_2 \mid \text{Salt, Acid, } Cl^- \mid \text{HgCl} \mid \text{Hg} \qquad (9.15)$$

which Grzybowski [29] has used for K_2 of phosphoric acid, and

$$\text{Pt} \mid \text{Quinhydrone, Salt, Acid, } Cl^- \mid \text{AgCl} \mid \text{Ag} \qquad (9.16)$$

whereby Parton and Gibbons [30] have determined K_1 of oxalic acid. Here, E^0 is the standard potential of the cell Pt | Quinhydrone, HCl | AgCl | Ag. This has been obtained by Harned and Wright.[31] By using deuterated quinhydrone and acid in D_2O in a micro form of cell (9.1), the dissociation constants of deuterated acetic and salicylic acids have been assessed as 5.8×10^{-6} and 2.5×10^{-4} at 25°C respectively.[32]

Bruckenstein and Kolthoff,[33] besides reviewing earlier potentiometric studies of cells in which the solvent is acetic acid, have obtained data in this medium with the cell

$$\text{Pt} \mid C_6Cl_4(OH)_2,\ C_6Cl_4O_2,\ HX(C_{HX}) \parallel \text{Ref. Electrode} \qquad (9.17)$$

The reference electrode was a form of the calomel electrode containing a solution of NaCl and $NaClO_4$ in acetic acid, with a connecting arm and sinter to the main body of the cell. The chloranil electrode functions as a H_2 electrode. Combination of

$$E_{HX} = E^0_{HX} + k' \log a_H + E(\text{liquid junction}) \qquad (9.18)$$

with (see Chapter 10),

$$K_{HX} = a_{H^+}a_{X^-}/(a_{HX} + a_{H^+X^-}) \qquad (9.19)$$

produces

$$E_{HX} = E^0_{HX} + \tfrac{1}{2}k' \log K_{HX} + \tfrac{1}{2}k' \log C_{HX} \qquad (9.20)$$

Since K_{HCl} is known from spectrophotometry (Chapter 10), E^0 can be derived and hence K_{HX} of other acids found; (p$K_{HCl} = 8.55$, $E^0 = 90.95$ mV, at 25°C). The obtained pK values are: $HClO_4 = 4.87 \pm 0.07$, $H_2SO_4 = 7.24 \pm 0.01$, $HTs = 8.46 \pm 0.02$ ($Ts = p.CH_3.C_6H_4.SO_3^-$). The approximate conductimetric values are 6.07, 8.13, and 9.19 respectively.[34]

The potentials of the concentration cell

$$H_2 \mid HA_1, m_1; NaA_1, m_1; NaCl, m - m_1 \,\|\, 3.5\text{ N KCl} \,\|\, HA_2, m_1; NaA_2, m_2; NaCl, m - m_2 \mid H_2$$

as Everett and Pinsent[35] have shown, may give K_{HA_2} if K_{HA_1} is known, since $E/k' = \log (K_{HA_2}/K_{HA_1})$ if the liquid junction potentials cancel (this was proved) and the ratio γ_A/γ_{HA} is the same for both acids. The chosen reference was acetic acid, and results between $I = 0.05$ to 0.20 were extrapolated to $I = 0$.

2. The Ionic of Water (K_w) and the Dissociation Constants of Hydroxides

A series of systematic studies of the cell

$$Pt \mid H_2 \mid \text{Cation}, OH^-, Cl^- \text{ or } Br^- \mid AgCl \text{ or } AgBr \mid Ag \qquad (9.21)$$

by Harned and his co-workers[9,30] had, as its main objective, the determination of K_w at various temperatures and the effect of salt concentrations upon this. Since

$$K_w = a_H a_{OH}/a_{H_2O} = m_H m_{OH} \gamma_H \gamma_{OH}/a_{H_2O} \qquad (9.22)$$

from (9.2),

$$(E - E^0)/k' + \log (m_X/m_{OH}) = -\log K_w a_{H_2O} - \log (\gamma_X \mid \gamma_{OH}) \qquad (9.23)$$

where $X^- = Cl^-$ or Br^-. If a_{H_2O} and γ_X/γ_{OH} were both unity, K_w as calculated from the left-hand side of (9.23) should be constant for all ratios of m_X/m_{OH} unless the cation associates with the OH^- and/or X^- ions. It is evident from the information which has been obtained[36] that one or more of these conditions is not fulfilled and very often marked variations occur which are related to the halide or hydroxide concentrations and these variations are largely characteristic of the cations. So far as alkali and alkaline earth metal cations are concerned, these do not associate with X^- ions (with the possible exception of [37] Cs^+). Also, by reference to known data [3,7] a_{H_2O} is very close to unity in dilute solutions and up to $I = 0.1$, γ_X/γ_{OH} is also very close [3,7] to one, and on these assumptions (but see further remarks in Ref. 104),

$$(E - E^0)/k' + \log m_X K_w^0 = \log m_{OH^-} \tag{9.21}$$

where $K_w^0 = K_w$ at $I = 0$. If m_1 and m_2 are the stoichiometric halide and hydroxide molalities, then $m_2 - m_{OH^-}$ and $m_M = m_1 + m_2 - m_{MOH}$ for the alkali metal hydroxides, and corresponding equations can be written for the alkaline earth series. The data have been analyzed in this way and values of

$$K = m_H m_{OH} \gamma_\pm^2 / m_{MOH} \tag{9.22}$$

derived, using a generalized activity coefficient expression, namely (9.9) with $Q'' = 0.2$, and $A'' = 0.485$ at 5°C, 0.49 at 15°C, 0.50 at 25°C, 0.51 at 35°C, and 0.52 at 45°C. It was found that the values of K are sensitive to changes in K_w^0, and this provides a way of fixing the latter with precision. This is shown by the following example.

Example 9.4. Harned and Copson [36] obtained the potentials of cell (9.21) when this held 0.01 M LiCl and varying molalities of LiCl. The measurements at 25°C up to 0.10 M LiCl together with K_{LiOH} using three different values of K_w^0 are shown in Table 9.3. At the time of taking the measurements, k' was taken as 0.00019844 T where $T = 273.1 + t$°C $= 0.059155$ at 25°C, and E^0 was taken as 0.22239 V. It is seen that K_{LiOH} is nearest to constancy when $-\log K_w^0 = 13.997$, i.e., $K_w^0 = 1.007 \times 10^{-14}$.

TABLE 9.3.

Dissociation Constant of LiOH and the Ionic Product of Water at 25°C

	LiCl = 0.01	0.02	0.04	0.05	0.09	0.10
$-\log K_w^0$	E(V) = 1.04979	1.03175	1.01349	1.00755	0.99167	0.98883
13.996	K = 0.718	0.706	0.687	0.683	0.667	0.679
13.997	K = 0.644	0.655	0.653	0.654	0.649	0.661
13.998	K = 0.581	0.611	0.622	0.628	0.632	0.644

The average value of K for this and other hydroxides at 25°C are:[37] LiOH = 0.65, NaOH = 5.9, $CaOH^+ = 0.043$, $SrOH^+ = 0.150$, $BaOH^+ = 0.23$. They agree with answers derived by other methods.[37] The values of K_w^0 at various temperatures are given below, together with those obtained by Harned [36] (his full results extend from 0°C to 60°C). The NaOH—NaCl system in aqueous dioxan has also been examined.[38]

By using NH_4OH and NH_4I with a Ag—AgI electrode in cell (9.21), Owen [39] obtained $K_{NH_4OH} = 1.75 \times 10^{-5}$ at 25°C. NH_4Cl and Ag—AgCl electrodes introduce an extra factor in that allowance must be made [40] for the dissociation of $Ag(NH_3)_2^+$, which is largely avoided by taking NH_4A, NaCl solutions, A^- being the anion of a weak acid HA. At 25°C, $K_{NH_4OH} = 1.774 \times 10^{-5}$ by this method.[41] The relevant expressions and calculations are somewhat extensive.[17,41]

TABLE 9.4

Average values for the Ionic Product of Water ($\times 10^{15}$)

°C	5	15	25	35	45
Ref. 36	1.846	4.505	10.08	20.89	40.18
Ref. 37	1.84	4.50	10.06	20.75	39.77

With base B of concentration C_B instead of HX in cell (9.17) with acetic acid as solvent, $K_B = a_{BH^+}a_{Ac^-}/(a_B + a_{BH^+Ac^-})$ and $K_s = a_{H^+}a_{Ac^-}$ have been obtained.[33] At 25°C pK_B = 6.79, 6.58 and 6.15 for LiAc, NaAc, and KAc respectively, and pK_s = 14.45.

3. SALTS

Information on ion-association between metal cations and anions of acid which do not dissociate completely can be evaluated from the potentials of the cell [42]

$$Pt \mid H_2 \mid H^+, A^{n-}, M^{m-}, X^- \text{ solvent} \mid AgX \mid Ag \tag{9.23}$$

if the principle equilibria are

$$H^+ + A^{n-} \rightleftharpoons HA^{n-} \text{ and } M^{m+} + A^{n-} \rightleftharpoons MA^{(m-n)+} \tag{9.24}$$

and the dissociation constant of the first of these is known (together with those of any minor systems present). To take an example, K_{MgSO_4} is obtainable from [42]

$$H_2 \mid Pt \mid HCl, m_1; MgSO_4, m_2 \mid AgCl \mid Ag \tag{9.25}$$

since, from (9.2),

$$(E - E^0)/k' + \log \gamma_H\gamma_{Cl} + \log m_1 = -\log m_H \tag{9.26}$$

and since $m_1 - m_H = m_{HSO_4}$, from

$$\log m_{SO_4} = \log K_{HSO_4} + \log m_{HSO_4} - \log \gamma_H\gamma_{SO_4}/\gamma_{HSO_4} - \log m_H \tag{9.27}$$

the concentration of $MgSO_4$ ion-pairs can be found since

$$m_{MgSO_4} = m_2 - m_{SO_4} - m_{HSO_4} \tag{9.28}$$

Successive approximations are required, calculating each γ_i by means of (9.9). For this, $I = m_1 + 4m_2 - 2m_{HSO_4} - 4m_{MgSO_4}$, $B''a = 1.0$, $Q'' = 0.3$.

Example 9.5. At 25°C, with $m_1 = 0.007127$ M, $m_2 = 0.005222$, E is [22] 265.66 mV, and from (9.9), (9.26) and (9.27), and $K_{HSO_4} = 0.0105$, there is obtained $m_{HSO_4} = 0.001072$, $m_{MgSO_4} = 0.000830$, $I = 0.02555$, $K_{MgSO_4} = 0.0055$. In all, six solutions were studied, and are summarized by Table 9.5.

TABLE 9.5

Dissociation Constant of $MgSO_4$ in Water at 25°C from E.M.F.s

$10^3 m_1$	5.522	6.170	7.127	7.575	3.652	4.392
$10^3 m_2$	3.169	5.333	5.222	5.148	16.727	38.497
E(mV)	277.04	273.13	265.66	262.49	308.38	307.09
$10^3 I$	16.58	21.83	22.55	22.86	49.17	101.77
$10^3 K$	5.25	5.1	5.5	5.8	5.85	6.2

The average of the first four is 0.0054, and accepting that there is a slight drift with I, extrapolation gives $K = 0.0053$. Now the parameters $B''a = 1.0$, $Q'' = 0.3$ used for (9.9), on comparison with (9.9) when $Q'' = 0$, correspond to $a = 4.5$ Å at $I = 0.02$ and 4.84 Å at $I = 0.1$, and according to Table 8.7, $K = 0.0053$ from conductance when $a = 5.0$ Å, so within reasonable limits, the two methods give the same answer with the same choice of ion-pair radius.

King [43] has made some interesting observations concerning K values and ion sizes. From studies of sulphamic acid by e.m.f. and conductance methods it was found that as a is increased, K from conductance decreases while K from potentials increases. The two answers coincide at $K = 0.103$ when $a = 3.85$ Å. It is stressed that this may be coincidence [43] since solutions of the acid were used for the conductances while mixtures of acid and salt were used in the e.m.f. cell, but it is important to find ways of obtaining the correct radius if accurate dissociation constants are to be determined. It is also worth stressing that experimental work of the greatest precision is needed when K is about 0.001 or higher; for instance,[42] an error of $\pm$ 0.1 mV can change K_{MgSO_4} by 10 to 20 %!

Some other investigations by means of cell (9.23) are [44] (a) $LaSO_4^+$ by use of HCl and $La_2(SO_4)_3$ (b) Mg malonate via H_2Mal, NaOH, $MgCl_2$; (c) Various lactates and glycollates via HCl, lactate or glycollate; (d) Mg phosphate, glucose 1-phosphate and glycerol 2-phosphate via $MgCl_2$, NaHA, Na_2A (A^{2-} = anion); (e) Ca glucose 1-phosphate; (f) various triglycinates via MCl_2, NaOH and $Z^{\pm}$; (g) MSO_4(M^{2+} = Zn, Mn, Co, Ni); (h) Mg^{2+}, Ca^{2+} Sr^{2+} and Ba^{2+} complexes, MV^{2-} of ethylenediamine tetra-acetic acid, H_4V(EDTA), by using KCl, $MCl_2(m_2)$, $K_2H_2V = K_4V = m_3$, where $m_2 = 3$

to 3.5 m_3; (i) the same complexes of nitrilotriacetic acid, H_3T, using MCl_2, K_2HT, K_3T, KCl.

Some other examples of metal ion-pair or complex studies by e.m.f. cells are: Cell (4.24) where MX_2 is a halide such as CdX_2, which forms CdX^+. Writing

$$K_{CdX^+} = m^2\alpha_c(1 + \alpha_c)\gamma_M\gamma_X/(1 - \alpha_c)m\gamma_{MX^+} \tag{9.29}$$

where α_c is the degree of dissociation of CdX^+, from (4.29),

$$E^0_{AgX} - E^0_M = E + (3k'/2) \log m\gamma_\pm\{\alpha_c(1 + \alpha_c)\}^{\frac{1}{3}} \tag{9.30}$$

Various values of K_{CdX^+} are tried and $\gamma_\pm$ calculated from a general form of the D.H. equation or is obtained from the figures for a similar salt that is fully ionized (at the same $I = m + 2m\alpha_c$). The correct K gives constant values of $E^0_{AgX} - E^0_M$ (assuming that the activity coefficients are correct). Some figures obtained in this way are: (a) [45] $CdCl^+ = 0.011$, (b) [46] $CdBr^+ = 0.006$, $CdI^+ = 0.004$, (c) [47] $PbCl^+ = 0.03$. Perrin [48] has studied a series of ferric amino-acid complexes by means of the cell

$$\text{Bright Pt} \mid Fe^{3+}, Fe^{2+}, ClO_4^-, H^+, Z^\pm \mid \text{Calomel Electrode} \tag{9.31}$$

pH Methods

1. Acids and Bases

The ideal definition of pH is

$$pH = -\log a_{H^+} = -\log m_H\gamma_H \tag{9.32}$$

but since γ_H cannot be measured directly, a "practical" definition is

$$pH = -\log m_H\gamma_\pm \tag{9.33}$$

For many routine operations, and under conditions where the H_2 electrode is impracticable (for instance, in the presence of certain ions such as those of Pb, Cu, Ag, Au and [27] Cd, oxidizing agents, certain organic compounds, and colloids) another electrode which is reversible to H^+ ions has to be employed, e.g., quinhydrone, antimony-antimony oxide, and the glass electrode. The last of these consists of a thin glass bulb sealed to a stout glass tube and inside is a buffered solution and connecting wire. The glass of the bulb is one of a special series of relatively low resistance glass (recent ones contain lithium and operate up to a pH of 13), although the actual resistance is still in the megohm range so that a potentiometer with valve amplification is needed to generate enough current to operate a galvanometer. Good insulation and screening are other necessities, and if the cell is to be kept at a specified temperature, an earthed oil-bath should be used and the stirrers switched off when taking measurements. Many types of cells and reference electrodes have been devised; a simple system is shown in Fig. 9.2. The cell contains entrances for making additions and for passing in CO_2-free gas when the pH is above 7. One of the major sources of uncertainty is the liquid junction potential at the junction of the solution and the KCl of the reference electrode:

Solution inside glass electrode

| Glass electrode | Solution ¦ KCl | Calomel of AgCl (9.34)

This potential is small, but it may alter during the course of a pH titration. Speakman [48] has described a cell in which the liquid junction can be formed with care, and the cell potential becomes constant very quickly.

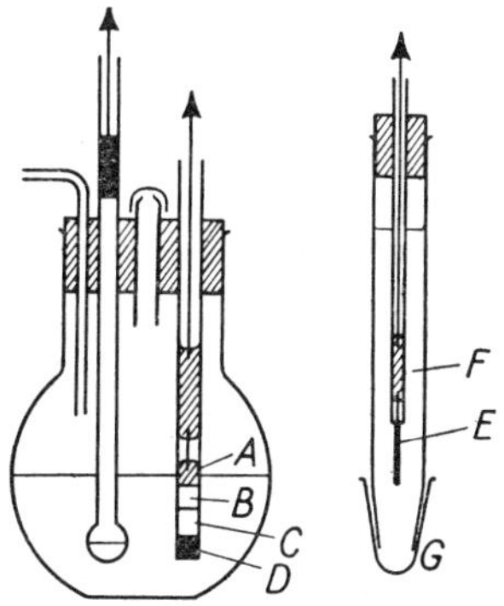

FIG. 9.2. pH titration vessel.
A = mercury; B = calomel mercury paste; C = sat. KCl; D = porous plug; E = AgCl | Ag electrode; F = KCl solution; G = ground glass cap.

The glass electrode is reversible only to H^+ ions; it is believed that this is due to the entry into or withdrawal of H^+ ions from the glass membrane. In theory, the potential developed across the membrane can be expressed by

$$E_g = k' \log (a_{\mathrm{H}}/a_{\mathrm{H'}}) \qquad (9.35)$$

where a_{H} and $a_{\mathrm{H'}}$ are the activities of the H^+ ions in the solutions separated by glass membrane. In practice this is not quite true for when the solutions are identical, E_g is not zero. A small "asymmetry" potential still exists which can be attributed to the extent of hydration of the silica in the surface layers, and to strains in the glass. This potential varies slowly with time but provided the electrode has been properly conditioned, the drift is not significant over a period of several hours. The best method of conditioning is to immerse the electrode in the type of solution that is to be studied. Further remarks on this point are made later on.

A lot of effort has gone into establishing standard pH solutions for calibration of pH meters, but for various reasons,[7,50] there have been several causes of ambiguity until recently. Nowadays, the accepted standards are founded on the potentials of the Harned cell (9.1) for studying weak acids. Most of this work has been carried out by Bates, Pinching, Bower, Acree and their collaborators.[51,52] From (9.2), (9.9) and (9.33),

$$\mathrm{pH} = (E - E^0)/k' + \log m_{\mathrm{Cl}} - A''\{\sqrt{I}/(1 + B''a\sqrt{I}) - Q''I\} \quad (9.36)$$

The values of $B''a$ and Q'' are chosen to fit the data over a range of concentrations, but it cannot be assumed that these apply when measuring unknowns. Robinson and Stokes[7] have given a very clear account on this and other aspects of pH standardization. Some of the accepted standards are[51,53] (Δ = approx. uncertainty).

TABLE 9.6

pH Values of Some Standard Solutions

	0°C	10°C	25°C	38°C	Δ
0.05M $KH_3(Ox)_2$	1.671	1.669	1.681	1.695	0.012
0.05M KHphthalate	4.012	4.001	4.005	4.025	0.009
0.01M NaHSuc + 0.01M Na_2Suc.	5.526	5.494	5.474	5.480	0.007
0.01M Borax	9.463	9.328	9.177	9.082	0.005

Some standards have also been given for methanol and water-methanol mixtures.[54]

A very wide range of acidic and basic electrolytes have been studied[8] by means of cell (9.34). It is particularly useful when only small amounts of material are available, such as peptides and proteins.[55] Mono-acids or bases are easy to deal with; the usual procedure is to carry out a pH titration and to calculate K values from the pH's of the partly neutralized solutions, e.g.,

$$\log K_{\mathrm{acid}} = -\,\mathrm{pH} + \log [\mathrm{A}^-]/[\mathrm{HA}] + \log f_\pm \quad (9.37)$$

where $[\mathrm{A}^-] = [\mathrm{Na}^+] + [\mathrm{H}^+]$ if NaOH is used, and $f_\pm$ is calculated by, e.g., (3.43).

Polybasic acids may require rather lengthy calculations. Speakman[56] has given a review of the methods which have been proposed for dibasic acids and has given another way which has been applied in a modified form by Pedersen[57] to tartronic and dihydroxytartaric acids. Hitchcock[58] has obtained the three K values of isocitric acid by finding the pH's at the mid-points of each neutralization stage; this is invalid when these pKs are of the same order.[59] However, all these methods are special cases of a more general analysis which is illustrated later on (see Example 9.11).

Within certain limits, the glass electrode can be used in mixed and non-aqueous solvents,[60] although for pH titrations of acids in the latter, NaOH and KOH are often unsuitable. A more satisfactory base is tetrabutyl ammonium hydroxide.[61]

Dunsmore and Speakman [49] have obtained the pK of benzoic acid in 0 to 50 % dioxan-water by conductance, e.m.f. and glass electrode pH methods. As the amount of dioxan increased the pKs obtained by the glass electrode were found to differ more and more from those given by the other two methods, which agree closely. Probably a partial explanation is that aqueous standard buffers were used, but the most likely cause is the relatively long time needed for the glass electrode to equilibrate in organic-rich solvents; this is dwelt on later.

Grunwald and his associates have made a series of measurements of the cell

$$\text{Glass electrode} \mid H^+, Cl^-, \text{other ions, solvent} \mid AgCl \mid Ag \qquad (9.38)$$

Part of their programme has been the obtaining of the K values of carboxylic acids in 0 to 80 % ethanol,[62] anilinium and ammonium salts in 0 to 95 % ethanol,[63] and both series in 0 to 95 % methanol.[63] A simplified version of their method and treatment is as follows. A known amount of HCl is added to a known amount of weak acid HA, and the potential E_1 found. Sufficient NaOH is now added to neutralize all of the HCl and some of the HA, and E_2 obtained. In both cases,

$$E^* = E_1 \text{ or } E_2 + k' \log c_{\mathrm{H}}c_{\mathrm{Cl}} - 2k'\{A''\sqrt{I}/(1 + B''a\sqrt{I})\} \qquad (9.39)$$

E^* being the potential of the AgCl electrode with reference to the glass electrode. This expression was tested by finding E^* was constant in 95 % ethanol with 0.001 to 0.024 HCl in the cell (taking $a = 5.0$ Å).

Letting c_1, c_2 and c_3 be the stoichiometric concentrations of HA, HCl and NaOH, and $u = (c_1 + c_2 - c_3)/c_1$, then

$$k_A = K_A f_{\mathrm{HA}}/f_{\mathrm{H}}f_{\mathrm{A}} = c_{\mathrm{H}}c_{\mathrm{A}}/c_{\mathrm{HA}} = c_{\mathrm{H}}\{c_{\mathrm{H}} + c_1(1 - u)\}/(c_1u - c_{\mathrm{H}}) \qquad (9.40)$$

whence

$$2c_{\mathrm{H}} = c_1(u - 1) - k_A + \{c_1^2(u - 1)^2 + 2k_Ac_1(1 + u) + k_A^2\}^{\frac{1}{2}} \qquad (9.41)$$

Writing $2Z$ for the right-hand side of (9.41), and c'_{H} and c''_{H} for c_{H} when only HCl, and when both HCl and NaOH have been added, then

$$c'_{\mathrm{H}}/c''_{\mathrm{H}} = Z'/Z'' \qquad (9.42)$$

A plausible first estimate of K_A is chosen, k'_A and k''_A calculated ($I = c_{\mathrm{H}} + c_3$) so that Z'/Z'' is found. If this is greater than $c'_{\mathrm{H}}/c''_{\mathrm{H}}$, the assumed value of K_A is too small, and the calculations are repeated until $c'_{\mathrm{H}}/c''_{\mathrm{H}}$ lies between two computed values of Z'/Z'', when K_A is found by interpolation.

Example 9.6. The following figures refer to propionic acid in 60 % methanol[64] (by vol.): $A'' = 0.885$, $B''a = 1.98$ (i) $10^4c'_1 = 78.02$, $10^4I' = 3.188$, $f' = 0.9655$, $u' = 1.0409$. (ii) $10^4c''_1 = 77.46$, $10^4I'' = 4.599$, $f'' = 0.9590$, $u'' = 0.9815$. $E_1 - E_2 = -946$ mV, $c'_{\mathrm{H}}/c''_{\mathrm{H}} = 8.64$. 10^7K_A assumed: $Z'/Z'' = 9.00 : 8.47$; $8.50 : 8.84$; $8.00 : 9.26$. By interpolation, $10^7K_A = 8.85$.

The data for acetic acid in aqueous methanol [63] agree [65] with those obtained by conductance.[66]

Kilpi [67] has also devised differential pH procedures for obtaining the pKs of acids and bases in aqueous and organic media, and has reviewed this work, which dates from 1935. It is shown that the most exact results emerge from pH changes at the inflection point of the titration, but the calculations are lengthy.

The treatments which are associated with the Harned cell, (9.1), are simpler and more suitable for deriving accurate pKs than any differential method, and during the last few years, Grunwald and his co-workers [68] have been developing suitable ways of using cell (9.38) for such purposes. This has been made possible by the introduction of several improvements. Firstly, pH measurements may now be measured with cell (9.38) to within 0.002 unit by certain commercial pH-meters, or by combination of a sensitive potentiometer and valve voltmeter.[76] Secondly, Purlee and Grunwald [68] have developed a silver mirror Ag—AgCl electrode (described in Chapter 4) which is very satisfactory in organic-rich solvents containing low H^+ and Cl^- concentrations; it equilibrates quite quickly, in contrast to the conventional thermal-electrolytic types. Thirdly, the glass electrode requires a relatively long time to equilibrate in solvents containing small concentrations of water,[69] e.g., in 70 % dioxan, about 24 hours are required; it is also important to equilibrate the electrode in freshly prepared solvent of the correct composition.

Instead of using standard buffers, E^* of (9.39) is found by means of dilute HCl solutions. In solvents of low dielectric constants allowance must be made for any HCl ion-pairs. When the solvent was 70 % dioxan, Purlee and Grunwald [69] took the value of K_{HCl} calculated from conductances [70] by the extended form of the limiting Onsager equation (Chapter 8), and a specialized form of the D.H. activity coefficient expression. At 25°C, this gave $E^* =$ 0.6318 V. Providing only dilute HCl is considered, standard methods are satisfactory; this is shown in the following example:

Example 9.6 As stated in Chapter 8, the Fuoss–Onsager conductance treatment gives [72] $K_{HCl} = 0.010$ in 70 % dioxan at 25°C. Also, $-\log f_i = 4.256\sqrt{I}(1 + B''a\sqrt{I})$ since $\epsilon = 19.07$ (see 3.35 to 37). By trial, $a = 4.5$ Å gives constant values of E^* in (9.39), calculating $c_H = c_{Cl}$ by approximations. The data are taken from those of Purlee and Grunwald:[69]

10^4c	1.037	1.660	2.372	3.510	4.151	5.013	6.047	7.445	8.596
$-E$(mV)	155.06	177.86	194.62	213.04	220.73	229.38	237.79	247.07	253.42
$-E^*$(mV)	631.88	631.93	631.72	631.84	631.81	631.86	631.80	631.57	631.85

The average is $E^* = -631.8$ mV.

Purlee and Grunwald [69] also compared the potentials of cell (9.38) with those of the cell: H_2 | 0.0014 to 0.035 N HCl in 70 % dioxan | AgCl | Ag: obtained by Harned and Calmon.[71] The difference is 698.8 mV so that $E^0_{AgCl} = 67.0$ mV compared with the original estimates of [71] 63.95 mV.

Grunwald has used cell (9.38) to obtain potentials of HCl + NaCl, and HCl + KCl in 70 % dioxan at 25°C. By taking [70] $K_{HCl} = 0.00667$, it was deduced that $K_{NaCl} = 0.00535$ and $K_{KCl} = 0.00253$. These would be higher if K_{HCl} were taken as 0.010, and would be in closer agreement with, e.g., $K_{NaCl} = 0.0083$ obtained [72] from conductance (see Chapter 8).

In further work, the dissociation constants of formic, acetic and propionic acids and of their sodium and potassium salts in 70 % dioxan at 25°C were estimated from potentials of mixtures of MA, HA and MCl in cell (9.38). The concentrations of HCl ion-pairs were small enough to ignore, and knowing K_{MCl}, K_{HA} and K_{MA} were estimated.

Prue [74] has explored the potentialities of a special type of low resistance glass electrode (0.3 to 0.6 megohm) which can be used with an ordinary potentiometer and sensitive galvanometer providing these, the thermostat bath and the screens of the leads are earthed. The electrode is open at the top so that the inside solution can be varied at will, and is then closed by a polythene stopper holding a Ag—AgCl electrode.

From the cell [74]

AgCl | 0.1111 M HCl inside electrode | Glass | 0.005 to 0.081 M HCl outside | AgCl (9.43)

$E^0 = E^*$, the standard potential of the electrode assembly was determined by application of (4.23), and thus $\gamma_\pm$ of HCl derived. The cell (cf. Fig. (4.5))

Glass electrode | HCl, m_1 ¦ HCl, m_2 | Glass electrode (9.44)

has been used for the same purpose, via (4.52). The activity coefficients of HNO_3 and of $HClO_4$ have been obtained in a similar fashion.[75] Up to 0.1 M, $B''a = 2.5$, $Q'' = -0.181$, -0.128 for HNO_3 and $HClO_4$ respectively, or $B''a = 3.0$, $Q'' = -0.303$, -0.242 in (9.9).

2. pH Methods for Salts

Glass electrode cells, although lacking the accuracy of the e.m.f. methods described in section C (p. 141), are one of the most important and widely used ways of following ion-association or complexing between metal cations and the anions of incompletely dissociated acids. The conventional system is cell (9.34) containing a mixed solution of a salt of the cation and the acid. This is titrated by small increments of NaOH and the pH found at each stage. Cell (9.38) has been used on a limited scale by Grunwald, as described in the previous section; ions which react with AgCl must be absent, and chloride ions must be present.

Example 9.7. Nancollas [76] has reported the pHs of mixtures of $MCl_2(C_1)$, $HA(C_2)$, and $NaOH(C_3)$, where M represents the alkaline earths and A represents formate and acetate. At 25°C, with $M = Ca^{2+}$, $A = Ac^-$, $C_1 = 0.00998$, $C_2 = 0.02014$, $C_3 = 0.00904$, the pH was 4.559. From electroneutrality, $[CaAc^+] = C_3 + [H^+] - [Ac^-]$. Also, $\log [Ac^-] = \log K_{HA} + pH + \log (C_2 - [Ac^-] - [CaAc^+]) + \phi(I)$, where $\phi(I)$ is given by (3.43) when $z_i = z_j = 2$, and $I = 3C_1 - C_3 - [H^+] + 2[Ac^-]$. For the first approximation, $[CaAc^+]$ is ignored, $[H^+]$ calculated and a first value of $[Ac^-]$ found; ($H_{HA} = 1.75_1 \times 10^{-5}$). The final figures are: $[CaAc^+] = 0.00066$, $I = 0.0377$, $K_{CaAc} = 0.058$.

Example 9.8. Dissociation constants of some MZ^+ and MZ_2 amino-acid complexes have been measured.[26,77] $NaOH(C_3)$ was added to a solution of $CoCl_2(C_1)$ and glycine (C_2). With $C_1 = 0.00476$, $C_2 = 0.06833$ and $C_3 = 0.00589$, the pH was 5.21 at 25°C. The required equations are, (a) $2[M^{2+}] + [MZ^+] + [Na^+] + [N^+] + [HZ^+] = [Cl^-] + [Z^-]$, (b) $[M^{2+}] + [MZ^+] + [MZ_2] = C_1$. $K_A = 4.47 \times 10^{-3}$, $K_B = 6.04 \times 10^{-5}$ are also needed,[24] and (3.43). First values of $[Z^-]$ and $[HZ^+]$ are calculated, and a first value of $[MZ^+]$ found (MZ_2 is negligible). A preliminary value of $K_1 = a_M a_Z / a_{MZ}$ is calculated.

With $C_1 = 0.00427$, $C_2 = 0.06833$, and $C_3 = 0.00589$, the pH was 6.85. Under these conditions, $[M^{2+}]$ is negligible, and a preliminary value of $K_2 = a_{MZ} a_Z / a_{MZ_2}$ is calculated. These preliminary values of K_1 and K_2 are used to calculate $[MZ_2]$ and $[M^{2+}]$ for further approximations in order to calculate the final values of K_1 and K_2. The final figures are (a) $[M^{2+}] = 0.00383$, $[MZ^+] = 0.00091$, $[MZ_2] = 0.00002$, $-\log [Z^-] = 5.642$, $I = 0.01335$, $-\log K_1 = 5.22$: (b) $[M^{2+}] = 0.00024$, $[MZ^+] = 0.00225$, $[MZ_2] = 0.00178$, $-\log [Z^-] = 4.091$, $I = 0.00886$, $-\log K_2 = 4.07$.

The majority of pH studies of metal complexes are made with media of constant ionic strengths by applying the methods described in the next section.

Consecutive Complexity or Stability Constants at Constant Ionic Strength

When more than one species of complex or ion-pair forms, the establishment of the related equilibrium constants often requires different processes of calculations to those given in the preceding sections. The only exceptions are when $K_1 \gg K_2 \gg \ldots$ so that the different species may be isolated by choosing the correct conditions.

Confining the discussion to the mononuclear species ML_n, where $0 \leqslant n \leqslant N$, N being the maximum number of "ligands," L, which the central ion M will accept, the problems are (i) to find N, (ii) to establish the values of the activity coefficients of the various species and (iii) to establish the values of the various equilibrium constants. Now by using solutions of constant ionic strength, in theory, the activity coefficients will remain constant, and under these conditions, the equilibrium constants can be resolved although, in general, I is so large that it is not possible to calculate the values of the constants at $I = 0$. According to Sillen [78] and Tobias,[79] who have given many references on the present topic, the idea of working with a constant salt "background" originated with Bodlander in 1904 and was first applied by Grossmann [80] in 1905 in his study of the mercuric thiocyanate system. The development of appropriate theories and the exploitation of experimental data has been mainly due to Bjerrum,[81] Leden,[82] Fronaeus [83] and other Scandinavians.[78,79] There are a number of other contributors, some of whom are referred to in subsequent paragraphs.

In order to form solutions of constant ionic strength or salt background, in practically every case high molarities of $NaClO_4$ are used. There are two main reasons for this choice and for using high molarities. Firstly, the Na^+ and ClO_4^- ions are among those that have the least tendency to associate with other ions or with themselves, and secondly it is often necessary to use high concentrations of the ligand to form the higher complexes of ML_n, so to make a complete study of the system, the concentration of ligand must cover a wide range. This means that the ionic strength would be high even if $NaClO_4$ were not added when forming the higher complexes. An important comment is that under certain conditions, both Na^+ and ClO_4^- can form ion-pairs, e.g.,[84] $NaSO_4^-$, and [85] $FeClO_4^{2+}$, and if M is small and ClO_4^- is very large, although the tendency to associate may be small, a relatively large fraction of M could be bound up as $MClO_4^{x+}$. Also, it has been observed by Biedermann and Sillen [86] that the activity coefficients of cations are mainly affected by variations in anion concentrations, and vice versa. They have suggested that it would be better to keep the ClO_4^- concentration constant rather than maintaining the ionic strength at a fixed level.

There are certain definitions common to all the treatments. These are

$$\beta_n = [ML_n]/[M][L]^n \tag{9.45}$$

where β_n is termed the stability or complexity constant. It is related to the concentration formation or association constant k_n, where

$$k_n = [ML_n]/[ML_{n-1}][L] \tag{9.46}$$

by

$$\beta_n = k_1 k_2 \ldots k_n$$

(k_n is the reciprocal of concentration dissociation constants, K_n').

Two other definitions are

$$C_M = \sum_0^N \beta_n[M][L]^n = \sum_0^N [ML_n] \tag{9.47}$$

$$C_L = [L] + \sum_0^N n\beta_n[M][L]^n \tag{9.48}$$

in which C_M and C_L are the total analytical or stoichiometric concentrations of the cation M and the ligand L respectively. Charges will be omitted for convenience. The summation is from $n = 0$, to the maximum of $n = N$. When $n = 0$, $\beta_0 = 1$, from (9.45). Another definition is [81]

$$\bar{n} = (C_L - [L])/C_M = \sum_0^N n\beta_n[ML_n]/\sum_0^N \beta_n[ML_n] \tag{9.49}$$

in which $\bar{n}$ is termed the "ligand number," and is the average number of ligands bound to each M. The right-hand side follows from (9.47) and (9.48).

The degree of formation of a given complex, ML_c, is given by

$$\alpha_c = [ML_c]/C_M = \beta_c[L]^c\Big/\sum_0^N \beta_n[L]^n \tag{9.50}$$

the right-hand side being derived from (9.45) and (9.47). Also, letting

$$-\log x = px \tag{9.51}$$

since

$$\mathrm{d} \log \alpha_c/\mathrm{d\,pL} = -\,\mathrm{d} \ln \alpha_c/\mathrm{d} \ln [L] = -\,[L]\mathrm{d}\,\alpha_c/\alpha_c\,\mathrm{d}[L] \tag{9.52}$$

and

$$\mathrm{d}\,\alpha_c/\mathrm{d}[L] = c\beta_c[L]^{c-1}\Big/\sum_0^N \beta_n[L]^n - \beta_c n[L]^c\Big/\sum_0^N \beta_n[L]^{n+1} \tag{9.53}$$

then

$$\mathrm{d} \log \alpha_c/\mathrm{d\,pL} = (\bar{n} - c) \tag{9.54}$$

The method of deriving the β_n values depends to some extent on the experimental analysis and on the relative values of the stability constants.[77]

1. [M] is Determined

(i) If all values of β_n are small, [L] can be found via $C_L \simeq [L]$.

(ii) If [98] $\beta_1 < 1{,}500$, values of $\log (C_M/[M])$ obtained with a fixed value of C_M are plotted against C_L. This is repeated with differing values of C_M, the family of curves being drawn on the same graph. Letting

$$C_M/[M] = X \tag{9.55}$$

it follows from (9.47) that if X is constant, then [L] is constant, and since, from (9.47) and (9.49),

$$\bar{n} = \sum_0^N n\beta_n[L]^n/X \tag{9.56}$$

then $\bar{n}$ is a function of [L] only, and is constant along a line of constant [L]. Lines are drawn at constant values of log X, and the C_L values at the intersection with the family of curves plotted against C_M. These plots are straight lines of slope $\bar{n}$ and [L] is the value of C_M at $C_L = 0$.

(iii) If [98] $\beta_1 > 1{,}500$, [L] can only be determined from the intercept of the C_L versus C_M curves when $C_L > C_M NX$. In such solutions, the highest complex is completely dominating and $C_M = [ML_n]$ and from (9.47), $C_M/[M] = \beta_N[L]^N$, so that β_N is obtained. To find β_1, etc., since differentiation of (9.47), together with (9.55) results in

$$dX = d[L]\sum_0^N (n\beta_n[L]^n)/[L] \qquad ([M] = \text{constant}) \qquad (9.57)$$

Then from (9.56),

$$\bar{n} = (dX/X)/(d[L]/[L]) = d \log X/d \log [L] \qquad (9.58)$$

whence

$$\bar{n} \log ([L]_a/[L]_f) = \int_{L_f}^{L_a} (d \log X) \qquad (9.59)$$

If $[L]_a$ and the corresponding value of X are chosen in the range where $[L]_a$ can be determined as the intercept of a C_L versus C_M line, then $[L]_f$ corresponding to the chosen X can be obtained by graphical integration from a plot of $1/\bar{n}$ against log X, finding the area from log X at $[L]_a$ to the chosen X_f. The best $[L]_a$ to take can be found from β_N, calculating values of [L] and X in the range where $X = \beta^N[L]_N$ holds. This is illustrated in example (8.12).

2. [M] AND [L] ARE DETERMINED

(i) Leden's method [82] can be used. This is based on the expression resulting from an expansion and rearrangement of (9.47):

$$A = (C_M - [M])/[M][L] = \beta_1 + \beta_2[L] + \ldots = \sum_1^N \beta_n[L]^{n-1} = g/[L] \qquad (9.60)$$

where g is defined by the term $\sum_1^N\beta_n[L]^n$. The intercept of the plot of A against [L] gives β_1; β_2 is the intercept of the plot of $(A - \beta_1)/[L]$ against [L], and so on.

(ii) A similar method, due to Olerup,[88] defines a function

$$B = C_M/[M] = 1 + \beta_1[L] + \ldots = \sum_0^N \beta_n[L]^n = 1 + g \qquad (9.61)$$

so that β_1 is the intercept of the plot of $(B - 1)/[L]$ against [L], and so on.

3. [ML$_c$] IS MEASURED

(i) Values of α_c can be calculated by (9.50), and $\bar{n}$ is found from the slope of the plot of log α_c against pL (see 9.54).

(ii) From (9.50), and above definition of g,

$$[L]^c/\alpha_c = \sum_0^N \beta_n[L]^n/\beta_c = 1/\beta_c + \beta_1[L]/\beta_c + \ldots = (1+g)/\beta_c \quad (9.62)$$

when β_c is the reciprocal of the intercept of the plot of $[L]^c/\alpha_c$ against [L], and β_1/β_c, β_2/β_c, etc., are obtained by successive extrapolations.

4. [L] IS DETERMINED

This is the most common procedure; $\bar{n}$ is calculated directly from (9.49).

(i) Bjerrum [81] has devised a number of methods, some of which give approximate answers which may then be used to derive accurate values. Carlson and others [89] have given a detailed account of Bjerrum's methods and some modifications to these.

If it is assumed that in a solution where $\bar{n} = n - \frac{1}{2}$ there are equal amounts of ML_n and of ML_{n-1}, then from (9.46),

$$k_n = (1/[L])_{\bar{n}=n-\frac{1}{2}} \quad (9.63)$$

The appropriate values of $\bar{n}$ are interpolated from a plot of $\bar{n}$ against pL (see 9.51).

A second approximate method is to find Δ from the mid-point slope of the plot of $\bar{n}$ against pL, where

$$\Delta = d\bar{n}/d \ln [L] = -0.4343\, d\bar{n}/d\, pL \text{ at } \bar{n} = \tfrac{1}{2}N \quad (9.64)$$

The average constant of the system, $\bar{k}$, is calculated from

$$\bar{k} = (1/[L])_{n=\frac{1}{2}N} \quad (9.65)$$

and k_n is calculated by means of

$$k_n = \{(N-n+1)/n\}\bar{k}x^{N+1-2n} \quad (9.66)$$

where x is derived from

$$\Delta = \left\{\sum_1^N n(n-\tfrac{1}{2}N)N^*x^{nN-n^2}\right\}\Big/\left(1+\sum_1^N N^*x^{nN-n^2}\right) \quad (9.67)$$

in which $N^* = N!/(N-n)!n!$. When $N = 2$, $x = (1-\Delta)/\Delta$, and when $N = 3$, $x = (9-4\Delta)/(12\Delta-3)$. For $N = 4$, etc., Δ is calculated for specific values of x, Δ is plotted against log x, and x is found by interpolation.

Irving and Rossotti [90] have described how precise values may be obtained when [97] $N = 2$, and Block and McIntyre [91] have dealt with systems when $N = 1$, 2 and 3 in the following fashion. From (9.45) and (9.49), [M] being constant,

$$\bar{n} = \sum_1^N (n-\bar{n})k_n[L]^n = \sum_1^N J_n k_n \quad (9.68)$$

where $J_n = (n-\bar{n})[L]^n$. The expressions for the individual values of k_n are obtained by solving the appropriate N sets of $\bar{n}$ equations. These give

$$\text{For } N = 1, \quad k_1 = \bar{n}/J_1 \quad (9.69)$$

For

$$N = 2,\ k_1 = (\bar{n}J'_2 - \bar{n}'J_2)/(J_1J'_2 - J_1J'_2);\ k_2 = (\bar{n}J_1 - \bar{n}J'_1)/(\bar{n}J'_2 - \bar{n}'J_2) \quad (9.70)$$

For $N = 3$, there are similar equations, but in addition there are a number of alternatives. Some metal amines are taken as examples.

Proceeding from (9.63), Carlson *et al.*[89] have shown that the accurate formation constants can be calculated via:

For $N = 2$, let $k_1 = 2y\sqrt{\beta_1}$, $k_2 = \sqrt{\beta_2}/2y$,

then from (9.45) and (9.49),

$$\bar{n} = (2y[L]\sqrt{\beta_2} + 2\beta_2[L]^2)/(1 + 2y[L]\sqrt{\beta_2} + \beta_2[L]^2) \quad (9.71)$$

When

$$\bar{n} = 1,\ \sqrt{\beta_2} = 1/[L]^2_{\bar{n}=1} \quad (9.72)$$

$$d\bar{n}/d\ln[L] = (2y[L]\sqrt{\beta_2} + 4[L]^2\beta_2 + 2y[L]^3\beta_2^{\frac{3}{2}})/(1 + 2y[L]\sqrt{\beta_2} + [L]^2\beta_2)^2 \quad (9.73)$$

since $\sqrt{\beta_2}[L]_{\bar{n}=1} = 1$, the slope of the formation curve is

$$d\bar{n}/d\ln[L]_{\bar{n}=1} = -\ 0.4343\ d\bar{n}/d\,pL = 1/(1 + y) \quad (9.74)$$

so that k_1 and k_2 are obtained.

For $N = 3$,

$$k_1 = 1/\{[L]_i(1 + 3k_2[L]_i + 5k_2k_3[L]_i^2)\} \quad (9.75)$$

$$k_2 = (1 + 3/k_1[L]_j)([L]_j + 3k_3[L]_j^2) \quad (9.76)$$

$$k_3 = (1 + 3/k_2[L]_k + 5/k_1k_2[L]_k)/[L]_k \quad (9.77)$$

where i, j and k refer to $\bar{n} = \frac{1}{2}$, $1\frac{1}{2}$ and $2\frac{1}{2}$ respectively. The values of k_1, k_2 and k_3 on the right-hand side of (9.75–7) are the approximate values given by (9.63). The k_1 and k_2 values of some diamine complexes are calculated.[89]

(ii) From (9.58),

$$\int_0^X (dX/X) = \int_0^L \bar{n}\ d[L]/[L] \quad (9.78)$$

whence

$$X = \exp \int_0^L \bar{n}\ d[L]/[L] = \int_0^N \beta_n[L]^n = 1 + g = C \quad (9.79)$$

This follows from (9.61), and the method of Leden, as modified by Fronaeus,[83] since C = B of (9.61), is to obtain β_1 by obtaining the intercept of the plot of (C — 1)/[L] against [L], etc. (See 2, i).

(iii) Olerup[88] has shown that complexity constants may be found by expanding

$$\sum_0^N (\bar{n} - n)\beta_n[L]^n = 0 \quad (9.80)$$

in the form

$$D = \bar{n}/[L] = (1 - \bar{n})\beta_1 + (2 - \bar{n})\beta_2[L] + \ldots = \sum_1^N (n - \bar{n})\beta_n[L]^{n-1} \tag{9.81}$$

Equation (9.80) is derived from (9.49) by re-arrangement.

β_1 is the intercept of the plot of D against [L], $2\beta_2$ is the intercept of the plot of $\{D - (1 - \bar{n})\beta_1\}/[L]$ against [L], and so on. The plots tend to be curved even at low values of [L], so extrapolation may be difficult.

(iv) Rossotti and Rossotti [87] has given a very straightforward method. Expanding (9.80) and rearranging,

$$\bar{n}/(1 - \bar{n})[L] = \beta_1 + \beta_2[L](2 - \bar{n})/(1 - \bar{n}) + \sum_3^N (n - \bar{n})\beta_n[L]^{n-1}/(1 - \bar{n}) \tag{9.82}$$

so a plot of $\bar{n}/(1 - \bar{n})[L]$ against $(2 - \bar{n})[L]/(1 - \bar{n})$ tends to a straight line of intercept β_1 and of slope β_2 as $[L] \to 0$. In general, an accurate value of any constant β_t may be obtained from

$$\sum_0^t \beta_n[L]^{n-t}(\bar{n} - n)/(t - \bar{n}) = \beta_t + \sum_{t+1}^N \beta_n[L]^{n-t}(n - \bar{n})/(t - \bar{n}) \tag{9.83}$$

where $0 < t < N$. If values of $\beta_1, \beta_2, \ldots$ have been previously calculated, the left-hand side of (9.83) is known so that a plot of this against $(t + 1 - \bar{n})/(t - \bar{n})$ gives β_t as the intercept and an approximate value of β_{t+1} from the limiting slope. Any errors in $\beta_1, \ldots \beta_{t-1}$ will accumulate in β_t and a check is to put $n = N$ in (9.80). Expanding and proceeding as before shows that a plot of $(N - \bar{n})[L]/(N - 1 - \bar{n})$ against $(\bar{n} - N + 2)/(N - 1 - \bar{n})[L]$ will have an intercept of β_{N-1}/β_N and a limiting slope of β_{N-2}/β_N as $1/[L] \to 0$.

(v) Hearon and Gilbert [92] have devised methods for calculating association constants when $N = 1$, 2 or 3, with particular reference to complexes with amino-acids and peptides obtained by pH titrations. Gilbert [93] has thereby studied the Co^{2+}-glycine, glycylglycine and glycylalanine systems.

Examples 9.9. Sonesson [94] has obtained the complexity constants of a number of lanthanide acetates and glycollates by means of the cell

$$\text{Au} \mid \text{Quinhydrone, } HClO_4 = m_1,\ NaClO_4 \text{ to } I = 2.0 \mid 2.0\text{ M } NaClO_4 \mid M(ClO_4)_3 = C_M,\ HClO_4 = \nu C_M,\ NaL = C'_L,\ HL = \delta C'_L,\ NaClO_4 \text{ to } I = 2.0, \text{ Quinhydrone} \mid \text{Au} \tag{9.84}$$

where M^{3+} = lanthanide, L^- = acetate or glycollate, $m_1 = 10.2$ mM (mM = millimolar), Table 9.6 gives some data at 20°C when $\delta = 0.5$ for lanthanum acetate. If E is potential of the cell, and E' its value when $C_M = 0$, then since I is constant,

$$E' - E = E_L = 58.16 \log ([H^+]/[H^+]') \tag{9.85}$$

It is assumed that the liquid junction potentials remain constant and that M^{3+} does not complex with quinhydrone or undissociated HL (these possibilities were ruled out by separate tests). Little change occurred when $NaClO_4$ was gradually exchanged for NaL (up to about 0.3 mV), but since complex formation reduced I from its theoretical value, further amounts of $NaClO_4$ were added, and this dilution changed E_L by an amount ΔE_L which was as much as 0.7 mV (the corrections are included in the E_L figures in Table 9.7.

From (9.85), with $C_M = 0$, $E' - 0 = 180.0$ mV $= -58.16 \log [H^+]'$, so $[H^+] = 2.77 \times 10^{-4}$ M. Then since

$$K_{HL} = [H^+]'(C'_L + [H^+]')/(C'_{HL} - [H^+]') = [H^+][L^-]/(C'_{HL} + C_H - [H^+]) \tag{9.86}$$

[L] can be calculated. Furthermore, $C_L = C'_L + [H^+] - C_H$, so $\bar{n}$ can be obtained by means of (9.49).

TABLE 9.7

Data for the Stability Constants of Lanthanum Acetate (Conc. in mM)

C'_L	C_M	E_L(mV)	C_H	[L]	$\bar{n}$
9.30	50.09	26.9	0.21	3.55	0.1228
19.61	49.62	24.8	0.21	7.50	0.2400
38.5	48.65	22.1	0.21	16.21	0.4253
56.6	47.74	20.1	0.20	25.7	0.6424
90.9	46.01	17.0	0.20	46.6	0.9601
122.8	44.39	14.7	0.19	68.9	1.213
152.5	42.89	12.9	0.18	91.8	1.414
200.0	40.49	10.6	0.17	131.7	1.686
285.7	36.15	7.7	0.15	210.9	2.066
375	31.63	5.7	0.13	299.5	2.386
500	25.3	3.8	0.11	430	2.752

According to the methods based on (9.79) and (9.81), β_1 is 36 ± 1, and from (9.82), $\beta_1 = 37$. To obtain β_2, the method of Fronaeus, from (9.79), is to plot $(X - 1 - \beta_1[L])/[L]^2$ against [L], and to extrapolate, X being obtained via $2.303 \log X$, this being the area under the plot of $\bar{n}/[L]$ against [L] from 0 to [L]. For instance, at $[L] = 0.030$ M, $\log X = 0.870/2.303$. Sonesson has given such data [94] some of which are in Table 9.8. The intercept corresponds to $\beta_2 = 290$. Similarly, from the plot of the last line of Table 9.7 against [L], $\beta_3 = 1{,}000$.

Taking Olerup's method, from (9.81), omitting the first two points of Table 9.7 the curve extrapolates to 560 for $2\beta_2$; by adjusting β_1 to 37.1, even the first two points fit the curve. Similarly, from a plot of

$\{D - (1 - \bar{n})\beta_1 - (2 - \bar{n})\beta_2[L]\}/[L]^2$ against [L], starting at $[L] = 0.0257$ gives $3\beta_3 = 3{,}100 \pm 100$.

With Rossotti's method for finding β_1, the slope of the plot gives an approximate value of β_2; this is about 330. The better method is to plot $\bar{n} + (\bar{n} - 1)\beta_1[\mathrm{L}]/(2 - \bar{n})[\mathrm{L}]^2$ against $(3 - \bar{n})[\mathrm{L}]/(2 - \bar{n})$. Starting at $[\mathrm{L}] = 0.0257$, the plot extrapolates to $\beta_2 = 300$, and the limiting slope gives $\beta_3 = 1{,}200$. From (9.83), when $t = 3$, $\beta_3 = 900$.

TABLE 9.8

Data for Calculating β_2 and β_3 by the Method of Fronaeus

[L]	0.030	0.070	0.140	0.200	0.300	0.400	0.500
X	2.384	5.365	14.87	29.2	71.1	144.9	266
$(X - 1 - \beta_1[\mathrm{L}])/[\mathrm{L}]^2$	341	369	447	523	657	810	986
$(X - 1 - \beta_1[\mathrm{L}] - \beta_2[\mathrm{L}]^2)/[\mathrm{L}]^3$	—	—	—	1160	1220	1295	1390

Example 9.10. Bjerrum [81] has made a number of studies of metal amines. For example, the formation constants of $Cu(NH_3)_4^{2+}$ were derived from the cell (9.34), i.e., Glass electrode | $Cu(NO_3)_2 = C_M$, $NH_3 = C_L$, 2 N NH_4NO_3 | 3.5 N Calomel. Some of the data at 30°C are shown in Table 9.9. $E - E_{st}$ is the potential difference between the cell and that when $C_M = 0$, $C_L = C_{st}$. It follows from the expressions (a) $E - E_{st} = k' \log ([H^+]/[H^+]_{st})$, where $k' = 60.13$ mV, and (b) $k_{NH_4} = [H^+][NH_3]/[NH_4^+]$, that when $[NH_4^+]$ is constant, $(E - E_{st})/k' = \log([\mathrm{L}]_{st}/[\mathrm{L}])$, where [L] is the value of $[NH_3]$ for each C_M and C_L. Having established [L], the methods illustrated in the previous example could be used. Bjerrum's methods are applied here.

TABLE 9.9

Data for the k_n Values of $Cu(NH_3)_4^{2+}$. (Concn. in mM)

C_M	20.27	20.57	10.32	10.32	20.57	20.57	20.57	125.4
C_L	5.00	10.02	10.02	19.90	59.56	79.32	99.7	749.5
$C_{L,st}$.	9.95	9.95	9.95	9.95	9.95	9.95	9.95	500
$-(E - E_{st})$(mV)	161.7	140.2	114.1	76.3	38.4	3.7	−21.5	18.3
pL	4.690	4.333	3.900	3.271	2.641	2.063	1.644	0.605
$\bar{n}$	0.243	0.486	0.959	1.876	2.785	3.437	3.744	3.996

From a plot of $\bar{n}$ against pL, and using (9.63), $\log k_1 = 4.45$, $\log k_2 = 3.51$, $\log k_3 = 2.85$, and $\log k_4 = 2.00$. Taking (9.64), Δ at $\bar{n} = 2$ is $1.53/0.4343 = 0.665$, and from (9.65), $\log \bar{k} = 3.20$. Interpolation from the plot of Δ against $\log x$ (see 9.67) gives $\log x = 0.132$, whence from (9.66), $\log k_1 = 0.602 + 3.20 + 3(0.132) = 4.20$. Similarly, $\log k_2 = 3.51$, $\log k_3 = 2.89$ and $\log k_4 = 2.20$.

Bjerrum [81] has obtained the formation constants of the ethylenediamine (*en*) complexes of some metals by the cell: H_2 or glass electrode | HCl = C_H, $MCl_2 = C_M$, *en* = C_L,N KCl | Calomel electrode. Defining

$\alpha_L = [L]/([L] + [LH] + [LH_2])$ and $\bar{n}_L = ([LH] + 2[LH_2])/([L] + [LH] + [LH_2])$

with $k_{LH} = [L][H]/[LH]$, $k_{LH_2} = [HL][H]/[LH_2]$, then

$$\alpha_L = k_{LH}k_{LH_2}/(k_{LH}k_{LH_2} + k_{LH_2}[H] + [H]^2) \tag{9.87}$$

$$\bar{n}_L = (k_{LH_2}[H] + 2[H]^2)/(k_{LH}k_{LH_2} + k_{LH_2}[H] + [H]^2) \tag{9.88}$$

and by elimination, since $C_L = [L] + [LH] + [LH_2] + \bar{n}C_M$,

$$\bar{n} = \{C_L - (C_H - [H^+])/\bar{n}_L\}/C_M \tag{9.89}$$

$$[L] = \alpha_L(C_H - [H^+])/\bar{n}_L \tag{9.90}$$

k_{LH} and k_{LH_2} are determined by the usual methods for dibasic acids. Bjerrum used $BaCl_2$ as the inert salt (at the same I) in the cell since this does not complex with *en*. Calvin and Wilson,[95] taking mixtures of $Cu(ClO_4)_2$, H_nL, $HClO_4$ in 50 % dioxan, and titrating with NaOH, have in an analogous manner, studied the cupric complexes of some diketones and *o*-hydroxyaromatic aldehydes.

Example 9.11. Speakman,[56] by titration of dibasic acids with NaOH in cell (9.34), has estimated the dissociation constants $K_1 = [H^+][HL^-]f_1f_2/[H_2L]$ and $K_2 = [H^+][L^{2-}]f_1f_3/[HL^-]f_2$. His method of obtaining K_1 and K_2 by a plotting procedure is a special case of (9.82). Substituting for β_1 and β_2, and rearranging, there is obtained

$$K_1K_2 + K_1(\bar{n} - 1)[L]f_3/\bar{n} = (2 - \bar{n})[L]^2f_1^2f_3/\bar{n} \tag{9.91}$$

so that a plot of $(\bar{n} - 1)[L]f_3/\bar{n}$ against $(2 - \bar{n})[L]^2f_1^2f_3/\bar{n}$ has a slope of K_1 and intercept of K_1K_2. If a is the molarity of acid and b is the added molarity of NaOH in the cell, then $C_L = 2a - b$, $[L] = [H^+]$, whence from (9.49), $\bar{n} = (2a - b - [H^+])/a$. I may be taken as $b + [L]$ when $b < a$, and as $2b - a$ when $b > a$, and f_1 and f_3 calculated by (3.43). Some figures for adipic acid [56] at 20°C are given in Table 9.10. From these, $pK_1 = 4.4_0$, $pK_2 = 5.4_5$. The pH titration of mixtures of acids and complex forming cations at low variable ionic strengths can be dealt with in this fashion, e.g., the titration of mixtures of oxalacetic acid and lanthanide chlorides [96] with HCl and NaOH (MA^+ and MA_2^- form).

Example 9.12. The silver complexes of some sulphonated anilines, ethers, substituted phosphines and arsines, which are very stable, have been investigated [98] by the cell:

Au | Quinhydrone, 0.01 M $HClO_4$, $NaClO_4 = I$ | $NaClO_4 = I$ | C_M, C_L, $NaClO_4$ to I | AgX | Ag

where C = $AgClO_4$, X = Cl^- or Br^-; $[Ag^+]$ is determined.

When L = e.g., diphenylphosphinobenzene-p-sulphonate, β_1 and β_2 are very large so that when C_L is very small, [L] cannot be determined by means of (9.56) and the associated plotting procedure. Instead, $E_M = E_0 - E$ is obtained with C_M small and C_L large (E_0 and E are the potentials of the cell with $C_M = 0$ and C_M respectively). Under these conditions, (9.56) is used to obtain [L], and β_N found from $C_M/[M] = \beta_N[L]^N$ (see 1, iii, p. 169). In this case $N = 3$. From the average β_N obtained in this way, an average value of $[L]_a$ is calculated from high C_L values (see 9.59). With β_3 established as 3×10^{19},

TABLE 9.10

Data for Deriving K_1 and K_2 of Adipic Acid at 20°C

pH	4.47	4.76	4.88	4.99	5.35	5.48
10^4a	7.75	7.66	7.63	7.60	7.50	7.47
10^4b	4.47	6.55	7.38	8.24	10.47	11.25
$\bar{n}$	1.371	1.101	1.016	0.904	0.598	0.490

and choosing $E_M = 600$ mV as standard, this means that from the above expression, $[L]_a = 0.77$ mM. The corresponding value of $\bar{n}$ is 3. Values of $\bar{n}$ at lower C_L values are calculated by the plotting procedure associated with (9.56), then $1/\bar{n}$ plotted against log X from the standard value. Thus any selected value of $[L]_f$ is found from (9.59) by graphical integration. Having obtained [L], any of the appropriate methods for obtaining β_1 and β_2 can be used. Some figures are:

E_M (mV)	600	500	450	350	280	230	200
log X	10.143	8.452	7.606	5.917	4.733	3.888	3.381
$\bar{n}$	3.0	3.0	3.0	2.89	2.68	2.42	2.22

For the last point, from the area under the plot of $1/\bar{n}$ against log X,

$$\log([L]_a/[L]_f) = 2.404, \text{ so } \log[L]_f = \log[L]_a - 2.404$$

The information obtained by the methods described in this section is very extensive. Up to 1956 it is summarized in the Tables of Schwarzenbach, *et al.*[8] Some recent studies are metal complexes of (a) amino-acids and peptides [99] (b) nucleotides [100] (AMP, ADP, ATP, etc.); (c) substituted and polyamines;[101] (d) polyphosphates [102] (e) N.N′ ethylene bis(2(o-hydroxyphenyl))glycine, ethylene diamine diacetic and aminomethylene phosphonic N.N′ diacetic acids.[103] Most of these are based on pH titrations. Examples where the measurements entail spectrophotometry, ion-exchange resins, solvent extraction, etc., are to be found in subsequent chapters.

REFERENCES

1. Harned, H. S. and Owen, B. B. *Chem. Rev.* **25,** 31 (1939). Ref. 2, Ch. 1.
2. Harned, H. S. and Ehlers, R. W. *J. Amer. chem. Soc.* **54,** 1350 (1932); **55,** 652 (1933).
3. Ref. 2, Ch. 1.
4. Nash, G. R. and Monk, C. B. *J. chem. Soc.* 4274 (1957).
5. Harned, H. S. and Robinson, R. A. *J. Amer. chem. Soc.* **50,** 3157 (1928).
6. Le Fevre, R. J. W. *Trans. Faraday Soc.* **34,** 1127 (1938).
7. Ref. 11, Ch. 1.
8. Ref. 14, Ch. 8.
9. Danyluk, S. S., Taniguchi, H. and Janz, G. J. *J. phys. Chem.* **61,** 1697 (1957).
10. Bates, R. G. *J. Amer. chem. Soc.* **70,** 1579 (1948).
11. (a) K_1, K_2: Darken, L. S. *ibid.* **63,** 1007 (1941). (b) K_2: Harned, H. S. and Fallon, L. D. *ibid.* **61,** 3111 (1939); Pinching, G. D. and Bates, R. G. *J. Res. nat. Bur. Stand.* **40,** 405 (1948).
12. (a) K_1: Harned, H. S. and Davis, R. *J. Amer. chem. Soc.* **65,** 2030 (1943); Harned, H. S. and Bonner, F. T. *ibid.* **67,** 1026 (1945). (b) K_2: Harned, H. S. and Scholes, S. R. *ibid.* **63,** 1706 (1941).
13. Hamer, W. J. *ibid.* **56,** 860 (1934).
14. Ashley, J. H., Crook, E. M. and Datta, S. P. *Biochem. J.* **56,** 190 (1954); Ashley, J. H., Clark, H. B., Crook, E. M. and Datta, S. P. *ibid.* **59,** 203 (1955); Clarke, A. B., Datta, S. P. and Rabin, B. R. *ibid.* **59,** 209 (1955); Datta, S. P. and Grzybowski, A. K. *ibid.* **69,** 218 (1958).
15. (a) K_1: Nims, L. F. *J. Amer. chem. Soc.* **56,** 1110 (1934). Bates, R. G. and Acree, S. F. *J. Res. nat. Bur. Stand.* **47,** 127 (1951). (b) K_2: Bates, R. G. and Acree, S. F. *ibid.* **30,** 129 (1943).
16. Bates, R. G. and Pinching, G. D. *J. Amer. chem. Soc.* **71,** 1274 (1949).
17. Berg, D. and Patterson, A. *ibid.* **75,** 5197 (1953). Wissbrun, K. F., French, D. M. and Patterson, A. *J. phys. Chem.* **58,** 693 (1954).
18. Carini, F. F. and Martell, A. E. *J. Amer. chem. Soc.* **75,** 4810 (1953).
19. Davies, C. W., Jones, H. W. and Monk, C. B. *Trans. Faraday Soc.* **48,** 921 (1951).
20. Klotz, I. M. and Singleterry, C. R. Theses, Chicago, (1940). Quoted by Robinson, R. A. and Stokes, R. H. Ref. 11, Ch. 1.
21. Kerker, M. *J. Amer. chem. Soc.* **79,** 3364 (1957).
22. Nair, V. S. K. and Nancollas, G. H. *J. chem. Soc.* 4144 (1958).
23. Harned, H. S. and Paxton, T. R. *J. phys. Chem.* **57,** 531 (1953).
24. Owen, B. B. *J. Amer. chem. Soc.* **56,** 24 (1934).
25. King, E. J. *ibid.* **73,** 155 (1951).
26. Datta, S. P. and Grzybowski, A. K. *Trans. Faraday Soc.* **54,** 1179 (1958).
27. Evans, W. P. and Monk, C. B. *ibid.* **51,** 1244 (1955).
28. (a) Alanine: Nims, L. F. and Smith, P. K. *J. biol. Chem.* **101,** 401 (1933). (b) Sarcosine and N-dimethylglycine, ref. 26. (c) Glycine in aqueous dioxan: Harned, H. S. and Birdsall, C. M. *J. Amer. chem. Soc.* **65,** 54, 1117 (1943). (d) Acetylglycine, propionylglycine, acetylglycine, acetyl amino-n-butyric acid, acetyl-β-alanine: King, E. J. and King, G. W. *ibid.* **78,** 1089, 6020 (1956). (e) Taurine: King, E. J. *ibid.* **75,** 2204 (1953). (f) γ-aminobutyric acid, King, E. J. *ibid.* **75,** 1006 (1954). (g) Sulphanilic acid: MacLaren, R. O. and Swinehart, D. F. *ibid.* **73,** 1822 (1951). (h) Metanilic acid: McCoy, R. D. and Swinehart, D. F. **76,** 4708 (1954). (i) Orthanilic

acid: Diebel, R. N. and Swinehart, D. F. *J. phys. Chem.* **61**, 333 (1957). (j) Glycine peptides: King, E. J. *J. Amer. chem. Soc.* **79**, 555 (1958).
29. Grzybowski, A. K. *J. phys. Chem.* **62**, 555 (1958).
30. Parton, H. N. and Gibbons, R. C. *Trans. Faraday Soc.* **35**, 542 (1939).
31. Harned, H. S. and Wright, D. D. *J. Amer. chem. Soc.* **55**, 4849 (1933).
32. Korman, S. and La Mer, V. K. *ibid.* **58**, 1396 (1936).
33. Bruckenstein, S. and Kolthoff, I. M. *ibid.* **78**, 2974 (1956).
34. Kolthoff I. M. and Willman A. *ibid.* **56**, 1001 (1934). Smith, T. L. and Elliot, J. H. *ibid.* **75**, 3566 (1953).
35. Everett, D. H., Landsman, D. A. and Pinsent, B. R. W. *Proc. roy. Soc.* **215A**, 416 (1952).
36. Harned, H. S. and others, *J. Amer. chem. Soc.* (a) CsOH, CsCl: **52**, 3892 (1930). (b) KOH, KCl: **55**, 2194 (1933). (c) LiOH, LiCl: **55**, 2206 (1933). (d) NaOH, NaBr and KOH, KBr: **55**, 4496 (1933). (e) NaOH, NaCl: **57**, 1873 (1935). (f) LiOH, LiBr: **59**, 1280 (1937). (g) $Ba(OH)_2$, $BaCl_2$: **59**, 2032 (1937). (g) $Sr(OH)_2$, $SrCl_2$: *J. phys. Chem.* **57**, 1280 (1953). Also Vance, J. E. *J. Amer. chem. Soc.* **55**, 2729 (1933).
37. Gimblett, F. G. R. and Monk, C. B. *Trans. Faraday Soc.* **50**, 965 (1954).
38. Harned, H. S. and Fallon, L. D. *J. Amer. chem. Soc.* **61**, 2374 (1939).
39. Owen, B. B. *ibid.* **56**, 2785 (1934).
40. Bates, R. G. and Pinching, G. D. *J. Res. nat. Bur. Stand.* **42**, 419 (1949).
41. Bates, R. G. and Pinching, G. D. *J. Amer. chem. Soc.* **73**, 1393 (1950).
42. Jones, H. W. and Monk, C. B. *Trans. Faraday Soc.* **48**, 929 (1952).
43. King, E. J. and King, G. W. *J. Amer. chem. Soc.* **74**, 212 (1952).
44. (a) Ref. 42. (b) Evans, J. I. and Monk, C. B. *Trans. Faraday Soc.* **48**, 934 (1952). (c) Davies, P. B. and Monk, C. B. *ibid.* **50**, 128, 132 (1954). (d) Clarke, H. B. Cusworth, D. C. and Datta, S. P. *Biochem. J.* **58**, 14, 146 (1954). (e) Clarke, H. B. and Datta, S. P. *ibid.* **64**, 604 (1956). (f) Ref. 27. (g) Ref. 22 and *J. chem. Soc.* 3934, (1959). (h) Carini, F. F. and Martell, A. E. *J. Amer. chem. Soc.* **76**, 2153 (1954). (i) Hughes, V. L. and Martell, A. E. *ibid.* **78**, 1319 (1956).
45. Harned, H. S. and Fitzgerald, M. E. *ibid.* **58**, 2624 (1936). Truemann, W. B. and Ferris, L. M. *ibid.* **80**, 5048 (1958).
46. Bates, R. G. and Vosburgh, W. C. *ibid.* **60**, 136 (1938).
47. Carmody, W. R. *ibid.* **51**, 2905 (1929).
48. Perrin, D. D. *J. phys. Chem.* **62**, 767 (1958); *J. chem. Soc.* 3120, 3125 (1958).
49. Dunsmore, H. S. and Speakman, J. C. *Trans. Faraday Soc.* **50**, 236 (1954).
50. Gold, V. "pH Measurements." Methuen and Co., Ltd., London; J. Wiley and Sons, Inc., New York, (1956).
51. Bates, R. G., Pinching, G. D. and Smith, E. R. *J. Res. nat. Bur. Stand.* **45**, 418 (1950); Bates, R. G., Bower, V. E., Miller, R. G. and Smith, E. R. *ibid.* **47**, 433 (1951); Bower, V. E., Bates, R. G. and Smith, E. R. *ibid.* **51**, 189 (1953); Bates, R. G., Bower, V. E. and Smith, E. R. *ibid.* **56**, 305 (1956); Bower, V. E. and Bates, R. G. *ibid.* **59**, 261 (1957).
52. Bates, R. G. "Electrometric pH Titrations." J. Wiley, and Sons, Inc., New York (1924)
53. (KHphthalate) British Standard Specifications, No. 1647, British Standards Institution, London (1950).
54. De Ligny, C. L. and Luyhx, P. F. M. *Rec. trav. Chim.* **77**, 154 (1958).
55. "Biophysical Chemistry." Academic Press, Inc., New York, Vol. I (1958). Cohen, E. J. and Edsall, J. T. "Proteins, Amino-acids and Peptides." Reinhold Publishing Corp., New York, (1943).

56. Speakman, J. C. *J. chem. Soc.* 855 (1940).
57. Pederson, K. J. *Acta Chem. Scand.* **9**, 1634 (1955).
58. Hitchcock, D. I. *J. phys. Chem.* **62**, 1337 (1958).
59. Monnier, D. and Kapetanidis, I. *Helv. Chim. Acta* **41**, 1652 (1958).
60. Schwabe, K. *Chem. Ing. Tech.* **31**, 109 (1959).
61. Harlow, G. A., Noble, C. M. and Wyld, G. E. *Anal. Chem.* **28**, 787 (1956); **30**, 69 (1958). Cundiff, R. H. and Markunas, P. C. *ibid.* **28**, 792 (1956).
62. Grunwald, E. *J. Amer. chem. Soc.* **73**, 4934 (1951). Grunwald, E. and Berkowitz, B. J. *loc. cit.* 4939.
63. Gutbezahl, B. and Grunwald, E. *ibid.* **75**, 559 (1953).
64. Bacarella, A. L., Grunwald, E., Marshall, H. P. and Purlee, E. L. *J. org. Chem.* **20**, 747 (1955).
65. Bacarella, A. L., Grunwald, E., Marshall, H. P. and Purlee, E. L. *J. phys. Chem.* **62**, 856 (1958).
66. Shedlovsky, T. and Kay, R. L. *ibid.* **60**, 151 (1956).
67. Kilpi, S. *J. Amer. chem. Soc.* **74**, 5296 (1952).
68. Purlee, E. L. and Grunwald, E. *J. phys. Chem.* **59**, 1112 (1955).
69. Purlee, E. L. and Grunwald, E. *J. Amer. chem. Soc.* **79**, 1366 (1957).
70. Marshall, H. P. and Grunwald, E. *J. chem. Phys.* **21**, 2143 (1958).
71. Harned, H. S. and Calmon, C. *J. Amer. chem. Soc.* **60**, 2130 (1938).
72. Nash, G. R. and Monk, C. B. *Trans. Faraday. Soc.* **54**, 1650 (1958).
73. Purlee, E. L. and Grunwald, E. *J. Amer. chem. Soc.* **79**, 1372 (1957).
74. Covington, A. K. and Prue, J. E. *J. chem. Soc.* 3696, 3701 (1955).
75. Covington, A. K. and Prue, J. E. *ibid.* 1567 (1957).
76. Nancollas, G. H. *ibid.* 744 (1956).
77. Monk, C. B. *Trans. Faraday Soc.* **47**, 297 (1951).
78. Sillen, L. G. *Acta Chem. Scand.* **1**, 461, 473 (1947); **2**, 828 (1948). *J inorg. nucl. Chem.* **8**, 176 (1958).
79. Tobias, R. S. *J. chem. Educ.* **35**, 592 (1958).
80. Grossman, H. *Z. anorg. Chem.* **43**, 356 (1905).
81. Bjerrum, J. "Metal Ammine Formation in Aqueous Solutions." P. Hasse and Son, Copenhagen (1941). *Chem. Abs.* **35**, 6527 (1941).
82. Leden, I. *Z. phys. Chem.* **188A**, 160 (1941). *Svensk. Kem. Tid.* **58**, 130 (1946).
83. Fronaeus, S. Thesis, Lund, (1948). *Acta. Chem. Scand.* **4**, 72 (1950).
84. Righellato, E. C. and Davies, C. W. *Trans. Faraday Soc.* **26**, 592 (1930).
85. Sutton, J. *Nature, Lond.* **169**, 71 (1952).
86. Biedermann, G. and Sillen, L. G. *Arkiv. Kemi.* **5**, 425 (1955).
87. Rossotti, F. J. C. and Rossotti, H. S. *Acta Chem. Scand.* **9**, 1166 (1955).
88. Olerup, H. Thesis, Lund (1944).
89. Carlson, G. A., McReynolds, J. P. and Verhoek, F. H. *J. Amer. chem. Soc.* **67**, 1334 (1945).
90. Irving, H. and Rossotti, H. S. *J. chem. Soc.* 3397 (1953).
91. Block, B. P. and McIntyre, G. H. *J. Amer. chem. Soc.* **75**, 5667 (1953).
92. Hearon, J. Z. and Gilbert, J. B. *ibid.* **77**, 2594 (1955).
93. Gilbert, J. B., Otey, M. C. and Hearon, J. Z. *ibid.* **77**, 2599 (1955).
94. Sonesson, A. *Acta Chem. Scand.* **12**, 165, 1937 (1958); **13**, 998 (1959).
95. Calvin, M. and Wilson, K. W. *J. Amer. chem. Soc.* **67**, 2003 (1945).
96. Gelles, E. and Nancollas, G. H. *Trans. Faraday Soc.* **52**, 98 (1956).
97. Schwarzenbach, G., Willi, A. and Bach, R. D. *Helv. Chim. Acta* **30**, 1303 (1947). Schwarzenbach, G. and Ackermann, H. *ibid.* **31**, 1029 (1948).

98. Ahrland, S., Chatt, J., Davies, N. R. and Williams, A. A. *J. chem. Soc.* 264, 276 (1958).

99. Li, N. C., Doody, E. and White, J. M. *J. Amer. chem. Soc.* **79,** 5859 (1957); **80,** 5901 (1958). Li, N. C. and Chen, M. C. M. *ibid.* **80,** 5678 (1958). Pelletier, S. *C.R. Acad. Sci., Paris,* **247,** 1996 (1958). Datta, S. P. and Grzybowski, A. K. *J. chem. Soc.* 1091 (1959). Datta, S. P. and Rabin, B. R. *Trans. Faraday Soc.* **52,** 1117, 1123, 1130 (1956); **55,** 1982, 2141 (1959).

100. Smith, R. M. and Alberty, R. A. *J. Amer. chem. Soc.* **78,** 2376 (1956). Namminga, L. B. *J. phys. Chem.* **61,** 1144 (1957). Walaas, E. *Acta Chem. Scand.* **12,** 528 (1958). Schwarzenbach, G. and Martell, A. E. *J. phys. Chem.* **62,** 886 (1958).

101. Bertsch, C. C., Fernelius, W. C. and Block, B. P. *ibid.* **62,** 444 (1958). Lotz, J. R., Block, B. P. and Fernelius, W. C. *ibid.* **63,** 541 (1959). Hoyer, E. *Z. anorg. Chem.* **297,** 176 (1958). Jonassen, H. B. *et. al. J. phys. Chem.* **61,** 504 (1957); *Acta Chem. Scand.* **79,** 4275 (1957).

102. Watters, J. I. and Lambert, S. M. *J. Amer. chem. Soc.* **78,** 4855 (1956); **79,** 3651, 4262, 5606 (1957); **81,** 3201 (1959). Wolhoff, J. A. and Overbeck, J. Th. G. *Rec. Trav. Chim.* **78,** 759 (1959).

103. Martell, A. E. *et. al. J. Amer. chem. Soc.* **80,** 530, 2351 (1958).

104. Bates, R. G., Bower, V. E., Canham, R. G. and Prue, J. E. *Trans. Faraday Soc.* **55,** 2062 (1959), have studied cell (9.21) from 0°C to 40°C with $Ca(OH)_2 + KCl$ and $Ca(OH)_2 + CaCl_2$ solutions, and have evaluated the corresponding values of $Ca(OH)^+$. They have shown that the answers are very dependent on the ratio γ_X/γ_{OH} and on the ion-size parameter a. The discussion preceding Table 9.2. and the r_h data of Table 14.4 support the assumptions made in Ch. 9.

CHAPTER 10

INCOMPLETE DISSOCIATION, PART III: SPECTROPHOTOMETRIC METHODS IN THE VISIBLE AND ULTRA-VIOLET REGIONS

MANY ionic species absorb radiations in the visible (4,000 to 8,000 Å, i.e., 400 to 800 mμ) and/or the ultra-violet (150 to 400 mμ) regions of the spectrum.[1–6,8] Among the cations, many of those of the transition elements, including the lanthanides and actinides [3] possess absorption bands, while the alkali and alkaline earth metals, La^{3+} and Zn^{2+}, are examples of ions which do not absorb down to at least 200 mμ. Where an absorbing cation has an anionic counterpart, this also absorbs somewhere in these regions, while many of the colourless anions such as CNS^-, Cl^-, Br^-, I^-, NO_3^- and $S_2O_3^{2-}$ show strong absorption [3] in the U.V. Even the ClO_4^- ion begins to absorb around 200 mμ, and this is doubtless true of many other ions which have not yet been examined at these wavelengths.

The study of ion-association or complexing by spectrophotometry depends on the fact that when these processes take place, many of the individual ion absorption bands are shifted to other frequencies, but this is not always so, and for certain shifts there are alternative explanations; these features are dealt with in the following paragraphs.

When radiation of intensity I_0 (intensity is a measure of energy) encounters matter, part of it is reflected with an intensity I_R, part is absorbed with an intensity I_A, and part, with intensity I_T, is transmitted. These intensities are related by

$$I_0 = I_R + I_A + I_T \qquad (10.1)$$

In spectrophotometry, the solutions are held in glass or fused silica cells with polished faces, so providing the solutions are optically clear, I_R is small and is largely balanced out by using a reference solution or solvent in a similar cell. Then the relationships between I_0 and I_T are described by the "laws" of Lambert or Bouguer, and of Beer. If the radiation is monochromatic, the Lambert–Bouguer law is expressed by $-\partial I/\partial d = \phi I$, where I is the intensity of the incident light on a thickness d of homogeneous transparent medium (d = pathlength in the cell), and ϕ is a constant. Letting $I = I_0$ when $d = 0$, and $I = I_T$ when $d = d$, then by integration, $\ln (I_0/I_T) = \phi d$. In terms of common logarithms, $\log (I_0/I_T) = \phi_1 d$; $\log (I_0/I_T)$ is termed the optical density, D (or absorption). E and A are also used instead of D.

Beer's law deals with the dependence of the absorption upon the number of absorbing particles which the radiation encounters in its path, i.e., upon the concentration. Combination with the Lambert–Bouguer law gives

$$\log (I_0/I_T) = \epsilon_s Cd = D(\text{or } E \text{ or } A) \tag{10.2}$$

where C is the concentration. If C is in molarities, and d is in cms., ϵ_s is termed the molar or molecular extinction coefficient. Other relationships are used in spectrophotometry, but these are mainly in connection with analysis.[5,6,8]

The essential features of a typical prism spectrophotometer suitable for the study of ionic equilibria and processes are shown in Fig. 10.1.

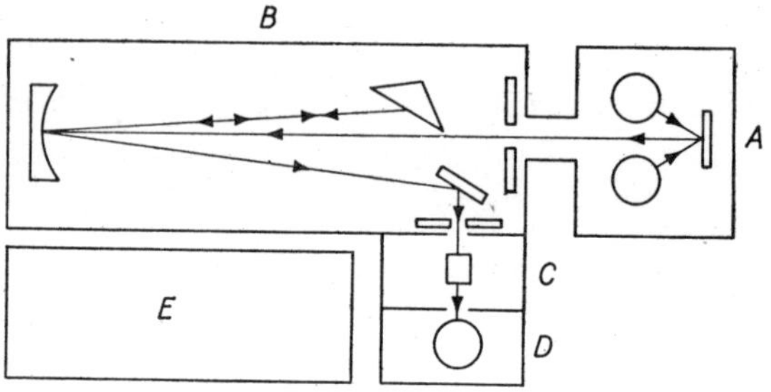

FIG. 10.1. Prism spectrophotometer.

A is the radiation source; it consists of a tungsten filament lamp for the visible and near U.V. regions (360 mμ upwards), and a hydrogen discharge lamp with a quartz envelope for radiations below 360 mμ. These lamps must be kept at a very constant level of illumination by means of well stabilized power packs (or large accumulators for the tungsten lamp). The radiation passes through a series of slits, mirrors and prism in the monochromator B so that a narrow parallel beam of nearly monochromatic radiation is directed on to the cells in C. Glass prisms and cells can be used in the visible region, down to about 400 mμ, but below this fused silica cells and a quartz (or better, fused silica) prism must be used. The radiation emerging from the cells is collected on the cathode of a photocell in D, the current of which is amplified in E by a D.C. amplifier in order to operate a galvanometer. Generally, there are two photocells, one with an Sb-Cs cathode that is sensitive up to about 650 mμ, and the second with a CsO cathode to operate above 650 mμ.

The current from the D.C. amplifier connected to the photocells produces a voltage drop across a resistance. This is balanced by an opposing potential from a potentiometer, a galvanometer acting as a null-point detector. Firstly, a balance is obtained with no light falling on a photocell, and then with light passing through the reference solvent or solution. The difference represents

100 % transmission so that when the test solution is moved into the light beam, the readings from the potentiometer controls on adjustment of the galvanometer to zero, give D directly.

Of the commercial instruments now available, some of the cheaper types are not sufficiently sensitive for the study of ionic equilibria.[6] On the other hand are those that have been specifically designed for research purposes; accounts of these are given in books by Gibson[7] (up to 1929), Mellon[6] (1930–47) and Kortüm[8] (up to 1953), and in articles such as those of Kortüm[10] and Zschiele.[11] Double beam instruments have also been devised; some of these are in commercial forms. These either have double detectors for reference and test solutions, or[12] a single beam which is chopped and diverted so that the detector is alternately exposed to the emergent light from the test and reference solutions. These instruments have the advantage that the cell carriage is not moved so as to place each cell in the beam, so there is no chance that the adjustments are upset during a comparison.

There are numerous sources of errors and anomalies in spectrophotometry. Most of these are described and discussed in the books already mentioned,[5–8] and in various articles. Goldring and his associates[9] have listed many of them and include in their classification the following:

(*a*) *Chemical and Photochemical Factors.* Turbidity can cause errors especially when coagulants are present; even laboratory dust can cause trouble. Centrifuging helps to disclose and eliminate this feature. In general, all solutions should be filtered even if they appear to be optically clear; forcing the solutions through a fine glass sinter is remarkably effective. Photochemical reactions will manifest themselves in the U.V. by the change in readings with time, and adsorption of material on the cell walls will be revealed by refilling the cell.

It is advisable to use covered or stoppered cells, especially when the solvent is volatile. A useful tip is to place a small open vessel of the solvent in the cell compartment to create a saturated atmosphere.

(*b*) *Instrumental Errors.* The wavelength setting can be set and checked in a number of ways. One method is to replace a cell by a small mirror at 45° so that the beam is diverted upwards; the two chief lines from the hydrogen lamp, red at 656.3 mμ and blue at 486.1 mμ, or the lines between 205 and 1,129 mμ from a mercury lamp[6] can then be used to line up the wavelength drum. There are also a number of useful solutions, especially (i) 0.04 g/l. of K_2CrO_4 in 0.05 N KOH which has maxima at 275 and 370 mμ, (ii) 14.5 g/l. of cobalt ammonium sulphate in 1% H_2SO_4, which has a maximum at 510 mμ.

Cell path lengths must be known accurately, especially when using small cells—the most common size is one cm. Also, the beam should be as parallel as possible for a serious source of error are multiple reflections from the cell walls, the exit slits and the outer face of the slit jaws.

Stray light can cause anomalies. It can be detected by means of glass filters of the sharp cut-off type [9,13] which absorb below the cut-off frequency. A possible source of stray light is light leakage into the cell or photocell compartments, but this is easily found and remedied. A more common type of what may be loosely termed stray light arises from using multichromatic light sources and finite slit widths so that a portion of the spectrum is passed on to the cell. The resulting stray light can be drastically reduced by using double monochromators (this is so in certain commercial instruments); any stray light entering the second stage is dispersed before the beam enters the cell. There are also other devices.[5] The theory relating stray light to I_0 has been stated by Zschiele and others.[11]

A practical way of correcting for stray light in the readings is suggested later on (see Table 10.3).

(*c*) *Cell Techniques*. Cells should have optically flat faces, and as far as possible, the cells for the reference and test solutions should match, i.e., the absorption difference should be zero when they are filled with the same solution. In practice, this is rarely so, and may vary slightly with usage, consequently an intercomparison should be made at frequent intervals. This provides a check on cleanliness, which can be critical, especially in the U.V. region, where traces of grease and other organic materials absorb strongly. Generally, washing with soapy water is sufficient, followed by distilled water, drying the outer faces by a soft cloth in such a way that fingering of the cell is avoided. A stream of alcohol vapour and HNO_3—H_2SO_4 solutions can remove obstinate contaminants.

To avoid handling the cells or altering their relative positions in the cell carriage after they have been placed in position (by forceps), a valuable technique [14,15] is to fill cells several times by pipette, emptying each time by suction jet.

The absorptions of many solutions are temperature-dependent, so a thermostatted cell holder [16] is required, this usually being water jackets with water circulation from a water bath.

It is easily shown [5,6,9] that when the solvent is the reference, the best value of D for the smallest relative error, $\Delta D/D$, is 0.4343, but there is little loss in accuracy if D is anywhere between 0.2 and 0.8. Greater precision is obtained by using a solution as reference such that $D_M = D_T - D_R$, where D_M is the measured absorption, and D_T and D_R are those of the test and reference solutions with respect to the solvent, is small. From (10.2), by differentiation, $\Delta C/C = D_M/D_T$, so that, in theory, the precision increases as D_T increases and D_M decreases. A practical limitation is that when D_T and D_R are high, the sensitivity of the spectrophotometer decreases unless the slits are opened wider, and this increases the spectral impurity.

The investigation of ionic processes by following the displacement or enhancement of absorption bands has its limitations. For example, the absorption band of the Cu^{2+} ion, as obtained with $Cu(ClO_4)_2$ solutions, is scarcely changed in the visible region on the addition of SO_4^{2-} ions (although there is a marked displacement to longer wavelengths of the U.V. band, which is ascribed to $CuSO_4$ ion-pair formation). The bands of the $Fe(CN)_6^{3-}$ ion are even more inert [50a] for neither the visible nor the U.V. bands are shifted by, e.g. La^{3+} ions, and spectral shifts are also absent in [19] tetraphenylammonium chloride solutions although conductance shows this to be a weak electrolyte. In contrast, Smith and Symons [20] have shown that the U.V. band of the iodide ion is shifted by e.g. salts which do not provide cations that associate with this anion (the major cause is probably an increase [20] in the polarization of the water molecules in the primary shell around the I^- ion).

The intense absorption bands of many anions in the U.V. are interpreted as being caused by charge-transfer,[2] the absorption of light by the hydrated ion giving rise to a free radical and a hydrated electron which remains in association with the hydration sphere. It can then either return to its origin or permanent chemical change can occur. Strong cation absorption bands are likewise ascribed to charge-transfer while the relatively weaker bands of the transition elements in the visible and near infra-red are due to internal $d - d$ or $f - f$ transitions. In complex ions electron transfer takes place between the co-ordinating groups and the central cation. The "crystal-field" branch of "ligand-field" theory [21–23] considers that the field of the (gaseous) ion is modified by associating groups or ions (ligands), causing a splitting in the energy levels. The bands then depend on transitions between these new levels (see Chapter 14). When ion-association takes place, it is usually the U.V. charge-transfer bands which are modified.

Although interest in the absorption of visible and U.V. radiations goes back for at least fifty years, pioneers such as Bjerrum [24] and Hantzsch [25] had to use rather tedious forms of visual and photographic methods and it was not until improvements in the latter and the introduction of photoelectric methods around 1920, that accurate and extensive investigations by workers such as von Halban and his associates,[26] Fromherz,[27] and Kortüm [28] were made possible, whereby dissociation constants could be derived from spectral shifts. With the marketing of commercial equipment, such information has accumulated rapidly.

The treatments which form the basis of expressions for calculating the formulae of ion-association products, and the corresponding dissociation constants, assume (with a few exceptions which are mentioned later) that the molar extinction coefficients of ions and ion-pairs or complexes are constant under constant physical conditions, and that, from (10.2), the measured absorptions are directly related to the concentrations of all the absorbing species.

The formula of the chief product of ion-association may, if it is necessary, be found by (i) Job's method [29] or (ii) the method of Bent and French.[30]

(i) For the equilibrium: $M + L \rightleftharpoons ML_n$, solutions of [31] M and L are mixed in such proportions that the sum of the moles of M and L is constant, C. In general, a solution contains $(1 - x)C$ molar M and xC molar L. The concentrations of M, L and ML_n, if they are C_1, C_2 and C_3, are related by $C_1 = C(1 - x) - C_3$ and $C_2 = Cx - nC_3$. The maximum value of C_3 is reached when $n = x/(1 - x)$, and a plot of D against x indicates this maximum; the sharpness of the maximum depends on the stability of ML_n and on the relative proportions of other association products. A good example is given by Moore and Anderson [32] with their study of ceric sulphate around 400 mμ, where SO_4^{2-} ions enhance the Ce^{4+} spectrum. The maximum occurs at $x > 0.5$, indicating that n = 1, but that some ML_2 also forms. By plotting $x/(1 - x)$ at D_{max} against C, and extrapolating, $x = 0.5$, i.e., $n = 1$ at $C = 0$.

(ii) Since $\log [ML_n] = \log [M] + n \log [L] - \log K'$, providing ML_n is not too stable, by keeping [M] constant and plotting log [L] against D, the slope of the line gives n. Job's method can be applied to other types of measurements, within certain limitations.[32]

If d is the path-length inside the cell, a the stoichiometric molarity of an absorbing ion M which has a molar extinction coefficient ϵ_M, then if the only other absorbing (mononuclear) species are M_1, ML_2, . . . ML_N formed with the non-absorbing ligand, L, the concentrations of which are $x_1, x_2, \ldots x_N$, and the molar extinction coefficients are $\epsilon_1, \epsilon_2, \ldots \epsilon_N$, then

$$D_1/d = \epsilon_M\left(a - \sum_1^N x_n\right) + \sum_1^N \epsilon_n x_n \tag{10.3}$$

$$D_2/d = a\epsilon_M \tag{10.4}$$

where D_1 and D_2 are the absorptions in solutions with and without L respectively. Also,

$$K_1 - \left(a - \sum_1^N x\right)\left(b - \sum_1^N nx_n - \sum_1^P\sum_1^M my_m\right)f_1f_2/xf_3) \tag{10.5}$$

where b is the stoichiometric molarity of L, and the double summation is to take account of P species of any non-absorbing species formed by L with other ions. This expression is for the dissociation constant K_1 of the equilibrium $M + L \rightleftharpoons ML_1$, and similar expressions can be written for K_2, etc.

Taking the simplest possible system, i.e., $n = 1$, $p = 0$, then from (10.3) and (10.4),

$$(D_1 - D_2)/d = x_1(\epsilon_1 - \epsilon_M) \tag{10.6}$$

and

$$K_1 = (a - x_1)(b - x_1)f_1f_2/xf_3 = K_1'f_1f_2/f_3 \tag{10.7}$$

where K_1' is the concentration dissociation constant. This will be constant if the ionic strength is kept constant, and this is commonly effected by addition of $NaClO_4$. By combining (10.6) and (10.7), on re-arrangement,

$$K'_1/(\epsilon_1 - \epsilon_M) = dab/(D_1 - D_2) - (a + b - x_1)/(\epsilon_1 - \epsilon_M) \quad (10.8)$$

$$= da(b - x_1)/(D_1 - D_2) - (b - x_1)/(\epsilon_1 - \epsilon_M) \quad (10.8a)$$

By plotting $dab/(D_1 - D_2)$ against $a + b - x_1$, ignoring x_1 in the first approximation, then the slope gives $1/(\epsilon_1 - \epsilon_M)$ and the intercept on the ordinate gives $K'/(\epsilon_1 - \epsilon_M)$. Having obtained a first estimate of $1/(\epsilon_1 - \epsilon_M)$, it is used in (10.6) to obtain first estimates of x_1, which in turn are used to derive better estimates of $1/(\epsilon_1 - \epsilon_M)$, and so on.

In order to obtain data at low ionic strengths, a second series of absorptions are obtained with mixtures of M and L without addition of $NaClO_4$, x_1 is found for each mixture via (10.6), and each K_1 via (10.7), using a general form of the D.H. equation (3.40–43).

Example 10.1. The NO_3^- ion has two absorption bands in the U.V. The minor one has its maximum at 300 mμ, and this is enhanced by Pb^{2+} ions due to $PbNO_3^-$ ion-pair [33, 18] formation; the Pb^{2+} ion itself does not absorb at this frequency. The absorptions of solutions containing *a* molar $NaNO_3$, *b* molar $Pb(ClO_4)_2$ and *c* molar $NaClO_4$, where *a* was fixed, *b* varied and *c* adjusted to keep *I* constant, were compared with a reference consisting of *a* molar $NaNO_3$. A trace of $HClO_4(e)$ was added to suppress hydrolysis. Some figures [18] are given in Table 10.1.

TABLE 10.1

Data for the Calculation of $1/(\epsilon_1 - \epsilon_M)$ for $PbNO_3^+$ at 25°C, 300 mμ (1-cm cells, $a = 0.0501$M, $I = 0.95$)

10^2b	10^2c	10^3e	$D_1 - D_2$	10^3x_1	I
30.75	—	7.7	0.228	19.4	0.94
26.90	11.6	6.8	0.208	17.7	0.955
23.06	23.3	5.8	0.187	15.9	0.95
19.22	34.9	4.8	0.167	14.2	0.955
15.37	46.5	3.9	0.141	12.0	0.96

According to Biggs *et al.*,[33] there is evidence, although it is mainly of a negative nature, that $Pb(ClO_4)_2$ dissociates completely; this, according to conductance data, is also true of $NaNO_3$. So from (10.8), the first estimate of $1/(\epsilon_1 - \epsilon_M)$, where $PbNO_3^+ = 1$, $NO_3^- = M$, is 0.085, and using this in (10.6) to obtain x_1, the final value of $1/(\epsilon_1 - \epsilon_M)$ is 0.088 ± 0.004, and K'_1 is 0.59.

The figures in Table 10.2 refer to dilute mixtures of $NaNO_3$ and $Pb(ClO_4)_2$. The individual values of K_1 were calculated by using (10.7), $I = a + 3b + d -$

$2x_1$, and $Q'' = 0.1$ in (3.42), this on trial, giving constant values of K_1, i.e., 0.066 ± 0.003. Including other series,[18] the overall average is 0.071 ± 0.009, which agrees well with 0.065 from conductance.[34]

TABLE 10.2

Data for $K_{PbNO_3^+}$ ($d = 1$ and* 4 cm.; 300 mμ)

10^2a	2.50	2.50	2.50	2.50	10.01	5.01	5.01	5.01
10^2b	0.96	1.44	1.92	2.40	19.22	30.75	38.43	48.04
10^3e	0.24	0.36	0.48	0.63	4.8	7.7	9.7	12.1
$D_1 - D_2$	0.063*	0.085*	0.097*	0.119*	0.346	0.228	0.267	0.291
10^2I	5.12	6.49	7.88	9.23	61.6	93.3	115.8	114.1
10^2K	6.1	6.3	6.9	6.5	6.5	6.9	6.4	6.9

The same methods of calculation have been used for (a) [18] K_{CuSO_4} by means of mixtures of $Cu(ClO_4)_2$, Na_2SO_4 or H_2SO_4, $NaClO_4$ and $HClO_4$, measuring shifts in the U.V. absorption of the Cu^{2+} ion at 260 mμ. Allowance was made for $NaSO_4^-$ and HSO_4^-, the dissociation constants of which are known [34, 35] ($m = 1$, $p = 1$ and 2 in (10.5); $x_1 - y_1 - y_2$ replaces x_1 in (10.8 and 8a) (Note: in Table 3 of this article the last two columns should be headed $10^5\beta$ and $10^3\alpha$); (b) [18, 36] $K_{Co(NH_3)_6SO_4^+}$ via shifts in the U.V. spectrum of the $Co(NH_3)_6^{3+}$ ion at 235 and 250 mμ (Note: in the second article,[36] α of Table I is $a(b - x - y)/(D_1 - D_2)$, in Table II, $1/(\epsilon_1 - \epsilon_M) = 0.00188$, and the 3rd row is 10^3y); (c) [37] various thiosulphates via the U.V. spectrum of the $S_2O_3^{2-}$ ion around 345 mμ; (d) [38] the first ion-pairs which the UO_2^{2+} ion forms with Cl^-, Br^-, CNS^- and SO_4^{2-} by enhancement of the U.V. cation band; (e) [39] the Cu^{2+}, N_3^- ion-pair, using NaN_3 at a pH of 5.0 to 5.5 and allowing for [40] undissociated HN_3.

McConnel and Davidson [41] obtained K_1' of $CuCl^+$ by an approximate version of (10.8), ignoring x_1, this being small when K_1' is large (Kruh [80] has extended such analyses to allow for CuX_2). This way of finding K_1' at specific ionic strengths has been applied to a number of systems such as (a) association of Fe^{3+} with [42] CNS^- and [43] Cl^-: (b) the combination of [44] Co^{2+} and [45, 46] Cr^{3+} complex ions with CNS^-: (c) [47] K^+, Mg^{2+} and Ba^{2+} with $Fe(CN)_6^{4-}$: (d)[48] Fe^{3+} and $Fe(CN)_6^{3-}$: (e) [46, 49] $Co(NH_3)_6^{3+}$ with Cl^-, Br^- and N_3^-.

Instead of using mixtures, Prue measured the absorptions of $CuSO_4$ solutions [50] with the solvent as reference. From (10.3),

$$D_1/ad = \epsilon = \epsilon_1 - (\epsilon_1 - \epsilon_M)\alpha_c \tag{10.9}$$

where a is the stoichiometric molarity of $CuSO_4$, ϵ_1 and ϵ_M refer to $CuSO_4$ ion-pairs and Cu^{2+} respectively, and $x_1 = (1 - \alpha_c)a$, α_c being the degree of dissociation. Of the various terms, ϵ_M can be determined from the absorption of $Cu(ClO_4)_2$, leaving two unknowns. By combining (10.9) with $K = \alpha_c^2 a f_{\pm}^2 / (1 - \alpha_c)$, calculating $f_{\pm}$ by (3.35) with $I = 4\alpha_c a$, K can be adjusted so that a plot of α_c against ϵ is linear. However, the values of K depends on the choice of the ion-size parameter; with this as 4.3 Å, $K = 0.008$ at 25°C and $\epsilon_1 = 353.9$, while with 10 Å, $K = 0.0040$ and $\epsilon_1 = 222.2$. Thus K cannot be decided unless the ion-size or ϵ_1 is known, which implies that methods of deriving ϵ_1 such as that based on (10.8) are invalid.

Panckhurst and Woolmington [50a] have used (10.9) and (10.11) for a study of ion-association between Tl^+, SO_4^{2-}; Tl^+, $Fe(CN)_6^{4-}$; and La^{3+}, $Fe(CN)_6^{4-}$, calculating each $f_{\pm}$ via (3.42). They conclude that $K_{TlSO_4^-}$ is between 0.087 and 0.25, $K_{TlFe(CN)^{3-}}$ is 0.00113, and $K_{LaFe(CN)_6^-}$ is $8.7 \pm 1.5 \times 10^{-6}$ (all at 25°C).

The same principles have been adopted for [51] deriving $K = [HCrO_4^-]^2 f_1^2 / [CrO_4^{2-}] f_2$ by means of slightly acid $K_2Cr_2O_7$, and [52] in a study of the PuO_2^{2+}, Cl^- system in which the enhancement of the PuO_2^{2+} absorption by NaCl addition to a 2 M $HClO_4$ solution of the cation was determined at eight frequencies between 294 and 837 mμ. Assuming that only PuO_2Cl^+ forms, with $K_1' = (a - x)[Cl^-]/x$, then from (10.8),

$$D_1/ad = \epsilon_M + (\epsilon_1 - \epsilon_M)[Cl^-]/(K_1' + [Cl^-]) \qquad (10.10)$$

where M refers to the cation of total concentration a, and 1 refers to the ion-pair of concentration x_1. The best values of K_1' and ϵ_1 were found at each frequency, and K_1 was found to be frequency dependent, which is attributed to PuO_2Cl_2 formation. From a lengthy analysis it is decided that $\beta_1 = 1.2$, and $\beta_2 = 0.3$ (see 9.45).

Prue [51] has aimed at greater experimental precision by using the spectrophotometer as a null-point instrument. When the test (T) and reference (R) solutions have almost identical absorptions,

$$\epsilon_R/\epsilon_T = D_R C_T d_T / D_T C_R d_R = (1 + \Delta D/D_T) C_T d_T / C_R d_R \qquad (10.11)$$

the smaller is ΔD, the more accurately is D_R found in terms of D_T. An exact balance can be found by varying the test solution, keeping the reference constant (or vice versa) so that ΔD varies from about 0.01 to -0.01, and to plot ΔD against C_T (or C_R) in this region. Then from (10.3) and (10.4), when $N = 1$,

$$\epsilon_M(a_2 - a_1)/(\epsilon_1 - \epsilon_M) = x_1 \qquad (10.12)$$

where a_2 and a_1 are the concentrations of M without and with L at the balance point.

Example 10.2. Li_2SO_4 of molarity b was added [38] to a fixed molarity a_1 (0.004860 M) of $UO_2(ClO_4)_2$ containing 0.00851 M $HClO_4$, and its absorption at 300 mμ was matched against those of $UO_2(ClO_4)_2 + HClO_4$ solutions in 1.00 cm cells at 25°C, by plotting the concentration of the latter against ΔD, and its matching concentration a_2 found by interpolation. The method described in example (10.1) was used to obtain $(\epsilon_1 - \epsilon_M) = 259$. To obtain ϵ_M, the absorptions of several concentrations of acidic $UO_2(ClO_4)_2$ were measured. These showed an apparent disagreement with the Beer–Bouguer–Lambert law, but since the addition of $NaClO_4$ did not affect the readings, it is assumed that the cause is a small stray light error. Allowing $D' = 0.006_3$ for this, ϵ_M is constant, as shown by Table 10.3, averaging $\epsilon_M = 59.9_3 \pm 0.10$.

TABLE 10.3

Data for $\epsilon_{UO_2^{2+}}$ ($d = 1$ cm)

$10^4 a$	29.16	48.60	68.77	85.05	105.7	123.0	127.3
D	0.181	0.298	0.420	0.515	0.640	0.741	0.769
$(D - 0.006_3)/a$	59.91	60.02	60.16	59.81	59.95	59.73	59.91

Some figures whereby $K_{UO_2SO_4}$ was calculated are shown in Table 10.4. For these, y_1 was obtained by approximations from $\log y_1 = \log (e - y_1)(b - x_1 - y_1) - \log K_{HSO_4} - 4\phi(I)$ (see Table 9.2, and 3.43), where $I = 3a_1 + 3b + e - 4x_1 - 2y_1$. Instead of obtaining a_2 by interpolation as described above, the values were calculated from $a_2 = (D - 0.006_3)/59.9_3$; (3.43) was also used in calculating K.

TABLE 10.4.

Dissociation Constant of UO_2SO_4 ($d = 1$ cm, $10^4 a_1 = 48.60$, $10^4 e = 85.1$)

$10^4 b$	D	$10^3 a_2$	$10^4 y_1$	$10^4 x_1$	$10^3 I$	$10^4 K$
10.28	0.416	6.84	1.76	4.59	23.93	11.5
19.99	0.515	8.49	3.58	8.44	25.00	11.3
35.02	0.641	10.59	6.3	13.25	27.03	11.8
50.06	0.743	12.29	9.2	17.25	29.4	11.8
56.15	0.778	12.88	10.2	18.60	30.4	11.9

Foley and Anderson [53] have applied the null-point procedure to some UO_2^{2+} and Cu^{2+} complexes of sulphosalicylate. They took as reference a solution containing, e.g., a 1 : 1 ratio of M to L at an ionic strength I_1, and compared with this the absorptions of a test solution containing a different ratio of M : L. The test was diluted until its absorption matched that of the reference, but keeping each at I_1 by an inert salt. K' was then found from $K' = (a_1 - x)(b_1 - x)/x = (a_2 - x)(b_2 - x)/x$ after deriving x from the two expressions.

Another way of handling absorption measurements has been described by Newton and Arcand.[54] From (10.3) and (10.4), when $N = 1$, $d = 1$, and writing $D_3 = a\epsilon_1$, then $D_1 - D_2 = x_1(\epsilon_1 - \epsilon_M) = (x_1/a)(D_3 - D_2)$. Also, $K' = (a - x_1)(b - x_1)/x_1$ so $K'/(b - x_1) = (a/x_1) - 1$, whence

$$D_1 = D_3 - K'(D_1 - D_2)/(b - x_1) \tag{10.13}$$

A plot of D_1 against $(D_1 - D_2)/(b - x_1)$ should be a straight line with a slope of $-K'$ and an intercept of D_3. When b is small, approximations are required to derive $b - x_1$. Newton and Arcand [54] used this method to study ion-association between Ce^{3+} and SO_4^{2-}; the Ce^{3+} spectrum consists of four bands in the U.V., the height but not the frequency range of the 296 mμ band is altered by associating ligands while the other three bands are scarcely changed. When b was large, the plot tended to curve upwards. This is attributed to $Ce(SO_4)_2^-$ formation. By taking the ion size parameter to be 5.73 Å, K_{CeSO_4} at $I = 0$ and 25°C $= 4.26 \times 10^{-4}$. Two similar investigations by this method are (a) [55] association of Fe^{3+} and SO_4^{2-}, for which $K' = 1/110$ at $I = 1.0$, and (b) [56] the association of $Co(NH_3)_5H_2O^{3+}$ and $Co(NH_3)_6^{3+}$ with SO_4^{2-} over a range of ionic strengths and frequencies. The result of $K = 4.76 \times 10^{-4}$ is considerably smaller than is given by (10.8) and by fresh measurements,[36] namely 13.0×10^{-4}.

Newton and Arcand [54] have drawn attention to an important aspect of spectrophotometry. They found that ϵ_{Ce} and $\epsilon_{CeSO_4^+}$ depend on the ionic strength; the former decreased from 19.5 to 16.8 as I increased from 0.2 to 2.0, while the latter increased from 49.2 to 56.8. This feature has also been observed with the UO_2^{2+}, Cl^- system,[38] $\epsilon_1 - \epsilon_M$ increasing as the ionic strength is increased by the addition of perchlorates (up to $I = 8.0$). Gates and King [57] have reported analogous effects concerning the absorption of the $Cr(H_2O)_4Cl_2^+$ ion, and so have Coll, Nauman and West,[43] with respect to the K' value of $FeCl^{2+}$; they attribute the effect as being due to the dehydrating action of perchlorates. Whatever explanation accounts for these medium effects, it is at least clear that it cannot be assumed that absorptions are independent of the ionic media, and allowances must be made if it is necessary.

A further feature has been emphasized by Biggs,[33] namely, that it cannot be assumed that activity coefficients remain constant in mixtures of constant ionic strength but of varying composition unless one of the components is always in large excess.

Weak Electrolytes

By adding an excess of the ligand L to a small concentration of the absorbing ion M, a limiting value of the absorption corresponding to that of ML will be obtained if ML is a weak electrolyte, and if further species ML_n $(n > 1)$ do not form. In this way, ϵ_1 of (10.6) is known, and K_1 can be obtained very easily. If the electrolyte is only moderately weak so that a large excess of L would be needed to convert all of M to ML, the corresponding limiting absorption may be found by extrapolating the plot of D against 1/[L]. This was done by Klotz and Ming [58] in a study of some metal chelates of pyridine-2-azo-p. dimethylaniline, and [37] for obtaining $K_{CdS_2O_3} = 0.0012$. An alternative approach is sometimes useful if more than one complex forms. For instance, Moore and Anderson [32] found that although the absorptions of solutions in which SO_4^{2-} was added to a small fixed concentration of Ce^{4+} ions did not tend to become constant owing to the formation of $Ce(SO_4)_2^-$, the reverse process of adding Ce^{4+} to a small fixed concentration of SO_4^{2-} led to a constant value of $\epsilon_{CeSO_4^+} = D/(\text{total Ce})$. This method has also been applied to Ni^{2+} complexes of amino-acids.[59]

If an electrolyte is very weak, it may be useless to apply direct spectrophotometry. Assuming the molecular form is ML, the most favourable conditions for dissociation are equal and very low concentrations of both M and L, but this may mean very small absorption differences. If L is the anion of a weak acid, indirect spectrophotometry whereby [L] is found via pH indicators and K_{HL} may be successful (or some of the methods described in Chapter 9).

Considering weak acids (or bases) which possess absorption bands, if

$$D_1 = \epsilon_L[L^-] + \epsilon_{HL}(C - [L^-]) \tag{10.14}$$

then

$$K_a = [H^+][L^-]f_\pm^2/[HL] = [H^+]f_\pm^2(D_1 - C\epsilon_{HL})/(C\epsilon_L - D_1) \tag{10.15}$$

One of the earliest applications is that of von Halban [26] to obtain K_a of 2 : 4 dinitrophenol; ϵ_L and ϵ_{HL} were obtained from measurements in the presence of excess NaOH and HCl respectively. He used solutions of the phenol alone or in the presence of salts, so that $[H^+] = [L^-]$.

A better way is to control $[L^-]$ and to have an accurate measure of $[H^+]$ by obtaining D_1 in the presence of non-absorbing standard buffers. This was first demonstrated by Hammett [60] with benzoic acid in acetate buffers. Edwards,[61] Biggs,[62] Robinson [63] and others have proved the possible precision of this method.

Example 10.3. Robinson and Biggs [64] found that for $C = 8.20 \times 10^{-5}$ benzoic acid at 25°C and 240 mμ, $D_1 = 0.581$ in N HCl, 0.241 in 0.01 N NaOH, and 0.309 in 0.1 M NaH succinate + 0.1 M NaCl where the pH = 4.684. Since $I = 0.20$, from (3.43), $f_\pm = 0.764$, and from (10.15), p$K_a = 4.199$.

If this method is used, it is important to see that there is no stray light error, e.g., by the method illustrated by Table 10.3. Another approach is to use the null-point method based on (10.12); this is illustrated by the study of

2 : 4 dinitrophenol is 20 % ethanol.[65] The absorptions of this in chloroacetate buffers were matched against a fixed concentration of the phenol in excess NaOH. It was necessary to obtain ΔD from $+ 0.02$ to $- 0.02$, plotting these against the molarity of phenol, since ϵ_{HL} is not zero at the chosen wavelength (446 mμ) and consequently the true balance-point is not $\Delta D = 0$. The approximate [L] derived from $\Delta D = 0$ gives a first estimate of [HL] and knowing ϵ_{HL}, a first value of ΔD where $[L^-]_T = [L^-]_R$ is obtained; several approximations of this sort are needed.

Ways of establishing K_1 and K_2 of a dibasic acid when these are relatively close together have been reviewed by Ang[66] in an article in which he has

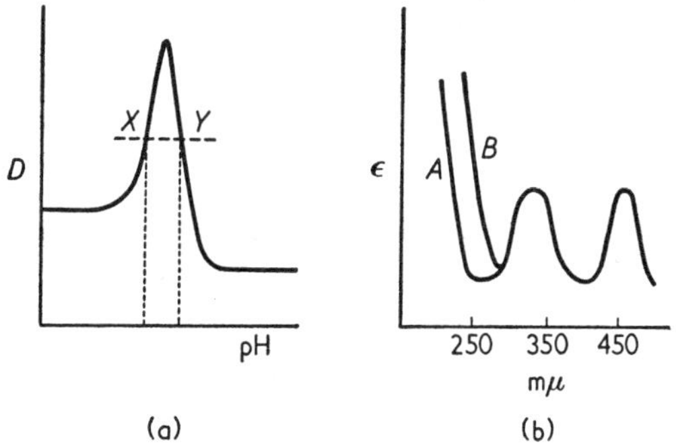

FIG. 10.2 (*a*). Variation of D with pH (Ang[66]); (*b*) Absorption spectra (A) $Co(NH_3)_6(ClO_4)_3$, (B) = (A)+Na_2SO_4.

described a way which can be seen from Fig. 10.2. On plotting D against pH, the curve shows a maximum or minimum. Consider the line XY which cuts the curve at a specific value of D. In general,

$$\epsilon C = D = \epsilon_1[H_2L] + \epsilon_2[HL^+] + \epsilon_3[R^{2-}] \quad (10.16)$$

$$K_1 = [HL^+]f_2a_H/[H_2L]f_1 = k_1f_2/f_1 \quad (10.17)$$

$$K_2 = [L^{2-}]f_3a_H/[HL^+]f_2 = k_2f_3/f_2 \quad (10.18)$$

whence

$$a_H^2 = (\epsilon - \epsilon_1) + a_Hk_1(\epsilon - \epsilon_2) + k_1k_2(\epsilon - \epsilon_3) \quad (10.18a)$$

Measurements in strongly acidic and alkaline solutions give ϵ_1 and ϵ_3. If now a'_H and a''_H refer to X and Y, and $\epsilon = \epsilon_2 - k_2P = \epsilon_1 + k_1Q$, where

$$P = (\epsilon - \epsilon_3)(a'_H + a''_H)/a'_Ha''_H \text{ and } Q = (\epsilon_2 - \epsilon)/(a'_H + a''_H) \quad (10.18b)$$

P is directly calculable, and a plot of ϵ against P gives k_2 from its slope and ϵ_2 at the intercept. Thus Q can be found, and a plot of ϵ against Q has a slope which gives k_1.

Example 10.4. The absorptions of a 1.608×10^{-4} M solution were measured at 25°C, 248 mμ and $I = 0.014$ over a pH range of 2.7 to 4.9 in formate and acetate buffers.[66] From excess acid and excess alkali solutions, $\epsilon_1 = 2193$, $\epsilon_3 = 2911$. In the Table, pH_1 and pH_2 refer to X and Y of Fig. 10.2

pH_1	2.90	3.00	3.10	3.20	3.30	3.40	3.50	3.60	3.70
pH_2	4.90_5	4.845	4.78	4.715	4.635	4.543	4.455	4.375	4.277
ϵ	2302	2323	2348	2374	2406	2442	2475	2504	2529

These give $pk_1 = 3.65$, $pk_2 = 4.44$, whence $pK_1 = 3.70$, $pK_2 = 4.60$ ($\pm$ 0.02). The pK_as of periodic acid [67] and [68] H_2S have also been obtained from the absorptions of their anions.

Using pH Indicators

Von Halban and Brüll,[69] to obtain K_{HIO_3}, matched solutions of the acid containing 2 : 4 dinitrophenol against solutions of HCl containing the same concentration of the indicator. This is a device that has general application when the acid anion does not have an absorption band. If a_1 and a_2 are the respective molarities of HIO_3 and HCl, b is the molarity of indicator, x is the undissociated HIO_3, and y_1 and y_2 are the In^- concentrations in the HIO_3 and HCl solutions, then

$$K^{HIn} = [H^+][In^-]f_\pm^2/[HIn] = (a_2 + y_2)y_2 f_\pm^2/(b - y_2) \qquad (10.19)$$

and at the balance-point,

$$a_1 - x + y_1 = a_2 + y_2 \qquad (10.20)$$

Since the ratio $y/(b - y)$ is the same for both solutions, $y_1 = y_2$, but $f_\pm$ is not necessarily the same. Equality can be assumed if the solutions of the two acids are dilute, but it is not true if buffer solutions are used for the reference or when the acid anions are of different valencies. Consequently $[H^+]$ is not of necessity identical in the two solutions.

Example 10.5. A solution of chloroacetic acid in 20 % ethanol [65] (a_1) at 25° C was balanced spectrophotometrically against HCl (a_2) in the same solvent, using 2 : 4 dinitrophenol (b) as the indicator. When $\Delta D = 0$, $10^4 a_1 = 24.76$, $10^4 a_2 = 10.11$, $10^5 b = 40.8$. Since [65] $K_{HIn} = 8.25 \times 10^{-5}$, with $\log f_\pm = 0.647\{\sqrt{I}/(1 + \sqrt{I}) - 0.2\, I\}$, and $I = a_2 + y$, from (10.19), $y = 1.23 \times 10^{-5}$, so from (10.20), $10^4 x = 14.65$, whence $K_a = 6.46 \times 10^{-4}$.

Sulphuric acid in water,[70,71] and sulphuric acid and propionic acid in 20 % ethanol [65] have been studied in a similar fashion.

It is also possible to study salt association with the aid of pH indicators if the acid of the anion or the hydroxide of the cation does not dissociate completely. For example,[38] using 2 : 4 dinitrophenol at 390 mμ in mixtures of

H_2SO_4 with $MgCl_2$ matched against mixtures of HCl and $MgCl_2$ (both in 20 % ethanol), from (10.19), $[H^+] = a_1 + y = (a_2 + y)f_2^2/f_1^2$, where f_1 and f_2 refer to the test and reference solutions, so $x = [HSO_4^-]$ is obtained from (10.20), and since $K_{HSO_4} = 0.00238$ in this solvent,[70] $[SO_4^{2-}]$ is obtained, and $[MgSO_4]$ is obtained by difference. In this way, $K_{MgSO_4} = 0.0024_5$ at 25°C. Since I is not known in the first instance, preliminary measurements are required to find the molarity of $MgCl_2$ to put in the reference solution. A possible source of uncertainty is association between Mg^{2+} and In^- (the indicator anion).

Spectrophotometry has proved to be of great importance for studying the dissociation of very weak acids and bases in acidic media.[76] Such solvents are termed "amphiprotic"; they can accept protons from acidic solutes, to form "lyonium" ions or can donate protons to basic solutes, to form "lyate" solvent ions.[72] Acidity in terms of basic indicator solutes has been formulated by Hammett [73] in terms of an acidity function, H_0, which is defined by

$$H_0 = - \log a_{H^+} f_B/f_{BH^+} = pK_a^B + \log (C_B/C_{BH^+}) \qquad (10.21)$$

where B is the indicator base.

The development and application of this subject are far too large to cover in the present book; Hammett [74] himself has given expositions, and so have others.[75]

A related topic which has been carefully examined by Kolthoff and Bruckenstein [77] is the ionisation of acids and bases in solvents of very low dielectric constant such as glacial acetic acid ($\epsilon_{d.c.} = 6.13$ at [78] 25°C). Electrolytes which are strong in water are weak in this solvent.

These authors differentiate between (a) ionization, and (b) dissociation; thus if HX is an acid and B a base,

$$HX + HAc \overset{a}{\rightleftharpoons} H_2Ac^+ . X^- \overset{b}{\rightleftharpoons} H_2Ac^+ + X^- \qquad (10.22)$$

$$B + HAc \overset{a}{\rightleftharpoons} BH^+ . Ac^- \overset{b}{\rightleftharpoons} BH^+ + Ac^- \qquad (10.23)$$

In common with other solvents, the extent of proton solvation may be unknown, so H^+ will be written for H_2Ac^+. Then the respective constants are

$$\text{(a)} \quad K_i^{HX} = a_{H^+X^-}/a_{HX}\,; \qquad K_i^B = a_{BH^+Ac^-}/a_B \qquad (10.24)$$

$$\text{(b)} \quad K_d^{HX} = a_{H^+}a_{X^-}/a_{H^+X^-}\,; \qquad K_d^B = a_{BH^+}a_{Ac^-}/a_{BH^+Ac^-} \qquad (10.25)$$

whence the overall dissociation constants are

$$K_{HX} = a_{H^+}a_{X^-}/(a_{XH} + a_{H^+}a_{X^-}) = K_i^{HX}K_d^{HX}/(1 + K_i^{HX}) \qquad (10.26)$$

$$K_B = a_{BH^+}a_{Ac^-}/(a_B + a_{BH^+Ac^-}) = K_i^B K_d^B/(1 + K_i^B) \qquad (10.27)$$

A suitable indicator for studying acids and bases in acetic acid is p.naphthol-benzein (PNB), which is a weak base (I) in this solvent. The equilibria

$$I + HX \overset{a}{\rightleftharpoons} IH^+ . X^- \overset{b}{\rightleftharpoons} IH^+ + X^- \qquad (10.28)$$

must be considered and confining conditions to low concentrations,

$$K_i^{\text{IHX}} = [\text{IH}^+\text{H}^-]/[\text{I}][\text{HX}] = K^{\text{HX}}K_{\text{I}}/K_d^{\text{IHX}}K_s \qquad (10.29)$$

where K_s is the autoprotolysis constant of acetic acid, $[\text{H}^+][\text{Ac}^-]$.

If HY represents a second acid, $R_1 = K_i^{\text{IHX}}/K_i^{\text{IHY}}$, and $R_2 = K_i^{\text{BHX}}/K_i^{\text{BHY}}$, then

$$R_1/R_2 = K_d^{\text{IHY}}K_d^{\text{BHX}}/K_d^{\text{IHX}}K_d^{\text{BHY}} \qquad (10.30)$$

In general, $R_1/R_2 \neq 1$, so when obtaining the relative strengths of acids in solvents of low dielectric constant by indicators, the reference indicator base must be stated. This reasoning shows that the Hammett acidity function is, as Hammett has stated, restricted to solvents of high dielectric constant, for in solvents such as acetic, it depends on the choice of indicator.

The equations for evaluating the various equilibria are derived from

$$[\text{IH}^+\text{X}^-] + [\text{IH}^+] = \sum[\text{IH}^+] \text{ (measured by spectrophotometer)} \qquad (10.31)$$

$$[\text{I}] = (C_{\text{ind.}})_t - \sum[\text{IH}^+] \qquad (10.32)$$

where $(C_{\text{ind.}})_t$ is the total indicator concentration. Likewise,

$$(C_{\text{HX}})_t = \sum[\text{IH}^+] + [\text{HX}] + [\text{H}^+\text{X}^-] + [\text{H}^+] = \sum[\text{IH}^+] + C^{\text{HX}} \qquad (10.33)$$

whence C_{HX} is found. For weak acids such as HCl and p . CH_3 . C_6H_4 . SO_3H (HTs), $C_{\text{HX}} = [\text{HX}]$, and in such solutions, from electroneutrality, $[\text{H}^+] + [\text{IH}^+] = [\text{X}^-]$, and from this and the preceding equations,

$$\sum[\text{IH}^+]/([\text{I}]\sqrt{C_{\text{HX}}}) = K_i^{\text{IHX}}\sqrt{C^{\text{HX}}} + K_i^{\text{IHX}}K_d^{\text{IHX}}(K_{\text{HX}} + K_i^{\text{IHX}}K_d^{\text{IHX}}[\text{I}])^{-\frac{1}{2}} \qquad (10.34)$$

Varying amounts of PNB were added to a solution containing a fixed $(C_{\text{HX}})_t$, $\sum[\text{IH}^+]$ plotted against [I], and $\sum[\text{IH}^+]$ values interpolated at selected values of [I]. This was repeated with several $(C_{\text{HX}})_t$ values, and a Table thereby constructed giving $\sum[\text{IH}^+]$ for a given [I] at the various values of $(C_{\text{HX}})_t$. Now when [I] is constant, (10.34) can be represented by $y = mx + b$, where y is the left-hand side, $m = K_i^{\text{IHX}}$, $x = \sqrt{C_{\text{HX}}}$, b is the right-hand term and is the intercept of the plot of y against x; its value depends on the chosen [I]. It follows that

$$[\text{I}]b^2 = -\, b^2K_{\text{HX}}/(K_i^{\text{IHX}}K_d^{\text{IHX}}) + K_i^{\text{IHX}}K_d^{\text{IHX}} \qquad (10.35)$$

which can be represented by $y' = m'x' + b'$, where $b' = K_i^{\text{IHX}}K_d^{\text{IHX}}$. Since K_i^{IHX} is known from the slope of the plot of y against x, then K_d^{IHX} and K_{HX} are derived from the slope and intercept of the plot of x' against y'.

Example 10.6. The basic form of PNB has one absorption maximum at 454 mμ, the acid form has two, at 464 and at 625 mμ. All measurements were made at 625 mμ to avoid overlap; $[\text{IH}^+]$ was found via $D = \epsilon_{\text{IH}^+}[\text{IH}^+]d$, where $\epsilon_{\text{IH}^+} = 2.86 \times 10^4$ from measurements of 1.30×10^{-5} M PNB in

0.02 M $HClO_4$. Other measurements in HBr, HCl, etc., gave the same ϵ_{IH}, showing the nature of X^- has no influence.
These give $K_{HX} = 2.8 \times 10^{-9}$, $K_i^{IHX} = 1.3 \times 10^2$, $K_d^{IHX} = 3.9 \times 10^{-6}$. By adding LiCl, the dissociation of IH^+Cl^- is repressed and by plotting $\sum[IH^+]/[I]$ against $-\log C_{LiCl}$, the minimum was found where IH^+Cl^- dissociation was

TABLE 10.5

Interaction of HCl with PNB in 99.965% Acetic Acid[77]; $10^6\Sigma[IH^+]$ Values

10^3HCl	10^6PNB = 1.82	3.64	5.46	7.28	9.08	10.87
0.663	0.427	0.798	1.11	1.44	1.73	2.00
1.326	0.596	1.16	1.66	2.11	2.58	2.98
2.05	0.792	1.56	2.26	2.98	3.61	4.24
5.31	1.07	2.07	3.05	4.00	4.92	5.81

TABLE 10.6

Interpolated Values of $10^6\Sigma[IH^+]$

10^3HCl	10^6[I] = 2.00	3.00	4.00	5.00	6.00
5.31	2.55	3.70	4.75	5.75	—
2.65	1.51	2.17	2.80	3.38	3.91
1.33	0.94	1.35	1.71	2.05	2.40
0.663	0.58	0.83	1.06	1.28	1.48
Slope of (10.34)	128	128	125	124	124
Intercept of (10.34)	9.00	7.50	7.15	6.80	6.45

negligible and ionic aggregates not yet formed. Using this value of $C_{LiCl} = 0.01$, a plot of $\log \sum[IH^+]/[I]$ against $-\log C_{HCl}$ should be a straight line of -1.00. This is so and from (10.29), the figures give $K_i^{IHCl} = 1.2 \times 10^2$, comparable with 1.3×10^2 derived in the absence of salt. The ionic strengths are different and these K values are based on concentrations.

With stronger acids like $HClO_4$ it was found the $\sum[IH^+]/[I]C_{HX}$ was independent of $(C_{HX})_t$ and of $NaClO_4$. This means that $K_d^{IHX} = K_{HX}$ and it can be shown that

$$K_i^{HClO_4} = (K_i^{IHX}K_d^{IHX}[I]C_{HX})/K_{HX}\sum[IH^+] \qquad (10.36)$$

where C_{HX} is given by (10.33). Since the K values on the right-hand side are known for HCl (and HTs), $K_i^{HClO_4}$ is found to be 0.8 via HCl, 1.0 via HTs. HBr proved to be intermediate between HCl and $HClO_4$, so no estimate of K_i^{HBr} could be made.

Colourless bases B of suitable strength can be studied. In the presence of $(C_B)_t$ and $HClO_4(= HX)$,

$$[BHX] + [BH^+] = \sum[BH^+] = (C_{HX})_t - C_{HX} - \sum[IH^+] \quad (10.37)$$

$$[B] = (C_B)_t - \sum[BH^+] \quad (10.38)$$

$$\sum[BH^+]/[B]C_{HX} = K_i^{BHX}/(1 + K_i^{HX}) = K_i^{BHX} \text{ since } K_i^{HClO_4} = 1.0 \quad (10.39)$$

Using HCl + 0.01 LiCl, $\sum[IH^+]/[I]$ gives [HX] and so is obtained $K_i^{BHX} = [BHX]/([B][HX])$

Acid (HX)	$HClO_4$	$HClO_4$	$HClO_4$	$HClO_4$	$HClO_4$	$HClO_4$	HTs	HTs	HCl	HCl
Base	PNB	Urea	H_2O	2 PrOH	EtOH	MeOH	PNB	Urea	PNB	Urea
K_i^{BHX}	2×10^5	1.6×10^5	68	17	15	8.8	370	1200	1300	490

Now $K_i^{IHClO_4}/K_i^{IHTs} = 540$, while $K^{BHClO_4}/K_i^{IHTs} = 140$, whence R_1/R_2 of (10.30) is 3.9. Similarly with HCl, $R_1/R_2 = 4.6$. These (and other comparisons) illustrate the remarks made earlier regarding acidity functions.

Evans and his associates [79] attributed the green-yellow colour which forms when triarylmethyl chlorides are dissolved in nitroalkanes to the partial change of $R_3C.Cl$ to $R_3C^+.Cl^-$ ion-pairs, and they have evaluated $K = [R_3C^+.Cl^-]/[R_3C.Cl]$ by spectrophotometry; the concentration of the ion-pair was evaluated by reference to the absorption of $R_3C.OH$ in concentrated H_2SO_4, in which the alcohol ionises completely. Pocker [80] has given a different interpretation. A solution of triphenyl methane chloride in nitromethane develops HCl, and the addition of Et_4NCl discharges the colour while Et_4NClO_4 has no effect. The ratio $[HCl]^2/[Ph_3C.Cl]$ was found to be constant, suggesting that $Ph_3C.Cl + CH_3.NO_2 = X + HCl$ occurs. The Cl^- is largely present as HCl_2^-, and the solutions contain free Ph_3C^+ ions.

Successive Complexity or Stability Constants

Ahrland [81] has obtained the stability constants of the complexes formed between the UO_2^{2+} ion and (a) $CH_2Cl.COO^-$, (b) CNS^-, (c) Ac^-, (d) SO_4^{2-}, (e) Cl^-, Br^-, NO_3^-, (f) $CH_2OH.COO^-$ ions by the spectrophotometric method devised by Fronaeus.[82] In terms of the nomenclature adopted for the treatments given in Chapter 9 and the present chapter,

$$D/d = \epsilon C_M = \epsilon_0[M] + \epsilon_1[ML] + \ldots = \sum_0^N \epsilon_n[ML_n] \quad (10.40)$$

where M refers to the UO_2^{2+} ion. Then from (9.47), with [M] constant,

$$D/C_M d = \epsilon = \sum_0^N \epsilon_n[ML_n] \Big/ \sum_0^N [ML_n] = \sum_0^N \epsilon_n\beta_n[L]^n \Big/ \sum_0^N \beta_n[L]^n \quad (10.41)$$

so that ϵ is some function of [L].

The practical work consists of obtaining value of ϵ with a fixed value of C_M and varying C_L, and drawing a smooth curve through the plot of ϵ against C_L. This is repeated with several values of C_M (which may mean using cells of different d values), all the curves being drawn on the same graph. Next, the curves are cut at selected values of ϵ, and the interpolated values of C_L are plotted against C_M. Extrapolation to $C_M = 0$ gives [L]; this follows from (10.41), for all solutions with the same ϵ must have the same [L]. The line drawn through these points is linear (if the complexes are mononuclear), and from (9.49), the slope of the line gives $\bar{n}$. Thus with [L] and $\bar{n}$ known the methods described in Chapter 9 can be applied for deriving β_n.

Example 10.7. The absorptions at 310 mμ of mixtures of $UO_2(ClO_4)_2$ and Na_2SO_4 containing sufficient $HClO_4$ to prevent hydrolysis and sufficient $NaClO_4$ to maintain $I = 1.0$ were measured by Ahrland.[81d] The C_L values were corrected for HSO_4^- via $[HSO_4^-] = [HClO_4] -$ measured $[H^+]$.

TABLE 10.7

Molar Extinction Coefficients of Uranyl Sulphate Solutions

10^3C_L for $C_M = 0.030$M	24.7	49.4	98.7	148.2	197.9	297.6	
(with $d = 0.1$ cm.)	102.5	136.5	175.1	195.8	207.4	219.9	
10^3C_L for $C_M = 0.010$M	9.8	24.6	49.3	98.8	148.5	198.4	298.1
(with $d = 0.3$ cm.)	87.1	120.8	154.8	187.5	203.3	214.2	224.4
10^3C_L for $C_M = 0.0030$M	9.8	24.6	49.3	98.9	148.7	198.5	298.3
(with $d = 1.00$ cm.)	93.7	129.0	161.0	191.3	206.0	216.0	226.3

The values of C_L for selected values of ϵ, together with values of [L], i.e., the extrapolated values of C_L at $C_M = 0$, and the values of $\bar{n}$ from slopes, are shown in Table 10.8.

TABLE 10.8

Interpolated Values of C_L, and Values of [L] and $\bar{n}$

$C_M =$	0.03	0.01	0.003	$0 = [L]$	$\bar{n}$
ϵ		10^3C_L			
105	26.3	16.6	13.6	12.1	0.47
130	44.2	30.0	25.1	22.9	0.71
160	74.2	55.1	48.0	45.2	0.97
190	131.6	105.1	95.6	91.6	1.33
205	186.0	155.5	144.2	139.6	1.57

In order to apply the method of Fronaeus, which is based on (9.79), $\bar{n}/[L]$ is plotted against $\bar{n}$, or preferably, $\bar{n}$ is plotted against ln [L], and values of C for values of [L] found by graphical integration. Since [L] does not extend to low concentrations, Ahrland evaluated $C/C_{0.01}$ from the graph areas, and by plotting these against [L] as abscissa, obtained C at [L] = 0, namely 0.62. Thus values of C for various values of [L] are known, and from a plot of (C — 1)/[L] against [L], $\beta_1 = 56$. Likewise a plot of $(C - 1 - \beta_1[L])/[L]^2$ against [L] leads to $\beta_2 = 480$. Some of the integrated areas are listed below:

10^3[L]	0	10	20	40	60	100	140
$C/C_{0.01}$	(0.62)	1.00	1.44	2.49	3.745	7.00	11.57

The method of Rossotti and Rossotti (see 9.82), when applied to Table 10.7, gives $\beta_1 = 50$, $\beta_2 = 550$, approx. Burns and Hume [85] have applied this treatment to derive the formation constants k_n (see 9.46) of the In^+, Br_n^- complexes, where $n = 1$ to 4 at $I = 4.0$.

Newman and Hume [83] have derived general equations for determining the successive formation constants of single and mixed ligand complexes. They have applied these to some bismuth and vanadium systems.[84] Space considerations prevent their treatments from being reproduced here, there being at least twenty-five different equations, according to the number of species and the number which have absorptions.

REFERENCES

1. "International Critical Tables." McGraw-Hill Book Co., Inc., New York (1929). "Landolt-Bornstein Tabellen." Springer-Verlag, Berlin (1927). "U.V. and Visible Absorption Spectra, 1930-1954," H. M. Hershenson. Academic Press, New York and London. *Ann. Rev. Phys. Chem.*, 1-10 (1950-9). Ann. Rev. Inc., Stanford, U.S.A.
2. Orgel, L. *Quart. Rev.* **8**, 422 (1954).
3. "The Transuranium Elements." National Nuclear Energy Series, Div. IV, **14B** (1954). McGraw-Hill Book Co., Inc., New York and London. W. C. Waggener, *J. phys. Chem.* **62**, 382 (1958) Onstott, E. I. and Brown, C. J. *Anal. Chem.* **30,** 172 (1958). Bayer-Helms, F. *Z. Naturf.*, **13A**, 161 (1958). B. Jezowska-Trzebiatowska, *et. al.*, *Nucleonika*, **3,** Spec. No. 39 (1958).
4. Moeller, T. and Brantley, J. C. *Anal. Chem.* **22,** 433 (1950). Hara, T. *Bunseki Kagaku*, **6,** 823 (1957).
5. Lothian, G. F. "Absorption Spectrophotometry." Hilger and Watts, Ltd., London (1958).
6. Mellon, M. G. "Analytical Absorption Spectroscopy." J. Wiley and Sons. Inc., New York; Chapman and Hall, Ltd., London (1950). E. Luscher, *Applied Spectroscopy*, **12,** 172 (1958).
7. Gibson, K. S. "Photoelectric Cells and their Applications." Cambridge University Press, Cambridge (1930).
8. Kortüm, G. "Kolorimetrie, Photometrie und Spektometrie." Springer-Verlag, Berlin, Band II (1955).
9. Goldring, L. S., Hawes, R. C., Hare, G. H., Beckman, A. O. and Stickney, M. E. *Anal. Chem.* **25,** 869 (1953).
10. Kortüm, G. and von Halban, H. *Z. phys. Chem.* **170A,** 212 (1934).

11. Hogness, T. R., Zschiele, F. P. and Sidwell, A. E. *J. phys. Chem.* **41,** 379 (1937). Zschiele, F. P. *ibid.* **51,** 903 (1947).
12. Hesketh, R. V. *J. sci. Instrum.* **35,** 371 (1958).
13. "Glass Color Filters." Corning Glass Works, U.S.A. (1948).
14. Davies, W. G. and Prue, J. E. *Trans. Faraday Soc.* **51,** 1045 (1955).
15. Thomas, G. O. and Monk, C. B. *ibid.* **52,** 685 (1956).
16. King, W. H. *J. opt. Soc. Amer.* **43,** 866 (1953). Campbell, H. and Simpson, J. A. *Chem. and Ind.* 887 (1953). Evans, R. F., Herington E. F. G. and Kynaston, W. *Trans. Faraday Soc.* **48,** 1284 (1953). Dixon, M. *Biochem. J.* **58,** 1 (1954). Grant, J. K. *Chem. and Ind.* 942 (1955) Eaborn, C. *J. chem. Soc.* 4858 (1956). Sweet, T. R. and Zehner, J. *Anal. Chem.* **30,** 1713 (1958).
17. Nasanen, R. *Acta Chem. Scand.* **3,** 179 (1949).
18. Bale, W. D., Davies, E. W. and Monk, C. B. *Trans. Faraday Soc.* **52,** 816 (1956).
19. Popov, A. I. and Humphreys, R. E. *J. Amer. chem. Soc.* **81,** 2043 (1959).
20. Smith, M. and Symons, M. C. R. *J. chem. phys.* **25,** 1074 (1956). *Disc. Faraday Soc.* **24,** 206 (1957); **54,** 338, 346 (1958).
21. Bethe, H. *Ann. Phys.* **3,** 133 (1929). Orgel, L. E. *J. chem. Soc.* 4756 (1952). Griffith J. S. and Orgel, L. E. *Quart. Rev.* **11,** 381 (1957).
22. Hamer, W. J. "The Structure of Electrolytic Solutions." J. Wiley and Sons, Inc., New York; Chapman and Hall, Ltd., London (1959). Article by J. F. Duncan and D. L. Kepert.
23. Linnett, J. W. *Disc. Faraday Soc.* **26,** 7 (1958) (and other articles in this). Sutton, L. E. *J. Inorg. and Nuclear Chem.* **8,** 23 (1958). Schaffer, C. E. and Jørgensen, C. C. *loc. cit.* 143, 149.
24. Bjerrum, N. *Z. anorg. Chem.* **63,** 140 (1909). See Guggenheim, E. A. *Proc. chem. Soc.* 104 (1960).
25. Hantzsch, A. and Clark, R. *Ber. dtsch. chem. Ges.* **41,** 1220 (1908). Hantzsch, A. and Garrett, G. S. *Z. phys. Chem.* **84,** 323 (1913).
26. von Halban, H. and others. *Z. phys. Chem.* **96,** 214 (1920); **100,** 208 (1922); 112, 321 (1924); **122,** 377 (1926); **132,** 401, 433 (1928); **170A,** 351 (1934); **173A,** 449 (1935); **181A,** 70 (1937). *Trans. Faraday Soc.* **21,** 620 (1926). *Helv. Chim. Acta,* **21,** 385 (1938); **27,** 1719 (1944).
27. Fromherz, H. and others. *Z. phys. Chem.* **7B,** 439 (1930); **153A,** 321 (1931); **167A,** 103 (1933); **171A,** 371 (1934); **178A,** 29 (1936); **43B,** 289 (1939).
28. Kortüm, G. and von Halban, H. *ibid.* **170A,** 212, 351 (1934). Kortüm, G. *ibid.* **33B,** 343 (1936); **42B,** 448 (1939).
29. Job. P. *C.R. Acad. Sci., Paris,* **180,** 928 (1925). *Ann. chim.* [10], **9,** 113 (1928); [11]. **6,** 97 (1936). Vosburgh, W. C. and Cooper, G. R. *J. Amer. chem. Soc.* **63,** 437 (1941), Katzin, I. I. and Gebert, E. *ibid.* **72,** 5455 (1950). Woldbye, F. *Acta Chem. Scand.* **9,** 299 (1955).
30. Bent, H. E. and French, C. L. *J. Amer. chem. Soc.* **63,** 568 (1941).
31. Jones, M. M. and Innes, K. K. *J. phys. Chem.* **62,** 1005 (1958).
32. Moore, R. L. and Anderson, R. C. *J. Amer. chem. Soc.* **67,** 167 (1945).
33. Hershenson, H. M., Smith, M. E. and Hume, D. N. *ibid.* **75,** 507 (1953). Biggs, A. I. Parton, H. N. and Robinson, R. A. *ibid.* **77,** 5844 (1955).
34. Righellato, E. C. and Davies, C. W. *Trans. Faraday Soc.* **23,** 351 (1927).
35. See Ch. 9.
36. Davies, E. W. and Monk, C. B. *J. Amer. chem. Soc.* **80,** 5032 (1958).
37. Gimblett, F. G. R. and Monk, C. B. *Trans. Faraday Soc.* **51,** 793 (1955). Thomas, G. O. and Monk, C. B. *ibid.* **52,** 685 (1956).

38. Davies, E. W. and Monk, C. B. *ibid.* **53,** 442 (1957). Bale, W. D., Davies, E. W., Morgans, D. B. and Monk, C. B. *Disc. Faraday Soc.* **24,** 94 (1957).
39. Saini, G. and Ostacoli, G. *J. Inorg. and Nuclear Chem.* **8,** 346 (1958).
40. Yiu, N. *Bull. Inst. phys. chem. Res., Tokyo* **20,** 390 (1941).
41. McConnell, H. and Davidson, N. *J. Amer. chem. Soc.* **72,** 3165 (1950).
42. Edwards, S. M. and Birnhaum, N. *ibid.* **63,** 1471 (1941). Lewin, S. Z. and Wagner, R. S. *J. chem. Educ.* **30,** 445 (1953).
43. Rabinowitch, E. and Stockmayer, W. H. *J. Amer. chem. Soc.* **64,** 335 (1942). Coll, H., Nauman, R. V. and West, P. W. *ibid.* **81,** 1284 (1959).
44. Senisse, P. and Perrier, M. *ibid.* **80,** 4194 (1958).
45. Phipps, A. L. and Plane, R. A. *ibid.* **79,** 2458 (1957).
46. Linhard, M. and others: *Z. Elektrochem.* **50,** 224 (1944). *Z. anorg. Chem.* **262,** 328 (1950). **264,** 321 (1951); **266,** 49, 73, 113, 121 (1951); **271,** 101, 131 (1952); **278,** 287 (1955). *Z. phys. Chem.* N.F., **5,** 20 (1955).
47. Cohen, S. R. and Plane, R. A. *J. phys. Chem.* **61,** 1096 (1957).
48. Ibers, J. I. and Davidson, N. *J. Amer. chem. Soc.* **73,** 476 (1951).
49. Evans, M. G. and Nancollas, G. H. *Trans. Faraday Soc.* **49,** 358 (1953). King, E. L., Espenson, J. H. and Visco, R. E. *J. phys. Chem.* **63,** 755 (1959).
50. Davies, W. G., Otter, R. J. and Prue, J. E. *Disc. Faraday Soc.* **24,** 103 (1957).
50a. Panckhurst, M. M. and Woolmington, K. G. *Proc. roy. Soc.* **244A,** 124 (1958).
51. Davies, W. G. and Prue, J. E. *Trans. Faraday Soc.* **51,** 1045 (1955).
52. Newton, T. W. and Baker, F. B. *J. phys. Chem.* **61,** 934 (1957).
53. Foley, R. T. and Anderson, R. C. *J. Amer. chem. Soc.* **71,** 909, 912 (1949).
54. Newton, T. W. and Arcand, G. M. *ibid.* **75,** 2449 (1953).
55. Whiteker, R. A. and Davidson, N. *ibid.* **75,** 3081 (1953).
56. Taube, H. and Posey, F. *ibid.* **78,** 15 (1956).
57. Gates, H. S. and King, E. L. *ibid.* **80,** 5011, 5015 (1958).
58. Klotz, I. M. and Ming, L. *ibid.* **75,** 4159 (1953).
59. Pelletier, S. *C.R. Acad. Sci., Paris,* **247,** 1187 (1958).
60. Flexer, L. A., Hammett, L. P. and Dingwall, A. *J. Amer. chem. Soc.* **57,** 2103 (1935).
61. Edwards, L. J. *Trans. Faraday Soc.* **46,** 723 (1950); **49,** 234 (1953).
62. Biggs, A. I. *ibid.* **50,** 800 (1954); **52,** 35 (1956).
63. Robinson, R. A. and King, A. K. *ibid.* **51,** 1398 (1955); **52,** 327 (1956). Robinson, R. A. Ref. 22.
64. Robinson, R. A. and Biggs, A. I. *Aust. J. Chem.* **10,** 128 (1957).
65. Bale, W. D. and Monk, C. B. *Trans. Faraday Soc.* **53,** 450 (1957).
66. Ang, K. P. *J. phys. Chem.* **62,** 1109 (1958).
67. Crouthamel, C. E. *et al. J. Amer. chem. Soc.* **71,** 3031 (1949); **73,** 82 (1951). Shiner, V. J. and Wasmuth, C. R. *ibid.* **81,** 37 (1959).
68. Ellis, A. J. and Golding, R. M. *J. chem. Soc.* **127** (1959).
69. von Halban, H. and Brüll, J. *Helv. Chim. Acta,* **27,** 1719 (1944).
70. Ref. 20, Ch. 9.
71. Ref. 2, Ch. 1.
72. Bjerrum, N. *Chem. Rev.* **16,** 287 (1935).
73. Hammett, L. P. and Deyrup, A. J. *J. Amer. chem. Soc.* **54,** 2721 (1932).
74. Hammett, L. P. *Chem. Rev.* **16,** 67 (1935). "Physical Organic Chemistry." McGraw-Hill Book Co., Inc., New York and London (1940).
75. Bell, R. P. "Acids and Bases." Methuen and Co., Ltd.; J. Wiley and Sons, Inc. (1952). Paul M. A. and Long F. A. *Chem. Rev.* **56,** 1 (1957). Bascombe, K. N. and Bell, R. P. *J. chem. Soc.* **158,** 935 (1957).

76. Bascombe, K. N. and Bell, R. P. *ibid.* 1096 (1959).
77. Kolthoff, I. M. and Bruckenstein, S. *J. Amer. chem. Soc.* **78,** 1, 10 (1956); **70,** 1 (1957).
78. Timmermans, J. "Physico-Chemical Constants of Pure Organic Compounds." Elsevier Publishing Corp., New York (1950).
79. Evans, A. G. and others, *Trans. Faraday Soc.*, **47,** 711 (1951); **50,** 16, 470, 568. *J. chem. Soc.*, 3803 (1954); 1020 (1957).
80. Pocker, Y. *ibid.* 240 (1958).
81. Ahrland, S. *Acta Chem. Scand.* (a) **3,** 783 (1949); (b) **3,** 1067 (1949); (c) **5,** 199 (1951); (d) **5,** 1151 (1951); (e) **5,** 1271 (1951); (f) **7,** 485 (1953).
82. Fronaeus, S. "Komplexsystem hos kopper." Thesis, Lund (1948).
83. Newman, L. and Hume, D. N. *J. Amer. chem. Soc.* **79,** 4571 (1957).
84. Newman, L. and Hume, D. N. *ibid.* **79,** 4576, 4581; **80,** 4491 (1958); **81,** 547 (1959).
85. Burns, E. A. and Hume, D. N. *ibid.* **79,** 2704 (1957).
86. Kruh, R. *ibid.* **76,** 4865 (1954).

CHAPTER 11

INCOMPLETE DISSOCIATION, PART IV: METHODS BASED ON SOLUBILITIES, FREEZING-POINTS, ION-EXCHANGE RESINS AND SOLVENT EXTRACTION

Solubilities

THE solubilities of sparingly soluble salts in certain solvent salt solutions, as compared with their solubilities in the pure solvent, may be much greater than can be accounted for in terms of activity coefficient effects (see Chapter 5). These enhanced solubilities are particularly noticeable when the solutions contain both high valence cations and anions, and when the sparingly soluble salt or the solvent salt contributes ions with tendencies to form ion-association products. Davies [1,2] has demonstrated that solubility measurements under these conditions can give quantitative information on the association equilibria, and this is in agreement with that obtained from conductances, e.m.f.'s, etc. The following example illustrates the method of calculation.

Example 11.1. The solubility of $Ca(IO_3)_2$ in water, NaCl and Na_2SO_4 solutions at 25°C (in mM) is:[2]

NaCl	0	12.50	25.0	50.0	100.0	Na_2SO_4:	6.25	12.50	18.75	25.00
Soly.	7.840	8.285	8.676	9.287	10.42		8.898	9.745	10.45	11.05

From (5.1) and (3.43), assuming that no ion-pairs form, extrapolation of a plot of log K_s against I gives

$$K_s = [Ca^{2+}][IO_3^-]^2 f_\pm^3 = 7.881 \pm 0.028 \times 10^{-7} \text{ g. ions}^3/l^3 \qquad (11.1)$$

The difference between the calculated values of $[Ca^{2+}]$ in the Na_2SO_4 solutions, and the measured values is attributed to $CaSO_4$ ion-pairs. Thus with the first Na_2SO_4 solution, the first approximation of $I = 3$ (Soly.) + $3(Na_2SO_4) = 0.04544$, whence $CaSO_4 = 0.008898 - 0.007861$. Thus a truer value of $I = 0.04139$, and by repetition, $CaSO_4$ (final) $= 0.001506$, whence $K_{CaSO_4} = 0.0054_4$. The values from the other data are 0.00546, 0.00533, 0.00524, and assuming the slight drift is due to various causes, at $I = 0$, $K = 0.0055$.

In a more accurate analysis, allowance should be made for [2] $K_{CaIO_3} = 0.13$, $K_{NaIO_3} = 3.0$, $K_{NaSO_4^-} = 0.2$; by doing this, $K_{CaSO_4} = 0.0053$.

A number of iodates have been used for such calculations; their solubilities are sufficiently low to permit a range of solvent salt concentrations to be covered while still keeping I below 0.1, the limit of the general form of the D.H. equation (3.43). In addition, they are rather strong electrolytes so that secondary ion-pairs are kept to low concentrations, and the solubilities can be determined accurately by titration. Examples are (a) [1, 3, 4, 5] $Ca(IO_3)_2$, $Ba(IO_3)_2$ and $Sr(IO_3)_2$ with sodium carboxylates (for ML^+), NaOH (for MOH^+) and $Na_2S_2O_3$ (for MS_2O_3); (b) $TlIO_3$, TlCl and TlBr[6] for ion-association of Tl^+ with OH^-, Cl^-, Br^-, NO_3^-, CNS^- and N_3^-; (c) $Cu(IO_3)_2$ with [7] amino-acid and carboxylates of sodium (for CuL^+ and CuL_2); (d) $AgBrO_3$ with [8] amino-acid anions (for AgL and AgL_2^-) and Ac^- in mixed solvents; (e) BaS_2O_3 with [9] MCl_2 solutions (for MS_2O_3); (f) $La(IO_3)_3$ with [1] Na_2SO_4 (for $LaSO_4^+$).

1. Successive Stability Constants

Since solubility measurements determine C_M, and [M] is deduced from the solubility product, Leden's method, which utilizes (9.60), can be applied.

Example 11.2. The β_n values of $CdCl_n^{n-2}$ where $n = 1$, 2 and 3 have been calculated at $I = 3.0$ from [10] the solubility of cadmium ferricyanide in NaCl, $NaClO_4$, $HClO_4$ solutions. The solubilities were found by spectrophotometric determination of $[Fe(CN)_6^{3-}] = 2S$, whence $C_M = 3S$. Also $[M]^3 = K_s/2S^2f_\pm^5$. From the solubility in 2.96 M $NaClO_4$ + 0.04 M $HClO_4$, $K_s/f_\pm^5 = 1.406 \times 10^{-19}$. Since $S \ll Cl^-$, [L] = [NaCl].

TABLE 11.1

Solubility Data for β_n Values of Cadmium Chloride

NaCl (M)	0	0.020	0.040	0.080	0.160	0.480	0.800
$S(\times 10^5)$	6.65	8.65	10.8	17.6	27.9	89	182
A	27.5	31.1	50.8	61.9	89	155	309
$(A - \beta_1)/[L]$	—	—	—	210	217	265	350
$(A - \beta_1 - \beta_2[L])/[L]^2$	—	—	—	—	240	240	250

From the data of Table 11.1, and (9.60), the plot of A against [L] extrapolates to $\beta_1 = 28 \pm 3$, and from the last two lines, $\beta_2 = 150$, $\beta_3 = 250$. Owing to scatter, the first three solubilities are unsuitable for finding β_2 and β_3.

Another example which makes use of Leden's method is [11] the determination of β_1, β_2 and β_3 of $PbAc_n$ from the solubility of $PbSO_4$ in NaAc, HAc solutions. Allowance was made for HSO_4^-, but not for undissociated $PbSO_4$. Tyrrell and Scaife [12] have also obtained formation constants at constant ionic strength; they have evaluated the constants of the $HgBr_3^-$ and $HgBr_4^{2-}$ ions from the solubility of $HgBr_2$ in NaBr and in NaBr, Br_2 solutions and the treatment of Rossotti and Rossotti (see 9.82).

The study of complex formation by solubilities may be followed by means of radioactive samples of sparingly soluble salts. Martin and his associates [13] have illustrated this by means of some lanthanide oxalates in buffered sodium oxalate solutions; $[M^{3+}]_{total}$ was measured by counting, $[Ox^{2-}]_{total}$ by $KMnO_4$, and the pH was measured. From K_1 and K_2 for H_2Ox, K_I and K_{II}, the dissociation constants of MOx^+ and MOx_2^- respectively, and $K_{SP} = m_{MOx^+}m_{MOx_2^-}f_1^2$, an expression was devised which equated $[M^{3+}]_{total}$ to the other quantities and constants, and K_I, K_{II} and K_{SP} derived by approximations; somewhat similar studies of [14] Y^{3+} and [15] Pu^{3+} oxalates have also been undertaken.

Freezing-points

Brown and Prue [16] have measured and analysed the freezing-points of dilute solutions of a number of divalent cation sulphates in a manner similar to that described in Chapter 5. The relevant equations are, with ϕ as defined by (5.22),

$$\phi = (1 - \alpha_c)/2 + \alpha_c\phi'; \; K = \alpha_c^2 m\gamma^2/(1 - \alpha_c) \qquad (11.2)$$

where ϕ' is related to γ, the mean ionic activity coefficient. Then from (5.30) and (5.31),

$$\theta(1 + b\theta) = 2\lambda m\phi = 2\lambda m\{(1 - \alpha_c)/2 + \alpha_c[1 - (2.303\,A'' \times 4\,\sigma\sqrt{I})/3 - \beta' I]\} \qquad (11.3)$$

If (3.43) is used, $B''a = 1.0$, $\beta' = 2.1$. In dilute solutions, $b\theta^2$ is negligible.

Values of K_{MSO_4} were calculated with different values of a and $\beta' = 0$, and since σ depends on x (see 5.29), I and therefore α_c involve approximation methods. It was found that constant, but different values of K were obtained as a is varied, e.g., for $MgSO_4$, with $a = 6.94$ Å, $K = 0.0041$, and with $a = 4.0$ Å, $K = 0.0105$. A close inspection of the $\theta - \theta_{calc.}$ values shows the deviations are least when $B''a = 1.0$, $\beta' = 2.1$ in (11.3), i.e., $a = 4.2$ to 4.4 Å (whence $K = 0.0090$). The next best set of results is with $a = 5.0$ Å, when $K = 0.0070$; these are around the figures associated with Table 9.5.

Kenttamaa's method of studying ion-association [17] by way of freezing-points has been to measure the depressions in saturated salt solutions ($KClO_4 = 0.454$ M, $KClO_3 = 0.258$ M, $KNO_3 = 1.15$ M) caused by MSO_4 (M = Cu, Zn, Ni, Mg). With I effectively constant, $\sum\overline{m}_i = [M] + [SO_4] + [MSO_4] = \theta/\lambda$ (see 5.20). The values of λ were found by means of HNO_3 in $KClO_4$ and in $KClO_3$, and by HCl in KNO_3, and plotting $\lambda' = \lambda + A$[acid] against [acid], where A is a constant. Letting m_s be the stoichiometric concentration ($= [M]+[MSO_4] = [SO_4] + [MSO_4]$), and $k_1 = [M][SO_4]/[MSO_4]$, then $1/k_1 = (2m_s - \sum\overline{m}_i)/(\sum\overline{m}_i - m_s)^2 = \overline{\psi}$ (say). The limit of $\sum\overline{m}_i$ as

$m_s \rightarrow 0$ is $\sum m_i$, and the corresponding value of $\bar{\psi}$ is ψ. This is obtained by linear extrapolation against m_s, so the intercept on the $\bar{\psi}$ axis gives $1/k_1$. Subsidiary ion-pairs such as KSO_4^- are ignored. To obtain k_1^0 at $I = 0$, $pk_1 = pk_1^0 - 3.9\sqrt{I}/(1 + B'a\sqrt{I}) + QI$ was used, leading to $k_1^0 = 0.0065$ for $MgSO_4$, $ZnSO_4 = CuSO_4 = 0.0055$, and $NiSO_4 = 0.0051$. Rossotti [79] has recently devised a "projection strip" method for analyzing cryoscopic data of this type.

Ion-exchange Resins

Ion-exchange materials, particularly those of the synthetic resin type are proving to be of increasing importance in elucidating the equilibria and the composition of the association products which form in ionic solutions. Much of the work reported up to 1951 has been reviewed by Schubert [18] in one of his many articles dealing with the applications of radioactive tracers and cation-exchange resins for deriving stability constants. For this work a sample of cation-exchange resin in its Na^+ form (or H^+ form) is prepared by saturating it with NaCl (or HCl), washing it with water, and drying at room temperature. A small weighed quantity m is shaken with a fixed concentration of tracer, M (usually carrier-free, or almost so) in a volume v, and the sodium salt of the anion L, which forms the ion-pairs ML, ML_2, etc., and sufficient NaCl to maintain a constant ionic strength. When L is the anion of a weak acid, it is necessary to buffer the solution. A number of samples, including $L = 0$ are equilibrated and samples taken for counting. There may be a correction to the volume due to the uptake of solvent and swelling of the resin. If v is the initial volume and $v\delta$ the volume at equilibrium, where δ is the swelling factor, this is found by measuring C_L before and after equilibrating with resin, with $C_M = 0$.

If the equilibrium distribution of tracer between resin and solution is defined by

$$K_d = (\% \text{ of M in resin})v\delta/(\% \text{ of M in soln.})m = M_s v\delta/M_1 m \qquad (11.4)$$

and by K_d^0 when $C_L = 0$, then if the cation on the resin is represented by MR,

$$[\mathrm{M}] + [\mathrm{ML}] + [\mathrm{ML_2}] + \ldots = [\mathrm{MR}]/K_d \qquad (11.5)$$

$$[\mathrm{M}] = [\mathrm{MR}]/K_d^0 \qquad (11.6)$$

so from (9.46),

$$1/K_d = 1/K_d^0 + k_1[\mathrm{L}]/K_d^0 + k_1k_2[\mathrm{L}]^2/K_d^0 + \ldots \qquad (11.7)$$

or

$$K_d^0/K_d = 1 + k_1[\mathrm{L}] + k_1k_2[\mathrm{L}]^2 + \ldots \qquad (11.8)$$

If only ML forms, or considering very low concentrations of L, extrapolation of the plot of $1/K_d$ against [L] gives $1/K_d^0$ at the intercept. This provides a useful check on the direct determination of this quantity. Plotting $(K_d^0/K_d) - 1$ against [L] gives a line passing through the origin and having a limiting slope equal to k_1. Plotting $\{(K_d^0/K_d) - 1\}/[L]$ against [L] gives k_1 from the intercept and has a limiting slope equal to k_1k_2. Plotting $\{(K_d^0/K_d) - 1 - k_1)\}/[L]$ against [L] has a limiting slope equal to k_1k_2, and so on.

Example 11.3. Schubert [18] obtained $k_1 = [SrCitrate^-]/[Sr^{2+}][Citrate^{3-}]$ at $I = 0.16$ by equilibration of the solutions listed below, each of which contained 100 mg of resin in 100 ml. of solution at a pH of 7.2. The initial count of the carrier-free Sr^{89} was 187.6 counts/min. ml. (cpm/ml.). Each solution contained 0.016 M sodium "veronal" buffer [19] (acetate + diethyl barbiturate); separate tests showed the buffer anions have negligible complexing action.

NaCl (mM)	144	134	126	120
NaCit (mM)	—	1.60	3.00	4.00
Sr in soln. (cpm/ml.)	51.2	81.7	98.9	107.3
K_d	2.664	1.296	0.897	0.748

$K_d = (187.6 - \text{cpm/ml.})/\text{cpm/ml.}$ By plotting, $K_d^0 = 2.70$, which checks the determined value. Plotting in the described manner, $k_1 = 2.45 \times 10^4$. Calculating $f_\pm$ by (3.45), k_1^0 at $I = 0$ is 2.21×10^4. Schubert obtained data from $I = 0.16$ to 0.0074 and by extrapolation found $k_1^0 = 1.29 \times 10^4$. Since K_3 for the acids is [20] 4.0×10^{-7}, at pH = 7.3, about 5 % of L is present as $HCit^{2-}$. Also, it has been shown that [21] k_1^0 for CaHCit is 1.0×10^3, and the corresponding value for SrHCit is probably larger. These two corrections would have a significant effect if taken into account.

If the buffer consists of anions different to L, corrections can be applied if these buffer anions associate with M by obtaining data in the absence of L; this has been demonstrated [22] in connection with the association of Ca^{2+} and Sr^{2+} with a series of organic acid anions. Similar studies of the Mn^{2+}, Co^{2+} and UO_2^{2+} systems [23] have also been made; to avoid hydrolysis, the pH was kept below 2 for the UO_2^{2+} series. With Mn Malonate, k_1 at $I = 0.16$ is 2.30, whence $k_1^0 = 3.31$; this agrees well with 3.29 obtained by pH colorimetry.[24] The k values of Co^{2+} and Zn^{2+} oxalates [25] and of Co^{2+} glycine peptides [26] (the results for which agree with those derived from pH titrations—see Chapter 9), the association of [27] Fe^{3+} and SO_4^{2-} ($\beta_1 = 95, \beta_2 = 900$ at $I = 1.0$), of [28] Ce^{3+} with Cit^{3-} and Ac^-, and [28a] of Gd^{3+} with Ac^- and Glycollate are other examples.

The MCl^{2+} complexes where M^{3+} = Pu, Am and Cm have been found [29] to have $k_1^0 = 14.7$ at $I = 0$ by studies of mixtures of the perchlorates and HCl at various ionic strengths, and a related series, ML^-, where M^{3+} = Am, Cm and Cf and L = $EDTA^{4-}$ has been studied by Fuger.[30] He equilibrated 2 to

10 mg of NH_4^+—resin with 0.001 M EDTA at a pH of 2 to 3.3 with trace amounts of these actinides and 0.1 M NH_4ClO_4. Writing $[EDTA]_{total} = [L^{4-}] + [HL^{3-}] + [H_2L^{2-}] + [H_3L^-] + [H_4L]$, it is easily shown from the dissociation constants of the acid,[31] that if

$$\theta = 1 + [H^+]/K_4 + [H^+]^2/K_3K_4 + [H^+]^3/K_2K_3K_4 + [H^+]^4/K_1K_2K_3K_4 \tag{11.9}$$

then from (11.8) since only ML^- forms,

$$K_d^0/K_d = 1 + k_1[EDTA]_{total}/\theta \tag{11.10}$$

whence $\log k_1 = 19.84$, 19.55, and 20.91 respectively.

Surls and Choppin [32] have given references to a number of studies dealing with elutions from resin columns of the lanthanides and actinides with citrate, lactate, α hydroxybutyrate and chloride in their account of the elution positions of the thiocyanate complexes; all of these should be suitable for quantitative studies of the related equilibria; this also applies to the metal phosphates which Salmon [33] has examined with cation exchangers.

A rather different way of finding association constants has been sought by Connick and Meyer [34] in connection with the association of Ce^{3+} with Cl^-, NO_3^- and SO_4^{2-}. For the exchange equilibrium $M^{3+} + 3N_R^+ \rightleftharpoons M_R^{3+} + 3N^+$, omitting charges,

$$K = [M_R][N]^3\gamma_{M_R}\gamma_N^3/[M][N_R]^3\gamma_M\gamma_{N_R}^3 = Q\gamma_{M_R}\gamma_{NX}^6/\gamma_{N_R}^3\gamma_{MX_3}^4 \tag{11.11}$$

Q is experimentally measurable. If [M] is small relative to $[N_R]$ and [N], then γ_{N_R} and γ_{M_R} should remain constant and γ_{NX} is independent of these and

$$\gamma_{MX_3} = k_N(Q\gamma_{XN}^6)^{\frac{1}{4}} \tag{11.12}$$

where k_N is a constant if the resin is nearly all in the N^+ form. Measurements were made with radioactive Ce^{3+} in $HClO_4$, $NaClO_4$, NaCl, HCl, HNO_3 and Na_2SO_4. Assuming that γ_{Ce} and γ_H are the same in HCl as in $HClO_4$, and that ClO_4^- does not complex, $[Ce^{3+}]$ is calculated via (11.11) and Q at the same ionic strength from the ClO_4^- system when studying the association with Cl^- and NO_3^-. Corrections for CeX^{2+} absorption were made via data for the Sr^{2+} system.

1. Successive Stability Constants

Fronaeus [35, 38, 78] has obtained the β_n values of e.g. cupric acetate by equilibrating solutions of $Cu(ClO_4)_2$, NaAc, HAc and $NaClO_4$ with Na^+ ion-exchange resins at $I = 1.0$ and 20°C. The total equilibrium concentration C_M was measured as $Cu(NH_3)_4^{2+}$ by spectrophotometer, which was also used to measure C_L by following the absorption change from CuL_n to Cu^{2+} on titration with $HClO_4$. The equilibria involving the resin (R) are $Cu^{2+} +$

$2NaR \rightleftharpoons CuR_2 + 2Na^+$ and $CuL^+ + NaR \rightleftharpoons CuLR + Na^+$. Carleson and Irving,[36] who have used H^+ ion resins in a study of the In^{3+} halides have broadened the original treatment of Fronaeus, and these additions are incorporated in the present account.

If the cations M^{x+} combine with L^- to form the species $\sum_0^N ML_n^{x-n}$, the general equilibria are

$$ML_n^{x-n} + (x-n)Na_R^+ \text{ or } H_R^+ \rightleftharpoons (x-n)H^+ \text{ or } Na^+ + (ML_n^{x-n})_R$$

and applying the law of mass action,

$$[ML_n^{x-n}]_R/[ML_n^{x-n}] = k_n[H^+ \text{ or } Na^+]_R^{x-n}/[H^+ \text{ or } Na^+]^{x-n} = l_n \qquad (11.13)$$

By keeping the load on the resin, C_{MR} (which $\ll H_R^+$ or Na_R^+), the pH, and the ionic strength of the solution constant, k_n and l_n are also constant.

The distribution of M^{x+} between resin and solution is given by

$$\phi = C_{MR}/C_M = l_0(1 + l_1'[L] + l_2'L]^2 + \ldots)/X \qquad (11.14)$$

where $l_n' = l_n\beta_n/l_0$, X is given by (9.55), which by (9.47) means

$$X = \sum_0^N \beta_n[L]^n \qquad (11.15)$$

In order to derive data at constant C_{MR} it is necessary to make several series of measurements over wide ranges of [L] in which C_M is varied slightly between each series at the same [L] and then to interpolate at a selected C_{MR} value. Fronaeus [35] studied four series and from the plots of ϕ against C_L and C_{MR} against C_L, interpolated to obtain ϕ and C_L at $C_{MR} = 1.25 \times 10^{-4}$ M. Carleson and Irving [36] studied three series of the same C_L but differing C_{MR}, and by plotting ϕ at constant C_L, obtained ϕ values at $C_{MR} = 1.00 \times 10^{-6}$ M. If (see 11.4)

$$vC_M' = v\delta C_M + mC_{MR} \qquad (11.16)$$

where C_M' is the starting concentration of M, this expression gives the experimental values of C_{MR} to use in (11.14) to obtain ϕ.

If $C_M \ll C_L$, $[L] \simeq C_L = C_L'/\delta$, where C_L' is the starting concentration of L. If this is not adequate then [35] a rough value of $\bar{n}$ is obtained from (9.58) and (11.14) by putting $l_n' = 0$ and $C_L \simeq [L]$, i.e., $X\phi \simeq l_0$, $dX/X \simeq d\phi/\phi$ so that $\bar{n} = [L]dX/X \,.\, d[L] \simeq -\, d\phi \,.\, C_L/\phi \,.\, dC_L$ since C_{MR} is constant, when [L] is given by (9.49), i.e., $[L] = C_L - \bar{n} \,.\, C_M$.

The treatment of Fronaeus is to differentiate ϕX twice with respect to [L], and taking l_2', . . . as zero in (11.14). This gives

$$\phi'' \,.\, X + 2\phi' \,.\, X' + \phi \,.\, X'' = 0 \qquad (11.17)$$

where $\phi' = d\phi/d[L]$, etc. Then substituting via (9.79), i.e., $X = \int_0^N \beta_n[L]^n$,

$$\phi'' + \sum_1^N (\phi'' \,.\, [L]^n + 2n[L]^{n-1}\phi' + \phi n(n-1)[L]^{n-2})\beta_n = 0 \qquad (11.18)$$

ϕ' and ϕ'' can be found from tangents of the plot of ϕ and then of ϕ' against [L]. In practice this method is only satisfactory for β_1, and β_2, etc., are best found by extrapolating the plot of $1/\phi$ against [L]; this, from (11.14), gives $1/l_0$. Then if a plot of $\phi_a = (1/\phi - 1/l_0)/[L]$ against [L] is extrapolated, since X for β_1 is $1 + \beta_1[L]$, from (11.14), l_1' can be determined by

$$(\beta_1 - l_1')/l_0 = \lim \phi_a \text{ as } [L] \to 0 \qquad (11.19)$$

With l_0 and l_1' determined, X can be found from (11.14), after which β_n values can be found by the methods described in Chapter 9.

Example 11.4. Fronaeus [35] has given the following information for cupric acetate at $C_{MR} = 1.25 \times 10^{-4}$ M/gm.

10^3[L]	0	20	40	70	100	150	200	250	300
$1/\phi$	(7.0)	10.8	15.3	20.4	26.5	39.0	56.5	80	104
ϕ_a	(195)	190	205	190	195	215	245	290	325
X		2.07	3.67	6.38	10.2	19.8	35.5	60.0	91

Obtaining ϕ' and ϕ'' by plotting, and using these at several values of [L] in (11.18), $\beta_1 = 45 \pm 2$. Taking the limit of $\phi_a = 195$, from (11.19), $l_1' = 17 \pm 2$, whence the given X values were obtained via (11.14). Then from (9.79), etc., $\beta_2 = 440 \pm 60$, $\beta_3 = 1{,}000 \pm 300$.

Carleson and Irving made use of two parameters:[36]

$$\phi_1 = \{(l_0/\phi) - 1\}/[L] \qquad (11.20)$$

$$f = (l_0/\phi)\{(\beta_1 - l_1')[L] - 1\}/[L]^2 \qquad (11.21)$$

For these, l_0 is found as already described or, more directly, from the fact that ϕ_0 = limit of ϕ as $[L] \to 0 = l_0$. In (11.21), $(\beta_1 - l_1')$ is derived from (11.25). Then from (11.14),

$$\phi_1 = \{\beta_1 - l_1' + [L](X_2 - l_2')\}/\{1 + l_1'[L] + l_2'[L]^2\} \qquad (11.22)$$

where, in general,

$$X_n = (X_{n-1} - \beta_{n-1})/[L] \quad (X_0 = X) \qquad (11.23)$$

and

$$f = [\beta_1(\beta_1 - l_1') + l_2' - \beta_2 + [L]\{X_2(\beta_2 - l_1') - X_3\}]/\{1 + l_1'[L] + l_2'[L]^2\} \qquad (11.24)$$

If ϕ_1^0 and f^0 are the limiting values as $[L] \to 0$,

$$\phi_1^0 = \beta_1 - l_1' \qquad (11.25)$$

$$f^0 = \beta_1(\beta_1 - l_1') - (\beta_2 - l_2') \qquad (11.26)$$

These are derived by graphical extrapolation of ϕ_1 and of f against [L].

By combination of (11.20), (11.21) and (11.22),

$$f = \beta_1\phi_1 - X_2 + l_2'(\phi_1[L] + 1) \qquad (11.27)$$

and if $\Delta f = f - f_0$, $\Delta\phi_1 = \phi_1 - \phi_1^0$,

$$\Delta f/[L] = \beta_1 \Delta\phi_1/[L] + l_2'(\phi_1^0 + \Delta\phi_1) - \beta_3 - [L]X_4 \qquad (11.28)$$

It was found that $[L]X_4 \simeq l_2'\Delta\phi_1$, so β_1 is the slope of the plot of $\Delta f/[L]$ against $\phi_1/[L]$, and l_1' is now known (from 11.25), so $\beta_2 - l_2'$ is known from (11.26).

An alternative expression for f can be derived from (11.20) and (11.27), i.e.,

$$f = \beta_1\phi_1 + \beta_2\phi_1[L] - X_3[L] - l_0(\beta_2 - l_2')/\phi \qquad (11.29)$$

and introducing

$$g = \beta_2\phi_1 - \beta_3 - X_4[L] = \{f - \beta_1\phi_1 + l_0(\beta_2 - l_2')/\phi\}/[L] \quad (11.30)$$

a plot of g against ϕ_1 has a slope of β_2 and intercept of β_3. With β_2 known, l_2' is found from (11.26), so that from (11.27), X_2 can be calculated for various values of [L], and a plot of [L] against $X_2 = \beta_2 + \beta_3[L] + \beta_4[L]^2 + X_5[L]^3$ has a limiting slope of β_3 and intercept of β_2. Alternatively, with l_0, l_1' and l_2' determined, X is found by (11.15) after which β_n values follow from (9.56).

Example 11.5. The following figures [36] for indium chloride refer to $C_{MR} = 1.00 \times 10^{-6}$ M/gm, at $I = 0.69$.

10^3Cl^-	0	9.49	23.72	47.45	71.17	142.35	237.25	427.04
$10^3\phi$	740	230	106.5	51.5	32.3	14.1	7.35	3.40
ϕ_1	(216)	233.6	250.8	281.7	307.8	361.7	420.1	507.3
$10^{-2}f$	(448.0)	486.1	527.0	594.7	652.1	771.0	898.9	1090
$10^{-2}g$	—	9694	10,615	11,870	12,970	15,170	17,730	21,420
X_2	—	4446	4286	4595	4843	5430	6275	7951

From these, $l_0 = \phi_0 = 740$, $\beta_1 - l_1' = 216$, $\beta_1 = 227 \pm 4$, $\beta_2 = 4{,}250 \pm 800$, $\beta_3 = 8{,}950 + 3{,}000$.

In a consideration of the systems $Th^{4+} + HSO_4^- \rightleftharpoons ThSO_4^{2+} + H^+$ and $Th^{4+} + 2HSO_4^- \rightleftharpoons Th(SO_4)_2 + 2H^+$, Zielen,[37] rather than employ the extrapolation procedures of Fronaeus, has taken advantage of electronic computing to obtain l_0, β_1 and β_2 in (11.14–15) by non-linear least squares methods (cf. Rydberg [76,80]), taking l_1' as zero in the first instance (this proved good enough in the final analysis).

A number of ionic equilibria and association processes have been studied by means of anion exchange resins. These include (a) [39,40] Cd^{2+}, Cl^-; (b) [41] Cu^{2+} and Cd^{2+} with SO_4^{2-}; (c) [42] Co^{2+}, Ox^{2-}; (d) [43] Zr^{4+} and Hf^{4+} with Ox^{2-} (e) [44] Zn^{2+} with Cl^-, Br^-, etc.; (f) [44a] Ca^{2+}, Mg^{2+}, Co^{2+} and Mn^{2+} complexes of some mononucleotides, e.g., AMP, ADP, ATP, using U.V. spectrophotometry to determine the nucleotide distribution.

Solvent Extraction

This is such an important and versatile method of extracting and purifying a very wide range of materials, including nuclear fuels and their fission products, that only a few aspects of its application to electrolyte solutions can be mentioned here. As Irving, Rossotti and Williams [45,46] have pointed out, the extraction of electrolytes into organic solvents can be classified into (a) inner complexes of metals with reagents such as dithiozone and acetyl acetone, (b) mineral acids, their salts and metal acido-complexes such as the well-known extraction of $FeCl_3$ by ether from aqueous HCl, and (c) certain salts or ion-pairs which incorporate bulky anions or cations, e.g., tetraphenyl arsonium per-rhenate and ferrous tris-1 : 10 phenanthroline in association with long-chain alkyl sulphates or sulphonates.[47]

The range of extractants is extremely wide and is being rapidly extended. Reviews on solvents which are suitable for extracting various cations have been published [46,48,51] and there are many articles dealing with particular aspects. Some of the more recent ones include those dealing with the extraction of Th^{4+} and UO_2^{2+} with polyamines [49,50] diluted with benzene (e.g., decylamine, octylamine), and with the extraction of [51,52] lanthanides, actinides, Zr^{4+}, Nb^{3+}, etc, with alkylphosphates, phosphoryl and phosphonic acids in kerosene, benzene, CCl_4 and other diluents. One of the most important of this group is tributyl phosphate (TBP); a review on its applications has been given by Irving and Edginton,[51] and there are numerous articles by McKay and his associates [53] and many others.

Taking class (a) mentioned at the beginning of this section, this is the most common application of solvent extraction for studying electrolytic dissociation. It depends [54] on finding the influence which a ligand L exerts on the distribution of a cation between an aqueous phase containing the ligand L, and an immiscible organic phase containing an agent which forms inner complexes with the cation. The essential principles can be seen by reference to some work of Day and Stoughton [55] with Th^{4+} systems. The extractant was a benzene solution of theonyl trifluoroacetone (TTA) equilibrated with dilute $HClO_4$ to convert most of the TTA to its enol form, HKe. Analysis shows that the Th^{4+} is extracted as $ThKe_4$, and the concentration equilibrium constant for the extraction process can be written as

$$K_T = [ThKl_4]_b[H^+]^4/[Th^{4+}][HKe]_b^4 \qquad (11.31)$$

where Ke^- is the enolate ion and b refers to the benzene phase. If the ligand L is present in the aqueous phase as a weak acid HL, and

$$k_n = [ThL_n^{4-n}][H^+]^n/[Th^{4+}][HL]^n \qquad (11.32)$$

while the extraction coefficient E_c is defined by

$$E_c = [\text{Th in organic phase}]/[\text{Th in aqueous phase}]$$
$$= [ThKe_4]_b/[Th^{4+} + ThL^{3+} \ldots] \qquad (11.33)$$

then if $E_c = E_c^0$ when [HL] (or L) = 0, it follows that

$$E_c^0/E_c = 1 + k_1[HL]/[H^+] + k_2[HL]^2/[H^+]^2 + \ldots \qquad (11.34)$$

so k_1, k_2, etc., can be found by determining E_c as a function of HL providing the activities of the various species are kept constant, which means keeping the ionic strength, the acidity and the concentration of TTA in the benzene phase constant (or making due allowance). When the ligand L is the anion of a strong acid, the above expressions are written accordingly; L replaces HL, $[H^+]$ is put as unity.

Example 11.6. The following figures have been obtained [55] at $I = 0.50$ with HNO_3, $HClO_4$ and HF, $HClO_4$ solutions.

HNO_3	0	0.10	0.20	0.30	0.40	0.50
$1/E_c$	0.63	0.94	1.24	1.54	1.84	2.14

The plot of $(E_c^0/E_c) - 1$ against $[NO_3^-]$ is a straight line, hence $k_2 = 0$. From the slope, $k_1 = [ThNO_3^{3+}]/[Th^{4+}][NO_3^-] = 4.8$.

10^5[HF]	0	1.00	2.00	4.00	6.00	8.00	20.0	30.0	100.0
$1/E_c$	0.63	1.22	1.83	2.94	4.08	5.50	14.6	24.0	133

The limiting slope of the plot of $(E_c^0/E_c) - 1$ against [HF]/0.5 gives $k_1 = 4.5 \times 10^4$. Inserting this in (11.34) for the last three solutions, $k_2 = 2.5$, 2.6 and 2.8×10^7; k_2 becomes constant at 3.1×10^7 if k_1 is taken as 4.3×10^4.

About 2 % of the TTA enters the aqueous phase, affecting K_T by about 9 %, but the effect on E_c^0/E_c should be negligible unless there is a marked salting-out effect. If the Th^{4+} or the ThL_n^{4-n} species form complexes with the Ke^- ion in the aqueous phase and, e.g., $k_t = [ThKe^{3+}]_{aq}[H^+]/[Th^{4+}][HKe]_{aq}$, this must be incorporated into (11.34). Day and Stoughton [55] estimated that k_t is about 7, which means k_1 and k_2 should be increased by about 7 % above the figures just given.

Zebrowski [56] has extended the study of Th complexes by means of TTA in benzene to its chlorides ($k_n = 1$ to 4), sulphates and phosphates; he and his associates estimated k_t to be 6.6 ± 5 %. A number of other extractants have also been used, including acetyl acetone,[57] TBP in benzene or kerosene,[58] and octylamine in benzene.[50] In oxygen-donor solvents such as TBP, metal nitrates are extracted as [59] neutral undissociated nitrates combined with TBP, $M'(NO_3)_3$,3TBP; $M''(NO_3)_4$,2TBP; $MO_2(NO_3)_2$,2TBP, where $M' = M^{3+}$ lanthanides and actinides, $M'' = Th^{4+}$, U^{4+}, Pu^{4+} and $MO_2 = UO_2^{2+}$, NpO_2^{2+}, PuO_2^{2+}. Metal chlorides are extracted in a similar way,[51, 60] e.g., $CoCl_2$,2TBP. Metal salts are probably extracted in amines as [50] $MX_x(R_3NH)_{x-2}$ where X is an inorganic anion. In an appendix, Hesford and McKay [53] have reviewed a number of k_n values of nitrates obtained with TTA:

Cation	I	k_1	k_2	k_3
Th^{4+}	6.0	2.8	1.4	—
Zr^{4+}	4.0	0.59	0.09	—
Pu^{4+}	6.0	4.7	0.96	0.33
PuO_2^{2+}	8.0	0.19	0.05	—

The association [61] of UO_2^{2+} with Cl^-, NO_3^-, SO_4^{2-}, and F^- at $I = 2.0$ by means of TTA (MKe_2 forms in the organic phase), of UO_2^{2+} with SO_4^{2-} by means of di-n-decylamine,[49] of Co^{2+} and Zn^{2+} with citrate and oxalate and of UO_2^{2+} with amino-acid anions [62] by means of radio-tracers and $CHCl_3$ solutions of cupferron and oxine, and of Zr^{4+} with NO_3^- and Cl^- (up to $k_n = 4$) with [63] TBP are other examples of the application of the solvent extraction method.

Example 11.7. Sullivan and Hindman,[64] in a review of methods of deriving stability constants, have reconsidered the data of Betts and Leigh [65] for the extraction of U^{4+} by benzene solutions of TTA from aqueous $H_2SO_4 + HClO_4$. Since the TTA concentration was varied, a new term is defined,[54]

$$E = E_c \gamma_{UKe_4}/[HKe]^4 \gamma_{HKe}^4 \tag{11.35}$$

the activity coefficients being almost unity. Some figures are given in Table 11.2.

TABLE 11.2

Data for the Stability Constants of USO_4^{2-} and $U(SO_4)_2$

$10^3[HSO_4^-] = C_L$	0	1.83	4.59	9.18	18.35	45.9	45.9	92.5
E_c	0.712	0.601	0.495	0.366	0.228	0.089	1.24	0.434
$10^4 E$	1.23	1.02	0.820	0.587	0.349	0.130	0.134	0.528
$\bar{n}$	—	0.15	0.35	0.55	0.85	1.15	1.15	1.35

Since [L] is not $\ll C_L$, $\bar{n}$ cannot be calculated directly. It is obtained by approximations by use of (9.47) and (9.58) since when C_M is constant, $\bar{n} = -\,d \log [M]/d \log [L] = -\,d \log E/d \log [L]$, so that $\bar{n}$ is given by the tangent of the log plot. For a first approximation, C_L is used, and [L] next calculated by (9.49). To do this, C_M is derived from $(U_{tot} - C_M)/C_M = E_c$, where the total $U^{4+} = 3.25 \times 10^{-3}$. By the Rossotti method (Chapter 9), $\beta_1 = [USO_4^{2-}]/[U^{4+}][HSO_4^-] = 140$, and $\beta_2 = [U(SO_4)_2]/[U^{4+}][HSO_4^-]^2 =$ 1,300. Sullivan and Hindman obtained 130 and 1,350, and Day,[66] who has repeated the work, together with studies of the Cl^- and CNS^- systems, reported $\beta_1 = 165$, $\beta_2 = 1{,}850$.

Some examples of class (b) extraction are included in a series of articles commencing with one by Irving, Rossotti and Williams [67] on a general treatment of solvent extraction. Assuming that hydrolytic species are absent, all the possible types which might exist in step-equilibria can be represented by $H_hM_mL_n(H_2O)_wS_s$ where S refers to solvent molecules. This is developed in terms of the possible equilibrium constants and partition coefficients, and is discussed in terms of the possible variables. The treatment is exemplified by studies of the extraction of [68] $FeCl_3$ by ethyl and iso-propyl ethers from aqueous HCl, where the organic phase holds $FeCl_3$, HCl and $HFeCl_4$, and the aqueous phase holds $FeCl_n^{3-n}$ ($n = 0$ to 4), and $HFeCl_4$, and by studies of the extraction of [69] In^{3+} by ketones, ethers, acetates, etc., from aqueous HI, HBr and HCl. The possible species are analogous to those assigned above to Fe^{3+}. Diamond [70] has reported on the extraction of In^{3+} from HCl and various chloride solutions by a wide range of solvents. The distribution of $HgCl_2$ between benzene and aqueous halides has been studied many times; Marcus [71] has tabulated earlier references and data in the first of his own contributions to this way of obtaining the complexity constants of $HgX_n^{-(n-2)}$ complexes, where $n = 1$ to 4.

The study of class (c) extractions is illustrated by Rydberg's work on the [72] partition of Th^{4+}-acetylacetone complexes between benzene and aqueous solutions containing $HClO_4$ and NaOH so that $I = 0.01$, and the pH was between 1.9 and 5.6.

If $\lambda_N = [ML_N]_{org}/[ML_N]_{aq}$, $q = [\text{total M}]_{org}/[\text{total M}]_{aq} = \lambda_N\beta_N[L]^N/\sum_0^N\beta_n[L]^n$, since, from (9.47) and (9.49), $\bar{n} = \sum_0^N n[ML_n]_{aq}/[\text{M total}]_{aq}$ and from (9.50) $\alpha_n = [ML_n]_{aq}/[\text{M total}]_{aq}$, then $q = \alpha_N\lambda_N$. From (9.54), $\bar{n} = n\, d \log \alpha_n/d\, pL$ so $\bar{n} = N + d \log q/d\, pL$ since λ_N is constant. Hence $\bar{n}$ can be obtained by plotting $\log q$ against pL. Then by taking $\bar{n}$ at 0.5, 1.5, 2.5, 3.5, and inserting into the expansion of (9.80), N equations are obtained from which β_1, β_2, β_3 and β_4 can be calculated. In this way $\beta_1 = 7 \times 10^7$, $\beta_2 = 3.8 \times 10^{15}$, $\beta_3 = 7.2 \times 10^{21}$, $\beta_4 = 7.2 \times 10^{26}$.

Now from (9.46) and (9.63), $k_n = \beta_n/\beta_{n-1} = 1/[L]_{(\bar{n}=n-0\cdot5)}$, and from this, $k_1 = 7.0 \times 10^7$, $k_2 = 5.4 \times 10^7$, $k_3 = 1.9 \times 10^6$, $k_4 = 1.0 \times 10^5$.

Dyrssen and others [73] have made a number of such investigations but have analyzed their results by two parameter equations devised by Dyrssen and Sillen [74] and Sillen and Rossotti.[75]

Rydberg and Sullivan [76, 80] have used an electronic computer to obtain the β_n and k_n values of the U^{4+}-acetylacetone and other complexes.[77] Putting $y = [M]_{aq}/[M]_{org}$, $x = [L]$, $t_n = \beta_n/\lambda_N\beta_N$, and $z = [M]_{aq}[L]^N/[M]_{org} = yx^N$ into the definition of q given earlier on, then $z = \sum_0^n t_n x^n$.

This can be solved for t_n if n combinations of $z(x)$ are available. Since the number, T, of experimental points is generally $\gg n$, the T equations for z can be solved by least squares methods, making due allowance for experimental errors in x and z. The computer was fed with the data [L], $[M]_{org} \pm \sigma_{org}$, $[M]_{aq} \pm \sigma_{aq}$ where σ is the estimated error, and t_n evaluated for a minimum value of n, the corresponding z, the deviations and the overall accuracy. The results are: (log values) $k_1 = 8.89 \pm 0.35$, $k_2 = 8.23 \pm 0.22$, $k_3 = 6.51 \pm 0.10$, $k_4 = 5.97 \pm 0.09$; $\beta_1 = 9.02 \pm 0.29$, $\beta_2 = 17.27 \pm 0.26$, $\beta_3 = 23.79 \pm 0.29$, $\beta_4 = 29.77 \pm 0.29$.

REFERENCES

1. Blayden, H. E. and Davies, C. W. *J. chem. Soc.* 494 (1930). Davies, C. W. *ibid.* 2410, 2421 (1930).
2. Wise, W. C. and Davies, C. W. *ibid.* 273 (1938).
3. Colman-Porter, C. A. and Monk, C. B. *ibid.* 1312, 4362 (1952).
4. Davies, C. W. and Hoyle, B. E. *ibid.* 233 (1951).
5. Bell R. P. and Waind, G. *Trans. Faraday Soc.* **49**, 623 (1953).
6. Bell, R. P. and George, J. H. *ibid.* **49**, 619 (1953). Nair, V. S. K. and Nancollas, G. H. *J. chem. Soc.* 318 (1957).
7. Monk, C. B. *Trans. Faraday Soc.* **47**, 285 (1951). Lloyd, M., Wycherley, V. and Monk, C. B. *J. chem. Soc.* 1786 (1951). Evans, W. P. and Monk, C. B. *ibid.* 550 (1954).
8. Monk, C. B. *Trans. Faraday Soc.* **47**, 292 (1951). Davies, P. B. and Monk, C. B. *J. chem. Soc.* 2718 (1951).
9. Denney, T. O. and Monk, C. B. *Trans. Faraday Soc.* **47**, 992 (1951).
10. King, E. L. *J. Amer. chem. Soc.* **71**, 319 (1949).
11. Burns, E. A. and Hume, D. N. *ibid.* **78**, 3958 (1956).
12. Tyrrell, H. J. V. and Scaife, D. B. *J. inorg. nucl. Chem.* **8**, 353 (1958).
13. Crouthamel, C. E. and Martin, D. S. *J. Amer. chem. Soc.* **72**, 1382 (1950); **73**, 569 (1951). Jonte, J. N. and Martin, D. S. *ibid.* **74**, 2052 (1952). Renier, J. J. and Martin, D. S. *ibid.* **78**, 1833 (1956).
14. Feibush, A. M., Rowley, K. and Gordon, L. *Anal. Chem.* **30**, 1610 (1958).
15. Matorina, N. N. and Moskvin, A. I. *Soviet J. At. Energy*, **3**, 1115 (1957).
16. Brown, P. G. and Prue, J. E. *Proc. roy. Soc.* **232A**, 320 (1955).
17. Kenttamaa, J. *Suomen Kemistileht*, **29B**, 59 (1956). *Acta Chem. Scand.* **12**, 1323 (1958).
18. Schubert, J. *J. phys. Chem.* **56**, 113 (1952). Also, Fomin, V. V. *Uspekhi Khim.* **24**, 1010 (1955).
19. Michaelis, L. *Biochem. Z.* **234**, 139 (1931).
20. Bates, R. G. and Pinching, G. D. *J. Amer. chem. Soc.* **71**, 1274 (1949).
21. Davies, C. W. and Hoyle, B. E. *J. chem. Soc.* 4134 (1953).
22. Schubert, J. and Lindenbaum, A. *J. Amer. chem. Soc.* **74**, 3529 (1952).
23. Li, N. C., Westfall, W. M., Lindenbaum, A., White, J. M. and Schubert, J. *ibid.* **79**, 5864 (1957).
24. Stock, D. I. and Davies, C. W. *J. chem. Soc.* 1371 (1949).
25. Schubert, J., Lind, E. L., Westfall, W. M., Pfleger, R. and Li, N. C. *J. Amer. chem. Soc.* **80**, 4799 (1958).
26. Li, N. C., Doody, E. and White, J. M. *ibid.* **79**, 5864 (1957).
27. Whiteker, R. A. and Davidson, N. *ibid.* **75**, 3081 (1953).
28. Seal, R. S. *J. inorg. nucl. Chem.* **10**, 159 (1959). Fronaeus, S. *Svensk. Kem. Tidskr.* **65**, 19 (1953).

28a. Sonesson, A. *Acta Chem. Scand.* **13**, 1437 (1959).

29. Ward, M. and Welch, G. A. *J. inorg. nucl. Chem.* **2**, 395 (1956).
30. Fuger, J. *ibid.* **5**, 332 (1958).

31. Schwarzenbach, G. and Ackermann, H. *Helv. Chim. Acta* **30**, 1798 (1947). Cabell, M. J. *Analyst* **77**, 859 (1952). Carini, F. F. and Martell, A. E. *J. Amer. chem. Soc.* **74**, 5745 (1952).
32. Surls, J. P. and Choppin, G. R. *J. inorg. nucl. Chem.* **4**, 62 (1957).
33. Salmon, J. E. and others, *J. chem. Soc.* 2316 (1952); 2644 (1953); 28, 4013 (1954); 360, 1444 (1955); 269, 360 (1956); 256, 959, 3239 (1957); 1128 (1958); 1459 (1959).
34. Connick, R. E. and Meyer, S. W. *J. Amer. chem. Soc.* **73**, 1176 (1951).
35. Froneaus, S. *Acta Chem. Scand.* **5**, 859 (1951).
36. Carleson, B. G. F. and Irving, H. *J. chem. Soc.* 4390 (1954).
37. Zielen, A. J. *J. Amer. chem. Soc.* **81**, 5022 (1959).
38. Froneaus, S. *Acta Chem. Scand* (a) Ni^{2+}, Ac^{-}: **6**, 1200 (1952). (b) Ni^{2+}, CNS^{-}: **7**, 21, (1955).
39. Fomin, V. V. and others, *J. phys. Chem. Moscou*, **29**, 2042 (1955).
40. Marcus, Y. *J. phys. Chem.* **63**, 1000 (1959).
41. Froneaus, S. *Acta Chem. Scand.* **8**, 1174 (1954).
42. Fomin, V. V. and others, *J. inorg. Chem. U.S.S.R.*, **1**, 2316 (1956).
43. Ermakov, A. N., Belyaeva, V. K. and Marov, I. N. *Trudy Kom. Anal. Khim. Acad. Nank, S.S.S.R.* **9**, 170 (1958).
44. Horne, R. A., Holm, R. H. and Meyers, M. D. *J. phys. Chem.* **61**, 1651, 1655, 1661 (1957).
44a. Walaas, E. *Acta chem. Scand.* **12**, 528 (1958).
45. Irving, H. *Quart. Rev.* **5**, 200 (1951).
46. Irving, H., Rossotti, F. J. C. and Williams, R. J. P. *J. chem. Soc.* 1906 (1955).
47. Friedman, H. L. and Haugen, G. R. *J. Amer. chem. Soc.* **76**, 2060 (1954). Powell, R. and Taylor, C. G. *Chem. and Ind.* 726 (1954).
48. Morrison, G. H. *Anal. Chem.* **22**, 1388 (1950). Kuznetsov, V. I. *Uspekhi Khim.* **23**, 654 (1954). Morrison, G. H. and Feiser, H. *Anal. Chem.* **30**, 632 (1958).
49. McDowell, W. J. and Baes, C. F. *J. phys. Chem.* **62**, 777 (1958). Allen, K. A. *J. Amer. chem. Soc.* **80**, 4133 (1958).
50. Carswell, D. J. and Lawrence, J. J. *J. inorg. nucl. Chem.* **11**, 69 (1959).
51. Irving, H. and Edginton, D. N. *ibid.* **10**, 306 (1959). Ferraro, J. R. *loc. cit.* 323.
52. Kennedy, J. *Chem. & Ind.* 950 (1958). Hardy, C. J. and Scargill, D. *J. inorg. nucl. Chem.* **10**, 323 (1959).
53. Hesford, E. and McKay, H. A. C. *Trans. Faraday Soc.* **54**, 573 (1958).
54. Connick, R. E. and McVey, W. H. *J. Amer. chem. Soc.* **71**, 3182 (1949).
55. Day, R. A. and Stoughton, R. W. *ibid.* **72**, 5662 (1950).
56. Zebrowski, E. L., Alter, H. W. and Heumann, F. K. *ibid.* **73**, 5646 (1951).
57. Peshkova, V. M. and Zozulya, A. P. *Nauch, Poklady Vysshei Shkoly, Khim i Khim Tekhol.* No. 3, 470 (1958).
58. Fomin, V. V. and Maiorova, E. F. *J. inorg. Chem. U.S.S.R.* **1**, 1703 (1956). A.E.R.E. Harwell, (U.K.), Lib./Trans. 802, Paper 1 (1958).
59. Alcock, K., Grimley, S., Healy, T., Kennedy, J. and McKay, H. A. C. *Trans. Faraday Soc.* **52**, 39 (1956). Healy, T. V. and McKay, H. A. C. *loc. cit.* 633.
60. Gal, I. J. and Ruvarac, A. *Bull. Inst. Nuclear Sci. Belgrade,* **8**, 67 (1958).
61. Day, R. A. and Powers, R. M. *J. Amer. chem. Soc.* **76**, 3895 (1954).
62. Schubert, J., Li, N. C. and others. *ibid.* **79**, 5864 (1957); **80**, 4799, 5901 (1958).
63. Solovkin, A. S. *J. inorg. Chem. U.S.S.R.* **2**, 611 (1957). A.E.R.E. Harwell (U.K.), Lib./Trans. 791 (1958).
64. Sullivan, J. C. and Hindman, J. C. *J. Amer. chem. Soc.* **74**, 6091 (1952).
65. Betts, R. H. and Leigh, R. *Canad. J. Res.* **28B**, 514 (1950).

66. Day, R. A., Wilhite, R. N. and Hamilton, F. D. *J. Amer. chem. Soc.* **77,** 3180 (1955).
67. Irving, H., Rossotti, F. J. C. and Williams, R. J. P. *J. chem. Soc.* 1906 (1955).
68. Chalkley, D. E. and Williams, R. J. P. *loc. cit.* 1920.
69. Irving, H. and Rossotti, F. J. C. *loc. cit.* 1927, 1938 (1946).
70. Diamond, R. M. *J. phys. Chem.* **63,** 659 (1959).
71. Marcus, Y. *Acta Chem. Scand.* **11,** 329, 599 (1957).
72. Rydberg, J. *ibid.* **4,** 1503 (1950).
73. Dyrssen, D. and others, *Svensk. Kem. Tid.* **65,** 43 (1953). *Acta Chem. Scand.* **9,** 1567 (1955); **10,** 107, 353 (1956); **13,** 50 (1959).
74. Dyrssen, D. and Sillen, L. G. *ibid.* **7,** 663 (1953).
75. Sillen, L. G. *ibid.* **8,** 299, 318 (1954); **10,** 186 ; (with Rossotti, F. J. and H.) 203 (1956).
76. Rydberg, J. and Sullivan, J. C. *ibid.* **13,** 186 (1959).
77. Rydberg, J. and Rydberg, B. *Ark. Kemi.* **9,** 81 (1956).
78. Ryabchikov, D. I., Ermakov, A. N., Belyaeva, V. K. and Morkov, I. N. *J. inorg. Chem. U.S.S.R.* 1814 (1959). (*Chem. Soc.*, English Translation, 818, (1959). This gives full references to the work of Froneaus and Russian workers, and also deals with Hf and Zr tartrates.)
79. Rossotti, F. J. C. and Rossotti, H. *J. phys. Chem.* **63,** 1041 (1959).
80. Sullivan, J. C., Rydberg, J. and Miller, W. F. *Acta Chem. Scand.* **13,** 2023, 2057 (1959).

CHAPTER 12

INCOMPLETE DISSOCIATION, PART V: POLAROGRAPHY. RAMAN SPECTRA. NUCLEAR MAGNETIC RESONANCE

Polarography

THE determination of trace concentrations of reducible and oxidizable substances (10^{-2} M to 10^{-7} M) by means of the polarography is described in a number of text-books on analytical chemistry, and in various articles [1, 2] and monographs.[3] Consequently only a short account will be given here of Heyrovsky's [4] device, indicating the main features by means of Fig. 12.1a.

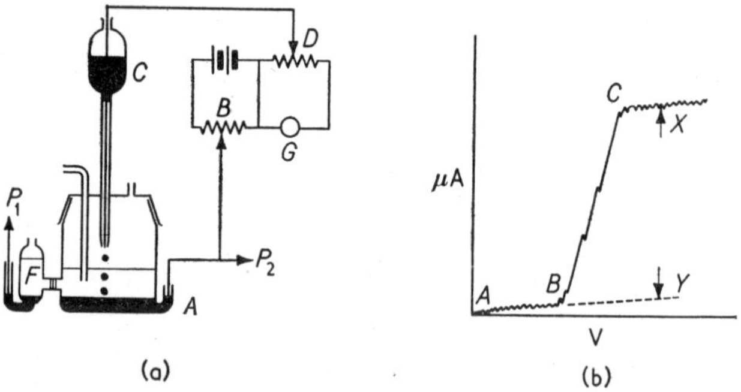

FIG. 12.1. (*a*) Polarograph circuit; (*b*) Polarogram.

The anode *A* is a pool of mercury while the cathode *C* consists of mercury falling as drops (2 to 4 per sec) from a capillary in order to provide a clean surface that is easily polarized but is free from products of oxidation, analgamation, etc. By adjusting *B*, a steadily increasing voltage can be applied to the cell; the corresponding current (in microamps, μA) is indicated by a sensitive calibrated galvanometer *G* which oscillates somewhat as the cathode drop grows and falls. The shunt *D* is to help damp down these oscillations. The electrolyte solution in the cell contains not only the minute concentration of material being determined, e.g., a cation M^{n+}, but also a fairly strong solution of a "base" or "supporting" electrolyte such as KCl, a "maximum suppressor" such as gelatin, and in addition the solution is freed from O_2. Under these conditions, as the voltage is increased a curve as that in Fig. 12.1b results. This is termed a polarogram. From *A* to *B* a small "residual" current flows due to movement of charge on the expanding cathode surface, while the sudden rise in current from *B* to *C* is due to discharge of e.g. M^{n+} ions, i.e., $M^{n+} + ne^- \rightarrow M/Hg$. The voltage at which this occurs is largely characteristic of the ion, and the vertical height *XY*, termed the diffusion current, i_d, is proportional to the concentration of M^{n+} provided certain factors and conditions are kept constant. These are described later on.

The voltage corresponding to halfway along *XY* is termed the "half-wave" potential, $E_{\frac{1}{2}}$, and is usually given by reference to a saturated calomel electrode (S.C.E.). The necessary calibration is effected either by obtaining the applied voltage using the calomel electrode *F* as anode, or by measuring the mercury anode potential against the calomel reference by potentiometer connected at P_1 and P_2. It is necessary to correct for the *iR* drop of the cell. *R* is measured by A.C. conductance bridge, *i* is the measured current at $E_{\frac{1}{2}}$.

The function of the supporting electrolyte is to supply sufficient ions to carry practically all of the current. Simple salts of the alkali and alkaline earth metals are suitable, and also the simple inorganic acids, since they have high $E_{\frac{1}{2}}$ values. With these ions present, M^{n+} ions can only reach the cathode by diffusion, and as they are discharged, [M^{n+} in the cathode surface] is reduced. It is assumed that the rate of diffusion is related to [M^{n+} in solution] —[M^{n+} in the cathode surface], and this will increase as the voltage is increased beyond *B* until [M^{n+} in the cathode surface] is zero. In other words, the ions are then discharged as they diffuse on to the cathode surface. When this condition is reached, the current has reached its maximum; this is termed the "limiting" current.

Any oxygen that is dissolved in the solution will give a two-step polarogram, the positions of which depend on the pH. The first step is due to the conversion of O_2 to H_2O_2, and the second to the reduction of H_2O_2 to H_2O or OH^-. It is in order to prevent these polarograms forming that the solution is freed from O_2 by saturating with N_2, or by other methods.

At times, instead of getting a normal polarogram, pronounced maxima occur. These are attributed to streaming of the diffusion layer around the cathode surface and, or, a flow of depolarizing ions towards the cathode drop as a result of a neighbouring heterogeneous field caused by a charging current developed during the growth of the drop. These maxima can usually be suppressed by adding a trace of surface-active material such as 0.01 % of gelatin or other colloids, or certain dyestuffs. These are absorbed on the cathode surface and they remove differences of interfacial tension.

An expression which relates the polarographic diffusion current i_d to the bulk concentration *C* of an ion, is that derived by Ilkovic,[3, 5] namely

$$i_d = 607\, nD^{\frac{1}{2}}m^{\frac{2}{3}}t^{\frac{1}{6}}C \qquad (12.1)$$

where i_d is in μA; 607 includes *F*, the surface area of a drop, and an allowance for the diffusion occurring on an expanding cathode surface; D is the diffusion coefficient of the ion in cm^2/sec; C is its concentration in g. ions/ml.; m is the mass of mercury in g dropping from the capillary in one sec; t is the time in sec for the formation of one drop; n is the number of electrons involved in the reduction stage. Providing the flow rate remains constant for a given capillary, and that the temperature, ionic strength, viscosity and medium (all of which influence D) remain constant, then (12.1) reduces to

$$i_d = k_s' I' C = k_s C \tag{12.2}$$

where k_s' is the "capillary constant" $m^{\frac{2}{3}}t^{\frac{1}{6}}$, and I' is the "diffusion current constant" $607nD^{\frac{1}{2}}$. To find k_s, i_d is measured with solutions of known C.

Instead of developing the type of polarogram shown in Fig. 12.1b, more convenient shapes can be produced and displayed on cathode ray oscilloscopes [3,6] with polarographs with potential sweeps of various frequencies of sinusoidal, saw-tooth and square-wave signals;[6,7] the latter instruments can handle concentrations down to 2×10^{-7} M.

The application of polarography to the evaluation of stability constants was first described by Lingane.[2] The procedure has been developed and generalized by De Ford and Hume,[8] who have shown that by obtaining $E_{\frac{1}{2}}$ and i_d values at constant ionic strengths with a fixed concentration of C_M and and varying C_L, Leden's method (see Chapter 9) can be used to obtain successive stability constants. Sartori [9] has given a general discussion on this topic.

A necessary preliminary is to derive a relationship between E_{de}, the potential at the dropping electrode, to $E_{\frac{1}{2}}$, i_d, and i, the current at any point on the wave.

At all points on the wave, E_{de} is related to E_a^0, the standard potential of the amalgam produced by $M^{n+} + ne^- \rightarrow M/Hg$ by (cf. 4.10)

$$E_{de} = E_a^0 - (k'/n) \log (C_a^0 f_a / a_{Hg} C_M^0 f_M) \tag{12.3}$$

where C_a^0 — [M/Hg] on the cathode surface, C_M^0 — $[M^{n+}]$ in the layer of solution on the surface of the drop, f_a and f_M are the corresponding activity coefficients and a_{Hg} is the activity of the mercury on the surface of the drop. Since C_a^0 is small, a_{Hg} is virtually constant so (12.3) can be written

$$E_{de} = E_a' - (k'/n) \log (C_a^0 f_a / C_M^0 f_M) \tag{12.4}$$

With the assumption already mentioned, i.e., the rate of diffusion is proportional to $[M^{n+}]$ in solution — C_M^0, the current i at any point on the wave, from (12.2) is given by

$$i = k_s(C_s - C_s^0) \tag{12.5}$$

C_s being $[M^{n+}]$ in solution. Now when $i = i_d$, $C_s^0 = 0$, i.e., $i_d = k_s C_s$, whence

$$C_s^0/C_s = (i_d - i)/i_d \tag{12.6}$$

Also, C_a^0 is proportional to i so that $i = k_M C_a^0$, where k_M depends on the diffusion coefficient of the M atoms in the amalgam. So from (12.4–12.6),

$$E_{de} = E_a' - (k'/n) \log (f_a k_s/f_s k_M) - (k'/n) \log \{i/(i_d - i)\} \quad (12.7)$$

Since $E_{\frac{1}{2}}$ is the value of E_{de} when $i = \frac{1}{2} i_d$,

$$E_{\frac{1}{2}} = E_a' - (k'/n) \log (f_a k_s/f_s f_M) \quad (12.8)$$

and

$$E_{de} = E_{\frac{1}{2}} - (k'/n) \log \{i/(i_d - i)\} \quad (12.9)$$

A plot of E_{de} against $\log \{i/(i_d - i)\}$ should be a straight line of slope k'/n and the value of E_{de} at the intercept gives $E_{\frac{1}{2}}$ (this is the standard procedure).

If ion-pairs or complexes, ML_n, are present, taking L to be a monovalent ligand, the reactions at the electrode will be $ML_n \rightleftharpoons M^{n+} + nL^-$; $M^{n+} + ne^- \rightleftharpoons M/Hg$, and if the dissociation of the complexes is sufficiently rapid to maintain equilibrium, then from (9.45), at constant ionic strength,

$$\beta_n [L]^n = [ML_n]/[M] = C_{ML_n}^0 / C_M^0 \quad (12.10)$$

where C_{MLn}^0 is the concentration of the nth complex on the cathode surface, and $[L] \gg [M]$. By summation,

$$C_M^0 = \sum_0^N C_{ML_n}^0 \Big/ \sum_0^N \beta_n [L]^n \quad (12.11)$$

so from (12.4),

$$E_{de} = E_a' - (k'/n) \log \sum_0^N \beta_n [L]^n - (k'/n) \log (C_a^0 f_a / f_s \sum_0^N C_{ML_n}^0) \quad (12.12)$$

Also see (12.6, 12.7),

$$i = k_c \sum_0^N ([ML_n] - C_{ML_n}^0) = k_M C_a^0 \quad (12.13)$$

$$i_d = k_c \sum_0^N [ML_n] = k_s [M] \text{ when } [L] = 0 \quad (12.14)$$

where k_c is the value of k_s in (12.2) when L is present. Then from (12.9),

$$(E_{\frac{1}{2}})_c - (k'/n) \log (k_M/k_c) = E_a' - (k'/n) \log \sum_0^N \beta_n [L]^n - (k'/n) \log (f_a/f_s) \quad (12.15)$$

and from (12.8),

$$(E_{\frac{1}{2}})_s = E_a' - (k'/n) \log (f_a/f_s) - (k'/n) \log (k_s/k_M) \quad (12.16)$$

where $(E_{\frac{1}{2}})_c$ and $(E_{\frac{1}{2}})_s$ are the $E_{\frac{1}{2}}$ values in the presence and absence of complexes. These last two expressions give

$$\log \sum_0^N \beta_n [L]^n = (n/k') \{(E_{\frac{1}{2}})_s - (E_{\frac{1}{2}})_c + \log (k_s/k_c)\} \quad (12.17)$$

In the last term of this, from (12.14), $k_s/k_c = i_d/i_d'$ where i_d and i_d' are the diffusion currents in the absence and presence of ML_n respectively. Then from

(9.60) or (9.61), a plot of $\{\sum_0^N \beta_n[L]^n - 1\}/[L] = (B - 1)/[L] = A$ against [L] gives β_1, etc.

Example 12.1. Some data for cadmium thiocyanate [10] at 30°C, where I was kept constant at 2.0 by KNO_3 are given below; C_M at trace (unknown) concentration.

KCNS(M)	0	0.1	0.2	0.3	0.4	0.5	0.7	1.0	1.5	2.0
$-E_{\frac{1}{2}}$ (mV)	572.4	584.7	594.3	601.1	607.1	614.2	628.5	634.3	651.6	664.6
$i_d(\mu A)$	7.56	7.34	7.29	7.24	7.22	7.08	7.08	7.07	7.06	7.03
B	—	2.64	5.54	9.40	14.89	26.24	53.3	122.2	459	1250

From (9.61), the intercept of the plot of $(B - 1)/[L]$ against [L] gives β_1, the intercept of the plot of $(B-1-\beta_1[L])/[L]^2$ against [L] gives β_2, and so on. The results are $\beta_1 = 10.5$, $\beta_2 = 55$, $\beta_3 = 6$, $\beta_4 = 60$ (only the last four points are suitable for β_3 and β_4).

The complexes of [11] Cd, Zn and Fe^{2+} with Cit^{3-}, $HCit^{2-}$ and H_2Cit^-, of [12] Zn with N_2H_4, of Pb with [13] NO_3^- and with [14] Ac^- (where $(E_{\frac{1}{2}})_s$ and i_d were obtained by extrapolation of the plot of $(E_{\frac{1}{2}})_c$ and i'_d against $[Ac^-]$, of some M^{2+} complexes with [15] formates, amino-acid and peptide anions [16] and with [17] CNS^- have all been studied by means of polarography. In some of the work the log (i_d/i'_d) term has been omitted; it is a significant correction when [L] is small. Some other recent examples are studied of the complexes of [18] Cu^{2+} with tri-*en*, tetra-*en* and penta-*en*, the complexes of [19] Cd, Pb with thiourea, of [20] Pb, Zn and Cd salicylates, of [21] Cd with *en*, N_2H_4, and [21a] $S_2O_3^{2-}$, Rh chlorides,[22] and [23] Pu^{3+} and Pu^{4+} with Ox^{2-}.

Raman Spectra

One consequence of the collision or interaction of radiation with matter is that some of the light may be scattered. If the radiation source is monochromatic, part of the scattered light may consist of weak radiations of different frequencies to that of the source. These new frequencies are related to the vibrations and rotations in the atoms of the molecules or ions of the irradiated substance, and apart from the special case of fluorescence, constitute what is termed the Raman effect.[24] It differs from fluorescence in that the latter is only produced by radiation of specific wavelengths, and is emitted after a very short but measurable time ($\sim 10^{-9}$ sec); in addition, Raman lines are often polarized. The polarization characteristics, the numbers and intensities of Raman lines, being related to molecular structure, are primarily used for such studies,[25] but the second and third of these characteristics are also useful for the study of ionic interactions, especially in acidic media.

It was found by Rao [26] that the integrated intensity of a Raman line of a particular species (molecule, ion, ion-pair, complex) is proportional to the concentration of the species, and according to the theory of Placzek,[27] the intensity as such depends on the change of polarizability with the change of internuclear distance associated with the vibration producing the line. For electrostatic binding, the polarizability should be independent of the internuclear distance and no Raman effect should appear. Woodward and his associates [28] do not accept this; the polarizability of an ion-pair is not merely the sum of the polarizabilities of the separate ions since the dipole produced in either ion by an external field will be modified by the field of the dipole produced simultaneously in the other ion. This mutual effect depends on the interionic distance, so the Raman effect will not be zero.

On applying a radiation of frequency ν_0, if a Raman line of frequency ν_{nm} arises from a transition between the vibrational levels n to m, the intensity I_{nm} is given by

$$I_{nm} = KM_n(\nu_0 - \nu_{nm})[\mathrm{P}]^2_{nm} \qquad (12.18)$$

where K is a universal constant, M_n is the molecular concentration of the scattering species in the n level and $[\mathrm{P}]_{nm}$ is the transition moment. Woodward [28] has developed this, and from tentative calculations concluded that $[\mathrm{P}]^2$ for covalent bonding is 100 to 10^5 times that for electrostatic bonding, i.e., I_{nm} from an ion-pair is that much weaker than that of a covalent bond.

The essential parts of a Raman spectrometer (full details are given elsewhere [25]) are indicated in Fig. 12.2. S is the sample tube which is irradiated

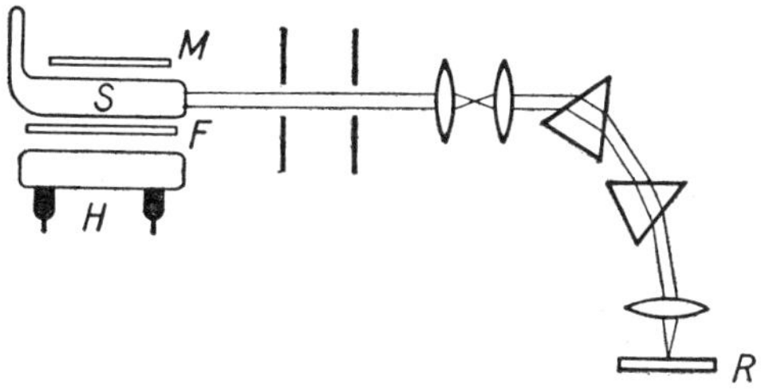

FIG. 12.2. Raman spectrometer.

by a mercury arc tube H; F is a filter for rendering the radiation as monochromatic as possible (it is often a cylindrical tube with focusing properties), and M is a plane mirror for increasing the light intensity. The Raman radiations which are scattered at 90° to the incident beam are focused by slits and lenses on to the recorder R which is either a photographic plate, or better, a photomultiplier tube with D.C. amplification of the anode so that a pen-recorder can be actuated. The frequencies of the lines are found by comparison with those of a standard (iron arc). One of the major difficulties about lines on photographic plates is that the intensity of a line is not proportional to the blackness of the line on the plate since the width increases as the intensity grows. The relative intensities of the lines obtained by photography are measured by microphotometer [29] or other devices.[30] The integrated intensities are obtained from pen-recorder charts by graphical integration of the area under the curve.

The Raman spectrum of HNO_3 has been subjected to a very thorough examination. The pure acid has eight Raman lines [31] and one band. Six of these lines and the band are due to vibrational modes in the molecule,[32] the line at 1,049 cm^{-1} is attributed to the NO_3^- ion and that at 1,394 cm^{-1} to the NO_2^+ ion. $NaNO_3$ solutions give three lines, that at 1,049 cm^{-1} being strong and sharp. There are various other features such as N_2O_5 in HNO_3; discussions on these are given elsewhere.[33, 34]

TABLE 12.1

Data and Calculations for K_{HNO_3} from Raman Spectra

C	4.51	6.60	8.90	10.30	11.89	14.23
M	5.2	8.25	12.4	15.8	20.1	28.8
$\log a_2$	1.51	2.09	2.62	2.92	3.22	3.66
(a) $1 - \alpha_c$	0.177	0.327	0.51	0.61	0.68	0.86
(a) $\log K_m\gamma_u$	1.55	1.66	1.82	1.94	2.08	2.27
(b) $1 - \alpha_c$	0.16	0.29	0.42	0.51	0.60	—
(b) $\log K_m\gamma_u$	1.59	1.71	1.90	2.01	2.14	—

The most recent investigation of HNO_3 solutions is that of Krawetz [35] who used the method of Redlich and Bigeleisen [36] to estimate K_{HNO_3}. Each acid solution (below 10 M) was compared with at least two $NaNO_3$ solutions of such concentrations that the intensity of the 1,049 line in the acid fell between those of the $NaNO_3$ solutions. In this way, $[NO_3^-]$ in the acid was found by interpolation. Then defining

$$K_m = a_2/(1 - \alpha_c)m\gamma_u \tag{12.19}$$

where

$$a_2 = a_H a_{NO_3} = m^2\gamma_\pm^2 = (m_i/\alpha_c)^2(\gamma_i^2\alpha_c)^2 = a_i^2 \tag{12.20}$$

in which γ_u is the activity coefficient of the undissociated HNO_3; m_i, γ_i and a_i refer to the ions, values of $\log K_m\gamma_u$ were estimated with the aid of a_2/M data obtained by vapour-pressure and freezing-points [37] of HNO_3 solutions. K_m was obtained by plotting $\log K_m\gamma_u$ against M and extrapolating. Some data are given in Table 12.1. Here (a) refers to Redlich and Bigeleisen, and (b) to data based on interpolations taken from a plot [34] of the unpublished results of Krawetz.[35]

Except for the last point of (a), the plots are linear, leading to $K_m = 25 \pm 1.5$. Plotting against M is an empirical procedure, and McKay,[38] who has derived $K_m = 23.5 \pm 0.5$ from the data of Krawetz, has considered the electrolyte as exerting a salting effect on the non-electrolyte part, and expressed this by

$$\log \gamma_u = \lambda_c[r + (1 - r)\alpha_c]m \qquad (12.21)$$

where $r = \lambda_u/\lambda_c$, and λ_u and λ_c are the salting coefficients of the molecules and ions respectively. He plotted $\log K\gamma_u$ against $[r + (1 - r)\alpha_c]m$, and adjusted r till the plot was linear. This gave $r = 0.7$, $\lambda_c = 0.048$ and $\lambda_u = 0.035$ (approx.).

H_2SO_4 solutions have also been subjected to detailed investigations.[33, 39] Four of its lines are common to those of the alkali metal sulphates, and of these, that at 980 cm^{-1} is used to estimate $[SO_4^{2-}]$. By comparison with NH_4HSO_4, the line at 1,040 cm^{-1} can be used to estimate $[HSO_4^-]$. From the available figures, K_{HSO_4} at 25°C is about 0.01, and although this agrees with other estimates (see Chapters 8, 9, 10), the measurements for low acid concentrations are not good enough for a critical assessment. Above 15 M, $[SO_4^{2-}]$ is negligible, and a rough estimate of $K_{H_2SO_4}$ is possible, namely 50 to 2,000, depending on how the activities are interpreted.[40]

Within the limits of the photographic method, $HClO_4$ is completely dissociated up to at least [29, 41] 6 M; for this work, the 935 line of the ClO_4^- ion was used. The Raman evidence for HCl indicates that this also is very strong,[42] but that molecules exist in solutions above 6 M. Iodic acid solutions show bands with Raman shifts [43, 44] of 330, 630, and 780 cm^{-1} and a shoulder indicating a band near 825. On dilution of a concentrated solution, the 780 band shifts to 800; this is attributed to $(HIO_3)_n \rightarrow n\ HIO_3$. On further dilution, the intensity at 800 increases relative to the shoulder at 825; this is attributed to $HIO_3 \rightarrow H^+ + IO_3^-$, for only the 800 band exists in dilute solutions and this is also found in $NaIO_3$. The ratio R of the Raman intensities at 800 and 825 were measured. After correcting for the background,

$$R \text{ corrected} = (\gamma_A C_A J_{A800} + \gamma_I C_I J_{I800})/(\gamma_A C_A J_{A825} + \gamma_I C_I J_{I825}) \qquad (12.22)$$

where $C_A = [HIO_3]$, $C_I = [IO_3^-]$ and the specific Raman coefficient J represents the intensity of the Raman emission from a solution of unit concentration of HIO_3 or IO_3^- (A or I). The γ term represents the factor to reduce the different intensity scales from which the J's are derived to a common scale, i.e. that of the solution under investigation. Since $C_A/C_I = (1 - \alpha_c)/\alpha_c$, then

$$(1 - \alpha_c)/\alpha_c = \gamma_I J_{I800} R_A(R_I - R)/\gamma_A J_{A800} R_I(R - R_A) \qquad (12.23)$$

R_I and R_A were obtained from solutions of $NaIO_3$ and of 1.85 M HIO_3 respectively, it being assumed that the latter is undissociated. The γ_I/γ_A ratios were based on the assumption that $\alpha_c = 0.88$ in a 0.034 M solution, this was derived from N.M.R., and by doing this the α_c values up to 1.3 M HIO_3 agree closely with those obtained by N.M.R. An analogous study by Siebert [45] with $NaIO_4$ and H_5IO_6 is claimed to give $K = 0.023$ for the equilibrium $H_5IO_6 \rightleftharpoons IO_4^- + H^+ + 2H_2O$.

Up to the present time only qualitative information about ionic interaction in non-acidic electrolyte solutions has been obtained. Woodward and his associates [28] made a careful examination of TlOH but could find no Raman line or band attributable to the undissociated form whereas K_{TlOH} from solubilities and kinetics [40] is below 0.4. This means that the ion-pairs are electrostatically bound; the bonding is non-covalent. It was remarked earlier on that Woodward [23] considers Raman effects in such ion-pairs to be of very low intensity. This, for the study of acids is a fortunate accident since it is assumed that the salt references for such work are completely dissociated whereas there is evidence from conductance that this is not strictly correct for e.g. $NaNO_3$ and $(NH_4)_2SO_4$, etc.

A number of Raman effects have been observed with mixtures [46, 67] of the type $MX_n + mKX \rightarrow MX_{n+m}^{m-}$ where M = Cd, Hg, Ga, In, Sn and Ti and X = Cl, Br, I, CN. These are of help in deciding structural forces, and related features. Possibly the study of acid + salt mixtures would be a profitable way of obtaining quantitative information about the acids.

Nuclear Magnetic Resonance

Atomic nuclei, excluding those with even numbers of both neutrons and protons, are regarded as having spin, and as a consequence, possess magnetic moments. These views originate from a suggestion of Pauli in 1924 that the hyperfine structure of certain line spectra are due to magnetic coupling between nuclei and electrons. Since then, a number of other methods have been found and developed for studying and measuring nuclear magnetic moments [48] namely by investigation of the alternating line intensities of the band spectra of homonuclear diatomic molecules, atom-beam methods, microwave spectroscopy,[49] paramagnetic and nuclear magnetic resonance [50] (N.M.R.). The last of these is the outcome of contributions by both Purcell and Bloch,[51] but since the studies of the ionization of acids reviewed in this section have been made with apparatus based on the Bloch method, only this form will be described. The main parts of such NMR spectrometers, as indicated in Fig. 12.3a consist of an electromagnet *MM* and *SS* in which the pole faces (about one ft diameter) are wound so that the magnetic field H_0 is in the *Y* direction. The coils on *MM* are energized by a highly stabilized D.C. supply so that a field of up to about 10,000 gauss (oersteds) is produced. The coils on *SS* are also wound to give a field in the same direction and are connected to an A.C. "sweep generator" operating at low frequency (from about one in several secs up to about 60/sec); this provides a weak adjustable field so that the total field sweeps through a critical (resonance) value. T is a glass tube containing the sample under investigation, and lies in a coil OO wound in the Z, Y plane and connected to a high frequency oscillator which can be set at various frequencies between about 5 and 50 Mcs. T also lies in a coil RR lying in the X, Y plane and connected to a receiver tuned to the frequency of the transmitter. The receiver signals are fed to a pen-recorder.

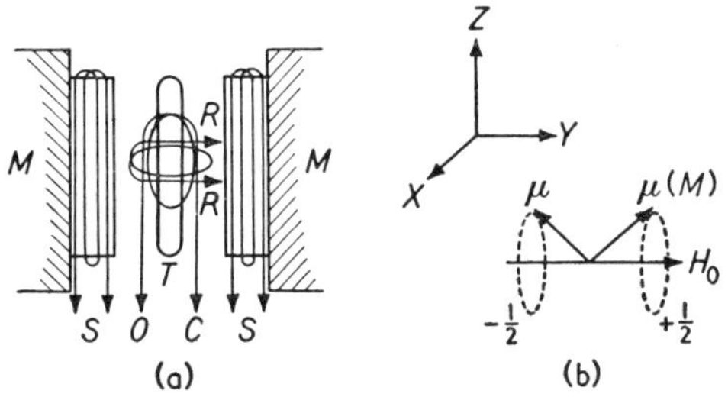

FIG. 12.3. N.M.R. spectrometer.

Nuclear spin values, I, are expressed as multiples of $\frac{1}{2}$. Examples of $I = \frac{1}{2}$ are ^{1}H, ^{19}F, ^{13}C, ^{15}N and ^{29}Si; in all of these the neutron : proton ratio is $n \pm 1 : n$ where n is an integer. Other examples are $^{2}H = 1$, $^{17}O = 5/2$, $^{107}Ag = 7/2$. When $I > \frac{1}{2}$, the charge on the nucleus behaves as if it were distributed over an ellipsoidal surface, and the nucleus has a nuclear quadrupole moment. This particular aspect is dealt with in books [40, 50] and reviews [50, 52] on NMR.

The magnetic moment of a nucleus, μ, is given by

$$\mu = g\mu_n[I(I + 1)]^{\frac{1}{2}} \tag{12.24}$$

where μ_n, the nuclear magneton, is defined by

$$\mu_n = eh/4\pi Mc^* = ehN/4\pi M'c^* = 5.0459 \times 10^{-24} \tag{12.25}$$

in which M is the proton mass, M' its weight on the atomic weight scale, and c^* is the velocity of light. The nuclear factor g is the ratio (magnetic moment)/(angular momentum). It can be determined if μ is measured and I is known; NMR provides one method.

Before H_0 is switched on, the magnetic dipoles are randomly distributed in all directions, and when H_0 is applied, the first thing that happens is that the dipoles precess (gyrate) about the H_0 axis. The permitted values of the μ vectors along H_0 are $I, I - 1, \ldots - I$, i.e., $\frac{1}{2}$ and $-\frac{1}{2}$ for ^{1}H. There will be equal numbers of each, so there is no resultant magnetization of the substance (see Fig. 12.3b). In the presence of H_0, $I = \frac{1}{2}$ is the favoured energy state, so the spin system is not in equilibrium although it follows from the Boltzmann distribution law that owing to opposing thermal agitation, only a slight excess of nuclei could assume the favoured orientation.

The equilibrium distribution is established as a consequence of interactions between atomic and molecular vibrations, and by lattice vibrations. The establishment of equilibrium takes time (time of longitudinal relaxation, T_1), the growth rate of which is expressed by

$$M_Y = M_0\{1 - \exp(-t/T_1)\} \tag{12.26}$$

where M_0 is the equilibrium magnetization of the substance along H_0, and M_Y its value at time t. While equilibrium is being established, energy is transferred from the spin system to its environment.

There is another type of relaxation termed transverse relaxation. Consider the overall magnetization M Fig. 12.3b, which is precessing around H_0. This has a component in the XZ plane. If the precession frequencies become different for different nuclei, the component in the ZY plane will change. The time taken for this to occur is the time of transverse relaxation, T_2. One of the major causes is that different nuclei in a solution may have different nuclear neighbours and therefore be in different magnetic fields.

Now the field from OO is in the ZX plane and as H_0 is altered, the precession frequency of M is altered. At resonance, the two frequencies are equal, some of the nuclei absorb energy from the oscillator signal and are raised to the higher energy state corresponding to $I = -\frac{1}{2}$. Simultaneously the direction of M is altered and a component is produced along the Z axis which precesses in the XY plane. This is the axis along which RR picks up signals. So as the field from SS is varied through resonance, a signal is received which rises to a maximum at the resonance field strength. The resonance frequency is given by

$$h\nu = g\mu_n H_0 = gehH_0/4\pi M \tag{12.27}$$

For the proton,[53] $g = 5.5855$, so if $H_0 = 10^4$ gauss, $\nu = 42.5$ Mcs. To some extent the resonance frequencies depend on the distribution or density of electrons around the nucleus, and consequently upon the nuclear environment. Thus the proton signal from water is at a different frequency from that of H in $-CH_3$, $-COOH$, $-CH_{2}-$, etc., which in turn differ among themselves. If a compound containing such groupings is studied, the frequency spacings are proportional to H_0, and to provide a comparative basis, a parameter s is introduced. This is defined by

$$s = 10^6\{(H_R/H_T) - 1\} + g';\ g' = (2\pi/3)(\chi - \chi_0)10^6 \tag{12.28}$$

where s is termed the resonance or chemical shift. The term H_R and H_T are the fields for resonance with reference (often water) and test solutions, g' is a correction term for the interaction of electrons with H; χ_0 and χ are the volume magnetic susceptibilities of the reference and test solutions.

By means of a high resolution NMR spectrometer, Hood and Redlich have obtained a number of accurate observations relating to the dissociation of acids. The instrument operated at 40 Mcs with $H_0 = 9{,}400$ gauss and a sweep of 150 milligauss. The resonance shifts were determined by superimposing an A.F. of 40 cps on the sweep field to produce side band peaks to the water reference resonance,[54] spaced 9.4 milligauss apart. The χ and χ_0 were obtained by a method [55] which uses the spectrometer. The samples were sealed into the upper compartments of pyrex capillary tubes of 3 mm i.d., each sample being several cms. in length. The lower compartment of all tubes contained water. After obtaining records for the latter, the tube was pushed down so that the sample was in the field. This change could be made in less than 1 sec (sweep periodicity = 5 sec). Gutowsky and Saika [56] and Masuda and Kanda [57] have also shown that proton resonance shifts (and ^{15}N in HNO_3) can provide information concerning the dissociation of acids.

Hood, *et al.*[58] have used the treatment of Gutowsky and Saika [56] to analyze the resonance shifts of HNO_3, $HClO_4$ and HCl solutions. The proton resonance in a free and a hydrogen bonded OH are different.[54] Since the presence of any solute shifts the association equilibrium, the proton resonance value is indirectly influenced by the solute; even salts have effects.[59] However, as a first approximation, it is assumed that the contribution of water to the proton resonance is a linear function of the mole fractions of the dissolved species, that is to say, the influence of the addition of a certain species is independent of the composition of the solution.

Representing the dissociation of a monoprotic acid by $HL + H_2O \rightleftharpoons H_3O^+ + L^-$, if N_1 and N_2 are the stoichiometric mole fractions of HL and H_2O, and p is the stoichiometric fraction of protons in the form H_3O^+, then

$$p = 3N_1/\{2(N_2 - N_1) + 3N_1\} = 3N_1/(2N_2 + N_1) \qquad (12.29)$$

Taking [53] HNO_3, for $p = 1$, corresponding to $H_3O^+NO_3^-$, $s = s_2 = 4.25$, and for $p = 3$, corresponding to $(HNO_3)_2$ (other evidence indicates that dimerization occurs), $s = s_3 = 7.00$. Within this range the plot of s against p is linear indicating that only these two species are present. From the limiting slope of the plot between $p = 0$ and $p = 0.466$, s_1 for $H_3O^+ = 11.82$. When $p < 1$, the degree of dissociation is given by

$$s/p = s_1\alpha_c + s_2(1 - \alpha_c) \qquad (12.30)$$

Some data are shown in Table 12.2, using interpolated activity coefficients [33] and, from (12.19–20),

$$K_m\gamma_u = m\gamma_\pm^2/(1 - \alpha_c) \qquad (12.31)$$

A plot of log $K_m\gamma_u$ against M extrapolates to $K_m = 30 \pm 5$. Taking other results into accounts,[58] $K = 22$.

In the same way,[58] $K_{HClO_4} = 38$, and [60] $K_{CF_3COOH} = 1.8$ by both proton and F^- resonance (deriving $\gamma_\pm$ from freezing-point data [57]). Some other results are [44] $K_{HIO_3} = 0.18$ and [61] 1.3 for heptafluorobutyric acid. By conductance,[62] $K_{CF_3COOH} = 0.59$, and $K_{HIO_3} = 0.18$.

In [63] the most extensive of several NMR studies of H_2SO_4, several experimental improvements were introduced, namely the use of a sample spinner,[64] a flux stabilizer [65] and a triangular wave sweep generator. The first of these improves the band widening by smoothing out nuclear spin effects from constituents in the glass tube holding the sample. The samples were sealed in tubes which were in turn sealed in larger tubes, filling the anulus with

TABLE 12.2

N.M.R. Data for K_{HNO_3}

C	4.0	6.0	8.0	10.0	12.0	14.1
M	4.7	7.4	10.5	15.4	20.3	28.5
p	0.119	0.187	0.265	0.355	0.466	0.612
g'	0.15	0.17	0.19	0.23	0.25	0.27
s	1.24	1.89	2.41	2.98	3.20	3.65
α_c	0.815	0.774	0.641	0.548	0.346	0.226
$\gamma_\pm$	1.05	1.29	1.53	1.82	2.02	2.34
$\log K_m\gamma_u$	1.45	1.74	1.83	2.05	2.10	2.30

water. The sine wave output of a second generator was superimposed on the sweep, producing side bands on the resonances from the sample and the water reference. The frequency of the second generator was adjusted to bring the first side band of the sample close to that of the water, and the amplitude of the applied voltage was adjusted until these two side bands were of equal intensity. The exact frequency separation was found by interpolation.

Assuming that s is a function of the mole fractions of hydrogen in H_3O^+, HSO_4^- and H_2SO_4,

$$2s = x(1 - \alpha_1)s_3 + x\alpha_1(1 - \alpha_2)s_2 + 3x\alpha_1(1 + \alpha_2)s_1 \quad (12.32)$$

$$s/x = s_3 + \alpha_1 s_4 + \alpha_1\alpha_2 s_5 \quad (12.33)$$

with $s_4 = 1.5s_1 + 0.5s_2 - s_3$; $s_5 = 1.5s_1 - 0.5s_2$, where s_1, s_2 and s_3 are the shifts of H_3O^+, HSO_4^- and H_2SO_4, while α_1 and α_2 are the degrees of dissociation of H_2SO_4 and HSO_4^-. The value of s at 100 % acid was taken for s_3 (= 6.15), and s_4 was obtained from plot of $(s/x - s_3)x/(1 - x) = 8.10$ at $x = 1$. A rough estimate of s_5 was derived from $K_{HSO_4^-} = 0.01$; it is between 20 and 30. These figures gave $s_1 = 13.1$. In general, calculation proved α_1 and α_2 to be close to figures obtained by Raman spectra, but no independent value for $K_{HSO_4^-}$ could be derived.

It can be seen from Table 12.3 that the values of s/C for both HCl and $HClO_4$ are constant, within experimental error, over the whole range of concentrations which have been studied (but see remarks in Ref. 72). This implies that the proton shifts are related to [H_3O^+], and agrees with the accepted views that these acids dissociate completely (see Raman spectra); both sets average $s/C = 0.313$. The figures for HNO_3 show a decrease down to 18C. Reasonable extrapolations from s/C against C are $s/C = 0.325$ at $C = 0$ and $s/C = 0.23$ at $C = 20$, so that $1 - \alpha_c = (0.325 - s/C)/0.095$. By means of this, and the $\gamma_\pm$ values of Table 12.2, a plot of the derived values of log $K_m\gamma_u$ against M gives $K_m = 32 \pm 8$.

TABLE 12.3

Values of s/C, and K Values of HNO_3 and HIO_3 from N.M.R.

$C(HCl)$	2.00	3.24	4.25	6.07	2.02	4.03	6.05	8.06	10.09	12.10
$s/C(HCl)$	0.31	0.330	0.301	0.310	0.322	0.318	0.301	0.312	0.322	0.304
$C(HClO_4)$	5.97	7.95	9.95	16.82	17.60	3.94	5.92	7.90	9.86	11.84
$s/C(HClO_4)$	0.30	0.303	0.328	0.339	0.302	0.28	0.30	0.316	0.314	0.345
$C(HNO_3)$	4.00	5.99	7.99	9.97	11.99	14.06	15.98	17.93	21.95	23.98
$s/C(HNO_3)$	0.313	0.315	0.301	0.299	0.267	0.259	0.241	0.240	0.248	0.296
$\log K_m\gamma_u^{HNO_3}$	1.59	(2.38)	1.96	2.25	2.11	2.33	2.30	—	—	—
$C(HIO_3)$	0.034	0.076	0.104	0.160	0.362	0.662	1.020	1.318	1.850	3.292
$s/C(HIO_3)$	0.35	0.31	0.295	0.300	0.249	0.211	0.196	0.190	0.168	0.173
$\gamma_\pm(HIO_3)$	—	0.63	0.58	0.50	0.35	0.25	0.19	0.15	—	—
$-\log K_c\gamma_u^{HIO_3}$	—	0.82	0.87	0.80	1.06	1.23	1.32	1.43	—	—

A plot of s/C against C for HIO_3 extrapolates to about 0.36 at $C = 0$, falls to a minimum of about 0.16 at $C = 2.5$, and then rises again. Taking these figures, interpolated [37] $\gamma_\pm$ values, and (12.31, with C for M), extrapolation gives $K_c = 0.18$, in good agreement with [44] the conclusions reached by other methods. Nevertheless there are some uncertain and obscure features. The extrapolations at both ends of the scale are not obtainable with much accuracy, particularly that corresponding to $\alpha_c = 0$; this may because of the complication of polymerization mentioned in the previous section. Some of the devices of spectrophotometry which have been described in Chapter 10 such as obtaining measurements with a small fixed concentration of acid and varying amount of anion added as salt, and work at constant ionic strength may prove useful in future applications of NMR.

With [63] H_2SO_4, s/C decreases from about 0.700 at $C = 0$ to about 0.45 between 4 and 8C after which there is a slow decrease to about 0.34 at 18.5C. It is not possible to calculate $K_{HSO_4^-}$ from the lower concentrations, where [HSO_4^-] is significant, owing to the scatter.

The NMR frequency of the F^- ion is shifted by Na^+, K^+ and Al^{3+} ions,[66] and those of the Tl^+ and Tl^{3+} ions are shifted by certain anions.[67] These shifts may be attributed to ion-association although Connick and Poulson [68] report that F^- shifts by metal cations are not related to the cation electronegativities or to the stabilities of the complexes; cations of high atomic number produce a strong decrease in the magnetic shielding which is roughly correlated with the term (atomic number/interatomic distance). The proton magnetic resonance of water shows shifts in the presence of diamagnetic salts which are interpreted [69] in terms of the breakdown of the hydrogen-bonded structures of water and of the polarization of water molecules. Polyvalent cations cause (positive) shifts which are proportional to the salt concentration. An exception to this is $ZnCl_2$, which may be because of its incomplete dissociation; this is supported by the finding that the weak salt $HgCl_2$ does not produce a measurable shift.

Some other elements with isotopes for which chemical shifts have been reported are B, ^{17}O, N, Si, ^{35}Cl, ^{81}Br, Co and ^{207}Pb; these are reviewed in an article by Brownstein on high resolution NMR and molecular structure.[70] It has also been found [71] that the absorption line of ^{23}Na in certain aqueous solutions is broadened and decreased in amplitude, notably by phosphates, hydroxy and keto acids, and EDTA. These effects are attributed to an interaction of the ^{23}Na nuclear quadrupole with an electric field gradient and are interpreted in terms of weak sodium complexes.

REFERENCES

1. Kolthoff, I. M. and Lingane, J. J. *Chem. Rev.* **24,** 1 (1939).
2. Lingane, J. J. *ibid.* **29,** 1 (1942).
3. Kolthoff, I. M. and Lingane, J. J. "Polarography." Interscience Pub. Ltd., New York & London, 2nd edn., I and II (1952). Milner, G. W. C. "The Principles and Applications of Polarography." Longmans Green and Co., Ltd., London. Meites, L. "Polarographic Techniques." Interscience Pub. Ltd. (1955).
4. Heyrovsky, J. *Chem. Listy.* **16,** 256 (1922); *Phil. Mag.* **45,** 303 (1923). Heyrovsky, J. and Shikata, M. *Rec. trav. Chim.* **44,** 496 (1925).
5. Ilkovic, D. *Coll. Czech., Chem., Commun.* **6,** 498 (1934).
6. Loveland, J. W. and Elving, P. J. *Chem. Rev.* **51,** 67 (1952).
7. Barker, G. C. and Jenkins, I. L. *Analyst* **77,** 685 (1952). Ferrett, D. J. and Milner, G. W. C. *ibid.* **79,** 731 (1954); **80,** 132 (1955). Hamm, R. E. *Anal. Chem.* **30,** 350 (1958).
8. DeFord, D. D. and Hume, D. N. *J. Amer. chem. Soc.* **73,** 5321 (1951).
9. Sartori, G. S. *J. inorg. nucl. Chem.* **8,** 196 (1958).
10. Hume, D. N., De Ford, D. D. and Cave, G. C. B. *J. Amer. chem. Soc.* **73,** 5323 (1951).
11. Meites, L. *ibid.* **73,** 3727 (1951).
12. Rebertus, R. L., Laitmen, H. A. and Bailar, J. C. *ibid.* **75,** 3051 (1953).
13. Hershenson, H. H., Smith, M. E. and Hume, D. N. *ibid.* **75,** 507 (1953).
14. Burns, E. A. and Hume, D. N. *ibid.* **78,** 3958 (1956).
15. Hershenson, H. M., Brooks, R. T. and Murphy, M. E. *ibid.* **79,** 2046 (1957).
16. Li, N. C., Doody, E., and White, J. M. *ibid.* **75,** 4184 (1952); **76,** 221 (1954); **78,** 5218 (1956); **80,** 5678, 5901 (1958).

17. Iwase, I. *J. chem. Soc. Japan* **78**, 1689 (1957). Kwang-hsien, H. and Hon-gee, T. *Acta Chim. Sinica*, **24**, 277 (1958).
18. Jonassen, H. B., Bertrand, J. A., Groves, F. G. and Stearns, R. I. *J. Amer. chem. Soc.* **79**, 4279 (1957).
19. Lane, T. J., Ryan, J. A. and Britten, E. F. *ibid.* **80**, 315 (1958).
20. Vinogradova, E. N. and Pedanova, V. G. *Motody Analiza Redkikh i Tsvet. Metal. Sbornik*, 25 (1956).
21. Gordienko, A. A. and Shamsiev, A. Sh. *Proc. Acad. Sci. Uzbek. S.S.R.* No. 4, 33 (1958).
21a. Migal, P. K., Grinberg, N. Kh. and Tur'yan, Ya. I. *Russ. J. Inorg. Chem.* [in English Chem. Soc., 833 (1959)].
22. Cozzi, D. and Pantani, F. *J. inorg. nucl. Chem.* **8**, 385 (1958).
23. Fomin, V. V., Vorob'ev, S. P. and Andreava, M. A. *Sov. J. At. Energy*, **4**, No. 1, 63 (1958).
24. Raman, C. V. and Krishnan, K. S. *Nature, Lond.* **121**, 501, 619 (1928). *Indian J. Phys.* **2**, 387 (1928).
25. Herzberg, G. "Molecular Spectra and Molecular Structure," II, "Infra-red and Raman Spectra of Polyatomic Molecules." Van Nostrand and Co., Inc., Princeton (1945). Hibben, T. H. "The Raman Effect and its Chemical Applications." Reinhold Publishing Corp., New York (1939). Partington, J. R. "An Advanced Treatise on Physical Chemistry." Vol. IV, Longmans Green and Co., London, New York (1953).
26. Rao, I. R. *Proc. roy. Soc.* **144A**, 159 (1934).
27. Placzek, G. "Handbuch der Radiologie," **6**, 205, 366 (1934).
28. George, J. H. B., Rolfe, J. A. and Woodward, L. A. *Trans. Faraday Soc.* **49**, 375 (1953).
29. Redlich, O., Holt, E. K. and Bigeleisen, J. *J. Amer. chem. Soc.* **66**, 13 (1944).
30. Tolansky, S. "High Resolution Spectroscopy." Methuen and Co., Ltd., London (1947).
31. Redlich, O. and Nielson, L. E. *J. Amer. chem. Soc.* **65**, 654 (1943).
32. Ingold, C. K. and Millen, D. J. *J. chem. Soc.* 2612 (1950), and further references quoted therein.
33. Ref. 11, Ch. 1.
34. Hamer, W. J. "The Structure of Electrolytic Solutions." J. Wiley and Sons, Inc., New York (1959). Article by Young, T. F., Maranville, L. F. and Smith, H. M.
35. Krawetz, A. A. Thesis, Chicago (1955). See ref. 34.
36. Redlich, O. and Bigeleisen, J. *J. Amer. chem. Soc.* **65**, 1883 (1943).
37. Landolt-Bornstein "Tabellen." J. Springer, Berlin, 5th edn., 2nd and 3rd supp. (1931, 1936).
38. McKay, H. A. C. *Trans. Faraday Soc.* **52**, 1568 (1956).
39. Millen, D. J. *J. chem. Soc.* 2589 (1950).
40. Deno, N. C. and Taft, R. W. *J. Amer. chem. Soc.* **76**, 244 (1954). Wyatt, P. A. H. *Disc. Faraday Soc.* **24**, 162 (1957).
41. Reilly, C. A. *J. chem. Phys.* **22**, 2067 (1954).
42. Karetnikov, G. S. *J. phys. Chem. Moscow* **28**, 1331 (1954).
43. Rao, N. R. *Indian J. Phys.* **16**, 71 (1942).
44. Hood, G. C., Jones, A. C. and Reilly, C. A. *J. phys. Chem.* **63**, 101 (1959).
45. Siebert, H. *Z. anorg. Chem.* **273**, 21 (1953).
46. Delwaulle, M. and others: *C.R. Acad. Sci., Paris* **206**, 187, 1108, 1965 (1938); **207**, 340 (1938); **208**, 184, 1002, 1818 (1939); **211**, 65 (1940); **212**, 761 (1941); **219**, 64

(1944); **220,** 173 (1945); **227,** 1229 (1948); **238,** 84 (1954); **240,** 2132 (1955). *J. Amer. chem. Soc.* **74,** 5678 (1952).
47. Woodward, L. A. and others, *Trans. Faraday Soc.* **50,** 1275 (1954). *J. Chem. Soc.* 1699, 2655 (1955); 3721 (1956); 1055 (1959).
48. Kopferman, H. "Nuclear Moments." translated by Schneider, E. E. Academic Press, Inc., New York (1958).
49. Gordy, W., Smith, W. V. and Trambarulo, R. F. "Microwave Spectroscopy." J. Wiley, and Sons, Inc., New York (1953). "Microwave and R. F. Spectroscopy." *Disc. Faraday Soc.* **19,** (1955).
50. Smith, J. A. S. *Quart. Rev.* **7,** 279 (1953) Wertz, J. E. *Chem. Rev.* **55,** 829 (1955). Andrew, E. R. "Nuclear Magnetic Resonance." Cambridge University Press, Cambridge, U.K., (1956). Roberts, J. D. "Nuclear Magentic Resonance." McGraw-Hill Book Co., New York, London (1959).
51. Purcell, E. M., Torrey, H. C. and Pound, R. V. *Phys. Rev.* **69,** 37 (1946). Bloch, F. Hansen, W. W. and Packard, M. E. *ibid.* **69,** 127 (1946). Mays, J. M., Moore, H. R. and Shulman, R. G. *Rev. sci. Instrum.* **29,** 300 (1958).
52. Wertz, J. E. *Chem. rev.* **55,** 829 (1955).
53. Hipple, J. A., Sommer, H. and Thomas, H. A. *Phys. Rev.* **76,** 1877 (1949); **80,** 487 (1950).
54. Arnold, J. T. and Packard, M. E. *J. chem. Phys.* **19,** 1608 (1951).
55. Reilly, C. A., McConnell, H. M. and Meisenheimer, R. G. *Phys. rev.* **98,** 264 (1955).
56. Gutowsky, H. S. and Saika, A. *J. chem. Phys.* **21,** 1688 (1953).
57. Masuda, Y. and Kanda, T. *J. phys. Soc., Japan* **8,** 432 (1953).
58. Hood, G. C., Redlich, O. and Reilly, C. A. *J. chem. Phys.* **22,** 2067 (1954).
59. Shoolery, J. M. and Alder, B. J. *ibid.* **23,** 805 (1955).
60. Hood, G. C., Redlich, O. and Reilly, C. A. *loc. cit.* 2229.
61. Hood, G. C. and Reilly, C. A. *ibid.* **28,** 329 (1958).
62. Henne, A. L. and Fox, C. T. *J. Amer. chem. Soc.* **73,** 2323 (1951).
63. Hood, G. C. and Reilly, C. A. *J. chem. Phys.* **27,** 1126 (1957).
64. Bloch, F. *Phys. Rev.* **94,** 497 (1954). Anderson, W. A. and Arnold, J. T. *ibid.* **94,** 497 (1954).
65. Lloyd, J. P. *Amer. phys. Soc.* Ser. II, **1,** 92 (1956).
66. Connick, R. E. and Poulson, R. E. *J. Amer. chem. Soc.* **79,** 5153 (1957). *J. chem. Phys.* **62,** 1002 (1958).
67. Gutowsky, H. S. and McGarvey, B. R. *Phys. Rev.* **91,** 81 (1953).
68. Connick, R. E. and Poulson, R. E. *J. phys. Chem.* **63,** 568 (1959).
69. Schoolery, J. N. and Adler, B. J. *J. chem. Phys.* **23,** 805 (1955).
70. Brownstein, S. *Chem. Rev.* **59,** 463 (1959).
71. Jardetzky, O. and Wertz, J. E. *J. Amer. chem. Soc.* **82,** 318 (1960).
72. Hood, G. C. and Reilly, C. A. *J. chem. Phys.* **32,** 127 (1960) have reported more accurate data for $HClO_4$ and HNO_3 solutions, which owing to improvements in experimental technique, extend to dilute concentrations. Plots of s/C fall off markedly at low values of C; this could be attributed to the g′ factor of (12.28) being slightly too small (it is a correction term).

CHAPTER 13

INCOMPLETE DISSOCIATION, PART VI: REACTION KINETICS

THE contents of the present chapter are mainly restricted to considerations of the effect of ion-association upon the rates of reactions involving ions, taking more or less for granted the more general and fundamental aspects of this subject as described in numerous books and articles.[1]

Ion-Ion Chemical Reactions

Although most chemical reactions between ions take place with extreme rapidity, there are a certain number which are slow enough to follow by the conventional analytical and physical methods of studying the rates and mechanisms of reaction processes. Under fixed physical conditions of the type $A + B \rightarrow X \rightarrow C + D$, where X is a transition state or activated complex with a charge equal to the sum of the charges on A and B, the velocity of reaction, v, is expressed by

$$v = k[A][B] = d[X]/dt = k_x[X] \qquad (13.1)$$

It is assumed that X is in equilibrium with A and B, and an equilibrium constant expression can be written,

$$K = [X]f_x/[A][B]f_A f_B \qquad (13.2)$$

whence

$$k = k_x K f_A f_B / f_X = k^\circ f_A f_B / f_x \qquad (13.3)$$

where $k^\circ = k_x K = k$ at $I = 0$ (the effect of ionic strength on rates of reactions is often termed the primary salt effect). Providing the observations are confined to dilute solutions, and assuming the absence of ion-pairs, it should be possible to correlate k and k° by the D.H. activity coefficient expression,

$$-\log f_i = A'z_i^2\{\sqrt{I}/(1 + B'a\sqrt{I}) - Q'I\} = A'z_i^2\phi(I) \qquad (13.4)$$

for since $z_x^2 = (z_A + z_B)^2$, (13.3) and (13.4) lead to

$$\log k^\circ = \log k - 2z_A z_B A'\phi(I) \qquad (13.5)$$

The above expressions summarize the theories of Bronsted [2] and Bjerrum [3] on ionic rate processes; they have been verified by the data obtained for a number of reactions, e.g., [4, 5, 6]

$$(z_A z_B = 4)\ 2[Co(NH_3)_5Br]^{2+} + Hg^{2+} + 2H_2O = 2[Co(NH_3)_5H_2O]^{3+} + HgBr_2 \qquad (13.6)$$

$(z_A z_B = -2)$ $[Co(NH_3)_5Br]^{2+} + OH^- = [Co(NH_3)_5OH]^{2+} + Br^-$ (13.7)

$(z_A z_B = -1)$ $H_2O_2 + 2H^+ + Br^- = 2H_2O + Br_2$ (13.8)

$(z_A z_B = 2)$ $CH_2BrCOO^- + S_2O_3^{2-} = CH_2S_2O_3COO^{2-} + Br^-$ (13.9)

Example 13.1. La Mer and Fessenden [6] followed the rate of reaction between equimolar solutions of CH_2Br . COOK and $K_2S_2O_3$ at 25°C. The values of k were obtained by the standard "bimolecular" formula $k = x/at(a - x)$ where a is the initial concentration of each reactant and x is the amount of each changed in time t (in min); some of the figures are given in Table 13.1.

TABLE 13.1

The $BrAc^-$, $S_2O_3^{2-}$ Reaction (K^+ salts)

10^3a	0	0.50	0.75	1.00	1.60	2.50	5.00	10.00	20.00	25.00
10^2I	0	0.20	0.30	0.40	0.64	1.00	2.00	4.00	8.00	10.00
$-\log k_{obs.}$	0	0.525	0.504	0.482	0.451	0.431	0.368	0.311	0.224	0.199
$-\log k_{calc.}$	0.605	0.520	0.502	0.487	0.455	0.429	0.371	0.302	0.227	0.205

To obtain $k°$, $B'a$ was taken as unity and Q' as zero in (13.4–5), and the plot of the resulting $\log k°'$ values against I extrapolated, giving $-\log k° = 0.605 \pm 0.005$. The slope of the line gives $2z_A z_B A'Q'$, whence, with $A' = 0.5092$, $Q' = 0.44$. These results were used in (13.4–5) to obtain the $\log k_{calc.}$ values of the Table. Guggenheim and Prue [7] have made similar calculations both of this and of reaction (13.8).

It has been assumed in the above calculations that no ion-pairs form between K^+ and the reactant and product anions; this is not quite true since [8] $K_{KS_2O_3^-} = 0.11$, but there is not much error by ignoring this at low concentrations.[9]

In the presence of polyvalent cations, the rate of reaction (13.9) is enhanced. Davies, Wyatt and Williams [9, 10] have shown that this can be explained in terms of ion-pairing. Considering the reaction when M^{2+} ions are present, the possible reactant ion-pairs are MS_2O_3 and $MBrAc^+$, so if allowance is made for all the reaction routes,

$$v = k_1^0[S_2O_3^{2-}][BrAc^-]f_2f_1/f_3 + k_2^0[MS_2O_3][BrAc^-]f_1/f_1$$
$$+ k_3^0[S_2O_3^{2-}][MBrAc^+]f_2f_1/f_1 + k_4^0[MS_2O_3][MBrAc^+]f_1/f_1 \quad (13.10)$$

where f_3, f_2 and f_1 refer to species of charges 3, 2 and 1 respectively. Since the contribution of the k_3 term turns out to be a minor one, that due to k_4 may be neglected since, on the basis of ionic types, $k_4 < k_3$, and the product $[MS_2O_3][MBrAc^+]$ is always very small. Then with

$$K_1 = [M^{2+}][S_2O_3^{2-}]f_2^2/[M_2S_2O_3], \text{ and } K_2 = [M^{2+}][BrAc^-]f_2f_1/[MBrAc^+]f_1 \quad (13.11)$$

there is derived

$$v = [S_2O_3^{2-}][BrAc^-]\{k_1^0 f_2 f_1/f_3 + [M^{2+}]f_2^2(k_2^0/K_1 + k_3^0/K_2)\} \quad (13.12)$$

Putting k' for $k_2^0/K_1 + k_3^0/K_2$, then since $v = k_{obs.}C_1C_2$, where C_1 and C_2 are the stoichiometric molarities of thiosulphate and bromoacetate, it follows that

$$\log (k_{obs.}C_1C_2 - k_1^0[S_2O_3^{2-}][BrAc^-]f_1f_2/f_3) - \log ([S_2O_3^{2-}][BrAc^-][M^{2+}]) + 8A'\phi(I) = \log k' \quad (13.13)$$

Example 13.2. Davies and Williams [9] have analysed the data of Von Kiss and Vass[11] where M = Mg^{2+}, taking [12] $K_1 = 0.0145$ and [13] $K_2 = 0.28$, and $B'a = 1.0$, $Q' = 0.2$. Some of the data are given in Table 13.2 using $-\log k_1^0 = 0.605$ from Example 13.1. Within possible experimental error and uncertainties in the above values, $k' = 112$.

TABLE 13.2

Derivation of k' for the MgS_2O_3, $Mg(BrAc)_2$ Reaction at 25°C

$10^3C_1 = C_2$	$k_{obs.}$	$10^3[MgS_2O_2]$	$10^3[MgBrAc^+]$	10^3I	$10^2\phi(I)$	$\log k'$
0.50	0.352	0.016	0.001	2.69	4.88	2.05
1.00	0.408	0.051	0.004	5.29	6.68	2.07_5
2.00	0.485	0.157	0.013	10.35	9.03	2.09
5.00	0.583	0.615	0.066	24.93	13.14	2.03
12.50	0.720	2.13	0.30	59.35	18.41	2.01_5
20.00	0.801	3.90	0.65	94.10	21.60	2.01

Similar calculations have been made [9] for the K^+, Ca^{2+} and Ba^{2+} systems. The ratios $\log (k'/k_1^0)$, if the enhanced rates are entirely due to the reduction in charge resulting in a greater probability of cation containing activated complexes, should be related to the free energy changes for the association of cations with the $(BrAc,S_2O_3)^{3-}$ complex. The similarity between the ratios and the pKs of the ferricyanide dissociation constants supports this hypothesis.[9]

From an examination of the rates of some ionic reactions in the presence of "inert" salts, Olson and Simonson [14] have expressed the view that the rates of reactions between anions are dependent on the character and concentration of the inert cation, those of reactant cations depend likewise on the inert anion, and cation, anion reactions are similarly influenced by both of the inert salt ions. In other words, it is their opinion that the rates are not dependent on the ionic strength as such. Thus with reaction (13.9), they found the plot of log k in the presence of Na^+ and K^+ salts against $\sqrt{M^+}$ gave a smooth curve (although data with Mg^{2+} present could not be fitted in). Similarly, reaction (13.7) gives two distinct curves on plotting log k against $\sqrt{I}$ when the added salts are NaBr and Na_2SO_4. However, Davies and Williams [10] have shown that their ion-association treatment can account for all these apparent deviations from (13.5); thus for reaction (13.9) in the presence of $MgSO_4$ and $Mg(NO_3)_2$, the data fall into line when (13.13) is applied after making allowance for $MgSO_4$ ion-pairs. In the same way, for reaction (13.7) in the presence of Na_2SO_4, it is necessary to allow for $NaSO_4^-$ and PSO_4 ion-pairs ($K_{PSO_4} = 0.0028$, $P^{2+} = [Co(NH_3)_5Br]^{2+}$) and for the $PSO_4 + OH^-$ reaction. Proceeding as before,

$$\log (k_{obs.}C_1C_2 - k_1^0[P^{2+}][OH^-]f_2) - \log [P^{2+}][SO_4^{2-}][OH^-]f_2^2 = \log k' \tag{13.14}$$

From the calculations [10], $\log k' = 3.8$, whence $\log k_2^0 = 1.3$, whereas $\log k_1^0 = 2.1$. This difference in k_2^0 and k_1^0 is of the expected order considering the difference in charge signs of the reactants.

Olson and Simonson [14] also studied reaction (13.6) and found the plot of log k against $\sqrt{I}$ with varying amounts of Hg $(ClO_4)_2$ to be considerably displaced from the theoretical line. Furthermore, in the presence of $NaClO_4$ and $La(ClO_4)_3$, the plot of log k against $[ClO_4^-]$ gives a smooth curve. This reaction is obviously complex but one of the factors which has a major influence is that $Hg(ClO_4)_2$ is a weak electrolyte [15] ($K_{HgClO_4} \sim 0.001$), and the concentration of Hg^{2+} is far from the stoichiometric value.

Another slow ionic reaction for which salt effects have been observed is

$$S_2O_8^{2-} + 2I^- = 2SO_4^{2-} + I_2 \tag{13.15}$$

where (M — H)[1] the rate determining step is probably $S_2O_8^{2-} + I^- = S_2O_8I^{3-}$. Indelli and Prue [16] have given references to previous work in their own report for which they used a polarized electrode to determine the time of reappearance of I_2 after the injection of $Na_2S_2O_3$. In their view, the association of the activated complex with cations cannot alone explain the results since many of the salt effects cannot be explained by the conventional theory and because many anions also cause specific effects. The dissociation constants of persulphates are unknown but are probably of the same order as those of the corresponding sulphates; if so, the enhanced rates due to the addition of $LaCl_3$ and $Co(NH_3)_6Cl_3$ have at least a partial explanation for the ion-pairs $MS_2O_3^+$ will react faster than $S_2O_8^{2-}$ alone. Also, ion-association can partly explain some of those rates which are less than expected, e.g., in solutions containing $Na_3P_3O_9$, for which [17] $K_{NaP_3O_9^{2-}} = 0.009$. On the other hand, the reduced rates caused by NMe_4Br and NEt_4Br probably have some other explanation.

Some other systems where the specific salt effects may be produced by ion-pairs are (a) [18] $Fe^{2+} + Co(Ox)_3^{3-} = Fe^{3+} + Co^{2+} + 3\,Ox^{2-}$, (b) [19] the reversible reaction $Fe(CN)_6^{3-} + I^- = Fe(CN)_6^{4-} + \frac{1}{2}I_2$ and (c) [20] $Fe^{3+} + I^-$

$= Fe^{2+} + \frac{1}{2}I_2$ (which is retarded by OH^-, Cl^-, Br^-, NO_3^-, and SO_4^{2-}), (d) [55] $Co(OH)^{2+} + CeX^{2+}(X^- = ClO_4^-, NO_3^-, F^-)$.

Many reactions of complex ions, particularly those of Co^{3+}, Cr^{3+} and Pt^{2+}, undergo substitution reactions (including solvolysis), cis-trans isomerization and racemization, oxidation-reduction and catalytic reactions at measurable rates. The many studies of these which have been made form the subject-matter of various books and articles by Basolo and Pearson,[21] Ingold [22] and others.[1, 23] Only a few aspects will be described here. Firstly, Brown, Ingold and Nyholm,[24] from studies of the kinetics of substitutions that displace Cl^- from *cis*$[Coen_2Cl_2]^+$ in CH_3OH, postulated that for $X^- = NO_3^-, Cl^-, Br^-$ and NCS^-, substitution proceeds according to a first order law. The rates depend on the concentration of $[Coen_2Cl_2]^+$ but not on $[X^-]$. This is termed a nucleophilic S_N1 reaction and the mechanism is taken to be

$$[Coen_2Cl_2]^+ \xrightarrow{\text{slow}} [Coen_2Cl]^{2+} \underset{X^-}{\xrightarrow{\text{fast}}} [Coen_2ClX]^+$$

With [24] LiCl at 35.8°C, the reaction rate, k_1, increases by about 50 % from $I = 0$ to 0.09. Somewhat larger increases occur with LiCNS and $LiNO_3$. Pearson, Henry and Basolo [25] have stated that *cis* $Coen_2Cl_2^+$ forms ion-pairs with Cl^- and Br^- ($K = 0.075$ and 0.3 at $I = 0.02$). These were derived from spectral shifts. On the other hand, the *trans* form does not associate with these anions. Also, from conductance,[26] LiCl, LiCNS and $LiNO_3$ are highly ionized in CH_3OH, so in general, there is not much evidence for ascribing rate increases with increasing salt concentration to ion-pairing. Brasted and Hirayama [27] have followed the isomerization of *cis* $Co\textit{en}_2Cl_2^+$ and $Co\textit{pn}_2Cl_2$ in CH_3OH, C_2H_5OH and C_3H_7OH in the presence of LiCl, NaCl and $NaClO_4$. Their main conclusions are that the ratio [Cl] : [complex ion] is significant; with NaCl and the *en* salt, k_1 rises to a maximum at a ratio of 69 and then falls as $[Cl^-]$ increases. Secondly, the isomerization of the *pn* salt is decreased slightly by the presence of the salts. Thirdly, in LiCl solutions, the rate is increased by traces of water. Clearly more work is required to clarify the position.

When $X_2^- = NO_3^-$, N^- and CH_3O^-, Ingold and his associates [24] consider that substitution proceeds partly or wholly in accordance with a second rate law; part at least of the rate depends on $[X^-]$, or what is termed a bimolecular nucleophilic substitution, S_N2:

$$X^- + Coen_2Cl_2^+ \rightarrow Coen_2ClX^+ + Cl^-$$

Pearson and his associates [25] have shown that the rates of these reactions are considerably reduced on buffering the solutions, the effect being the greater with the *trans* isomer. They suggest other mechanisms such as intervention of the solvent and ion-pairing of the complex; on the other hand they do not take into account reaction with undissociated HX, although this is not so important as the main reaction. In support of this is their finding [28] that such a reaction applies in the case of $Co(NH_3)_5H_2O^{3+}$ + acidic nitrite → $Co(NH_3)_5ONO^{2+}$, for which the results can be expressed by: rate = k[aquo complex][HNO_2][NO_2^-].

Nyholm and Tobe [29] have reported k_2 values for the S_N2 reaction OH^- + *trans* $Coen_2NH_3Cl^{2+}$ → *cis* and *trans* $Coen_2NH_3(OH)^{2+}$ + Cl^- at 0°C in the presence of increasing amounts of NaOH. The results trend from about 1.4 at $I = 0.005$ down to about 1.0 at $I = 0.025$. On applying (13.5), with $A' = 0.488$, k_2^0 averages 1.83 ± 0.05 l./sec mole (omitting two results in which $NaClO_4$ was added, which give $k_2^0 = 1.62$ and 1.68). This treatment omits any consideration of ion-pairing between the *trans* Cl complex and OH^-; its dissociation constant is unknown but may be about 0.5 since that for $Co(NH_3)_6^{3+}$, OH^- is [30] 0.014.

The marked effect of the SO_4^{2-} ion upon the rates of reactions of complex ions is illustrated by the work of Garrick [31] on the reaction $Co(NH_3)_5Cl^{2+} + H_2O = Co(NH_3)_5H_2O^{3+}$ and the study of Taube and Posey [32] of the reversible system $Co(NH_3)_5H_2O^{3+} + SO_4^{2-} \rightleftharpoons Co(NH_3)_5SO_4^+ + H_2O$. The latter recognized that ion-association accounts for such facts as that the rate of the forward reaction is almost independent of [SO_4^{2-}] in dilute solutions and it can influence the position of equilibrium. They also found the rate to be affected by [H^+], although they attribute this to reduction in SO_4^{2-} to HSO_4^- rather than to any reaction of the latter with the aquo complex ion. The influence of ion-association upon the equilibrium position of a reversible reaction is well illustrated by the findings of Pedersen [33] concerning the reaction $Co(NH_3)_4CO_3^+ + 2H^+ + H_2O \rightleftharpoons Co(NH_3)_4(H_2O)_2^{3+} + CO_2$, for this goes almost completely to the right when using glycollate buffers but reaches equilibrium long before this in acetate buffers. The explanation is due to the strong association of higher valent glycollate salts.[34]

There are a number of reactions involving organic compounds capable of ionization where the rate of reaction depends on the pH and on the nature and concentration of metal cations that are present. The decarboxylation of β-keto materials such as acetone dicarboxylic acid is a typical example; Prue [35] has made a thorough examination of this by measuring the pressure of evolved CO_2 by a manometric method [36] at 42°C and I = 0.6 (maintained constant by KCl or KNO_3). Most of the measurements were made with acetate buffered solutions (pH 4.6 to 3.8), calculating [H^+] via [37] [H^+][Ac^-]/[HAc] = 3.30×10^{-5}. Under the chosen conditions, a second stage of breakdown, from acetoacetic acid to acetone and CO_2 is unimportant since this acid is largely dissociated and the anion is more stable than the acid. If k is the velocity constant for the breakdown of acetone dicarboxylic acid to acetoacetic acid and CO_2, it can be shown that [35]

$$\log (r' - r) + A/2.303(r' - r) = B - kt/2.303 \qquad (13.16)$$

where r and r' are the manometer readings at times t and $t + \tau$ and A and B are unknown constants. From the slope of the plot of log $(r' - r)$ against t, an approximate value of k is found. Then a first value of A is obtained by plotting log $(r' - r) + kt/2.303$ against $1/(r' - r)$, whence a plot of the left side of (13.16) against t gives a better value of k and by continuation, the true value of k is reached. Prue expressed the results (omitting the charge signs) in terms of

$$k = k_A\alpha_A + k_{HA}\alpha_{HA} + k_{M.A}\alpha_M + k_{MAc.A}\alpha_{MAc} \quad (13.17)$$

where each α represents the fraction of acid present as A^{2-}, HA^- and metal cation present as M^{n+} or $MAc^{(n-1)+}$ (the last applied to Cu^{2+}). Apart from La^{3+} and $Co(NH_3)_6^{3+}$, it was shown that there is a good linear relationship between log $k_{M.A}$ and log $K_{M.Malonate}$. This supports a suggestion that the activated complex is a chelate formed between cations and an enolate form of the A^{2-} ion, but Pedersen [38], in an investigation of the cation catalyzed halogenation of ethyl acetoacetate, suggested that the complex involves the keto form and also adopted this idea to explain the cation catalyzed decarboxylation of oxaloacetic acid.[39] He has also pointed out in connection with the decarboxylation of acetosuccinic acid,[40] that a COO^- group must be free since the cation stabilizes it if bound to it, i.e., the =CO group and the other COO^- chelate the cation. Steinberger and Westheimer [41] have used a similar explanation to account for dimethyloxaloacetic acid but not its monoester being subject to cation catalysis (the acid not being able to form an enolic complex).

Gelles and his associates have aimed at clarifying the situation by studies with oxaloacetic acid. Firstly, they have shown that the results for the effects of some lanthanides [42] and transition-metal [43] cations at constant ionic strength can be interpreted in terms of

$$k_{obs.}C_a = k_0[H_2A] + k_1[HA^-] + k_2[A^{2-}] + k_3[MA^{(n-2)+}] + k_4[MAH^{(n-1)+}] + \ldots \quad (13.18)$$

(cf. 13.10), where C_a is the stoichiometric concentration of oxaloacetic acid, A^{2-} is its anion, and M^{n+} the metal cation. In the absence of the latter, if α_1 and α_2 are the fractions of acid present as HA^- and A^{2-}, then

$$\alpha_1/(1 - \alpha_1 - \alpha_2) = K'_{H_2A}/[H^+],\ \alpha_2/(1 - \alpha_1 - \alpha_2) = K'_{HA}K'_{H_2A}/[H^+]^2 \quad (13.19)$$

where K'_{H_2A} and K'_{HA} are the concentration dissociation constants of the acid (these are 6.19×10^{-3} and 1.37×10^{-4} at [40, 44] $I = 0.1$ and 37°C). The first step [43] was to establish $k_u = k_{obs.}$ in the absence of M^{n+} at various concentrations of HCl, adding KCl to keep $I = 0.1$, and to calculate $[H^+]$, α_1 and α_2 from K'_{H_2A} and K'_{HA}. Then in the presence of $[M^{n+}] = C_M \gg [A^{2-}]$ (keeping the pH low, i.e., 1.3 to 1.9), if

$$K'_{MA} = [MA^{(n-2)+}]/[M^{n+}][A^{2-}],\ K'_{MAH} = [MAH^{(n-1)+}]/[M^{n+}][HA^-] \quad (13.20)$$

$$R = (k_{obs.} - k_u)/K'_{HA}\alpha_1 C_M = k_3 K'_{MA}/[H^+] + k_4 K'_{MAH}/K'_{HA} \quad (13.21)$$

Plotting R against $1/[H^+]$ gives a line of slope $k_3K'_{MA}$ and intercept of $k_4K'_{MAH}/K'_{HA}$ (the latter proved negligible). The values of K'_{MA} (and K'_{MA_2} in the case of the lanthanides) were determined [45, 46] by pH titration (cell 9.34).

Example 13.2. With [43] $C_a = 0.03$, $I = 0.1$ (maintained by KCl) and 36.9°C, the following results were obtained with $ZnCl_2$ (C_M). The values of $[H^+]$, α_1, α_2 and k_u were obtained with similar studies in the absence of $ZnCl_2$; since $C_M \gg [A^{2-}]$, this causes little error; (k in sec^{-1} $mole^{-1}$):

10^3C_M	5.00	7.50	5.00	10.0	10.0	20.0	15.0
10^3HCl	5.00	5.00	10.0	10.0	25.0	25.0	50.0
$10^3[H^+]$	13.9	13.9	17.5	17.5	29.9	29.9	53.0
$10^5k_{obs.}$	19.8	26.5	14.8	21.8	10.8	18.0	6.72
10^5k_u	8.4	8.4	7.25	7.25	4.95	4.95	3.25
$10^2\alpha_1$	30.7	30.7	26.1	26.1	17.15	17.15	10.48
$10^2\alpha_2$	30.3	30.3	20.0	20.0	7.9	7.9	2.7
R	542	574	422	407	249	278	161

From the plot of R against $1/[H^+]$, $k_3K'_{MA} = 7.6$, and [43] since $K'_{MA} = 0.143K_{MA}$, while K_{MA} is [46] 1700, $k_3 = 0.031$.

It was found with the lanthanides [42] that a plot of log k_3 against log K'_{MA} was almost linear, but not so with the transition metals,[43] where a plot of log k_3 against log K_{MA} of the oxalates [47] is more satisfactory; this is reminiscent of Prue's findings [35] as already described, but the position is involved since both acid and metal complexes exist as mixtures of enolic and ketonic forms,[48] and from the optical densities of alkylated forms, only the ketonic form decarboxylates. In addition, it is not easy to get precise values of [46] K'_{MA}.

Among other decarboxylations, that of nitroacetate is in contrast to that of the two acids dealt with above since [49,50] the rate decreases when, e.g., Cu^{2+} ions are added while other ions such as Ba^{2+} have little effect; in general, the rate decreases in relation to the stability of the acid and its metal complexes—only the anion is unstable. A somewhat similar feature characterizes acetosuccinic acid [51] in that Cu^{2+} ions have little effect, which points to a six-membered ring chelate. Pedersen obtained k_0, k_1 and k_2 (see 13.18) by measurements under particular conditions. For k_2, the diethyl ester was dissolved in $Ba(OH)_2$, when the A^{2-} produced by hydrolysis undergoes slow decarboxylation. The rate was followed by titrating samples at timed intervals since $BaCO_3$ is precipitated. The rate increases slightly with increasing hydroxide, and is decreased by $BaCl_2$ and KCl. The reaction is kinetically rather complicated, but these salt effects could be attributed to the presence of $BaOH^+$ since this has a dissociation constant of about [52] 0.23. To obtain k_0, data were obtained with HCl + NaCl solutions of the ester by a manometric method [53] at $I = 0.3$. Now

$$-dx/dt = (k_0\alpha_0 + k_1\alpha_1 + k_2\alpha_2)x \tag{13.22}$$

$$k = k_0\alpha_{0,\infty} + k_1\alpha_{1,\infty} + k_2\alpha_{2,\infty} \tag{13.23}$$

where x is the molarity of acetosuccinic acid and α_0, α_1 and α_2 are the fractions as H_2A, HA^- and A^{2-} at a time t, while ∞ refers to the end of the process (acetosuccinic to β-acetopropionic). Under the chosen conditions, k_2, α_2 and their ∞ values can be neglected so only K'_{HAcSuc} is required, so if h is the molarity of added HCl,

$$\alpha_1(h + \alpha_1 x)/(1 - \alpha_1) = h\alpha_{1,\infty}/(1 - \alpha_{1,\infty}) \tag{13.24}$$

The above expressions, followed by integration lead to

$$-kt + \text{constant} = \ln \alpha_1 x + (2kk_0^{-1} - 1)\ln [h + \alpha_1 x h^{-1}(1 - \alpha_{1,\infty})] \tag{13.25}$$

Then, putting $P = \phi x$ where P is the pressure difference of the manometer at the end and at time t, and ϕ is a constant derived from the ester molarity and P for $t = 0$, an approximate solution suitable for handling most of the data is

$$-kt + \text{constant} = \ln P - \alpha_{1,\infty}^2 P/\phi h \tag{13.26}$$

(a more exact expression has been derived [51]) so the slope of the plot of the right-hand side against t gives k. Since $k_2\alpha_2 x = 0$ in (13.22), by allowing for the small term $\alpha_1 k_1 x$ (via k_1 from below), k_0 is found; another way is to extrapolate the plot of k against $\alpha_{1,\infty}$

In a third series, the decarboxylation was studied manometrically with NaAc, HAc, NaCl buffers at $I = 0.3$. Assuming α_0, α_1 and α_2 vary linearly with x,

$$-dx/dt = (k + \beta' P)x \tag{13.27}$$

where

$$x_0\phi\beta' = \sum_0^2 k_n(\alpha_{n,0} - \alpha_{n,\infty}) \quad (x_0 = x \text{ at } t = 0) \tag{13.28}$$

To calculate the α's, the following data are needed: $K'_1 = 1.56 \times 10^{-3}$, $K'_2 = 3.64 \times 10^{-5}$ for acetosuccinic acid, $K' = 5.47 \times 10^{-5}$ for β-acetopropionic acid and [54] $K' = 3.20 \times 10^{-5}$ for acetic acid. Successive approximations are needed to obtain β'. It was found that the final equation involved $[Ac^-]$, which was regarded as indicating acetate catalysis of H_2A. However, K_{HAc} determinations by e.m.f. methods are complicated by the variation of E^0_{AgCl} with [HAc], and the dimerization of the latter (see Chapter 9), which might explain this feature.

The following figures summarize some of the available information:

Acid	°C	I	k_0	k_1	k_2	
Oxaloacetic[39]	37	0.1	0.341	15.13	4.2	$\times 10^{-3}$ sec^{-1}
Oxaloacetic[44]	37	0.1	0.345	15.4	4.2	$\times 10^{-3}$ sec^{-1}
Dihydoxytartaric[56]	25	0.3	0.014	4.15	17.7	$\times 10^{-5}$ sec^{-1}
Acetosuccinic[51]	25	0.3	17.23	3.05	0.19	$\times 10^{-5}$ sec^{-1}

The k_2 figure for oxaloacetic acid appears to be anomalous, although this may be related to the enol-keto mechanism.

If the decarboxylation process is written as a bimolecular reaction, it could be expressed by (for a dibasic acid):

$$k_{\text{obs}}[\text{acid}][H_2O] = [H_2A](k_1[H^+] + k_2[H_2O] + k_3[OH^-]) + [HA^-](k_4[H^+]f_1^2 + k_5[H_2O] + k_6[OH^-]f_1^2/f_2) + [A^{2-}](k_7[H^+]f_2 + k_8[H_2O] + k_9[OH^-]f_2) \qquad (13.29)$$

Under specific pH conditions a number of terms could be ignored, and k_2, k_5 and k_8 may be negligible, but there are not sufficient data to examine the remaining terms.

Another aspect of metal ion catalysis is the effect of cations upon the rates of alkaline (KOH) hydrolysis of $KOOC\,.\,(CH_2)_n\,.\,COOC_2H_5$. This has been studied by Hoppé and Prue[57] who found that Ca^{2+}, Ba^{2+} and $Co(NH_3)_6^{3+}$ chlorides and Tl^+ nitrate accelerate the reaction when $n = 0$ and 1 (this is attributed to chelation), while Li^+, Na^+ and K^+ chlorides retard the hydrolysis slightly when $n = 2$ and 3 (this is attributed to the separation of the charges of the ester ion from the reaction centre). In terms of the main reactions, the appropriate expression is

$$k_{\text{obs}}C_1C_2 = k_1^0[\tfrac{1}{2}\text{Est}^-][OH^-]f_1^2/f_2 + k_2^0[\tfrac{1}{2}\text{Est}^-][MOH^{(n-1)+}]f_1f_{n-1}/f_{n-2} \qquad (13.30)$$

where C_1 and C_2 are the stoichiometric concentrations of KOH and $\frac{1}{2}$Ester, and M^{n+} is the salt cation. Besides the indicated ion-pair, others such as $M\,.\,\frac{1}{2}\text{Est}^{(n-1)+}$, $MCl^{(n-1)+}$ and $TlNO_3$ affect the concentrations of reactants and the total ionic strength. Hoppé and Prue[57] have listed the relevant available dissociation constants (Ba^{2+}, Ca^{2+}, $Co(NH_3)_6^{3+}$ oxalates, malonates, adipates and hydroxides; $Co(NH_3)_6^{3+}$,Cl^-; TlOH and $TlNO_3$). By means of (13.30) they derived the values of k_1^0 and k_2^0.

A totally different aspect of ion-ion kinetics is illustrated by the rate and mechanism of the decomposition of sulphonium salts, $R_3SX = RX + R_2S$, where R = e.g., $C_6H_5CH_2$ —(Bz) and CH_3, while X = Cl^-, I^-, ClO_4^-. Swain and Kaiser[58] have reached a different conclusion to that of earlier workers,[59] who regarded the reaction to be of the first order, and which was supported by Gleave, Hughes and Ingold[60] who specify the mechanism (R = CH_3,

$X = Cl^-$, Br^- and CO_3^{2-}) to be of the S_N1 type, the rate determining mechanism being the rate of ionization $(CH_3)_3SX = (CH_3)_3S^+ + X^-$. Swain and Kaiser studied the reaction in 90 % acetone, with and without added salts, and they measured the conductances of Bz_3S, Cl and $NaClO_4$ in the stated solvent. By application of the limiting Onsager equation, they concluded that the two substances are almost completely dissociated, but it can be seen from their data that some ion-pairing does occur although there is not enough information at low concentrations to apply the methods of Chapter 8 [some dissociation constants which have been derived from conductances in 90 % acetone are [61,62] $AgNO_3 = 0.0012$, $LiBr = 0.012$, $KBr = 0.013$, $HBr = 0.0063$ at 25°C].

With 0.016 M Bz_3S Cl alone, k_1, the apparent first order rate constant was practically constant, k_2, the apparent second order constant varied, while in 0.13 M $NaClO_4$, k_1 varied while k_2 was constant. On applying (13.3), k_2^0 of the first run was found to be sufficiently constant to substantiate the claim that the reaction is second order.

Example 13.3. With [58] 0.0165 M Bz_3SCl in 90 % acetone at 50°C ($\epsilon = 22$, $A' = 3.045$), the following figures and calculations were obtained with $\log k_2^0 = \log k_2 + 2A'\sqrt{I}/(1 + \sqrt{I})$ assuming complete dissociation.

% reaction	15.5	34.0	44.2	60.0	70.5	78.1	87.3	(0)
$10^2 I$	1.39	1.09	0.92	0.66	0.49	0.36	0.21	(0)
$10^2 k_2(\text{sec}^{-1})$	2.70	3.51	3.91	4.95	5.88	6.72	8.76	
k_2^0	0.119	0.132	0.133	0.142	0.147	0.149	0.162	(0.164)

The drift in k_2^0 is in the opposite direction to that explicable by incomplete dissociation but may be due to the form of the D.H. equation.

Electron Exchange Reactions

Reactions between simple ions of the type $M^{n+} + M^{(n+x)+} \rightleftharpoons M^{(n+x)+} + M^{n+}$ can usually be studied only in strongly acidic media owing to hydrolysis. This means that studies are confined to media of high ionic strength, making quantitative studies of salt effects difficult. This is a great disadvantage since there are a number of indications that the mechanisms involve products of ion-association. Examples are (a) [61] Fe^{2+}, Fe^{3+}; the rate is accelerated by Cl^-, F^- and at lower acidities where OH^-, Fe^{3+} association occurs; (b)[62] Ce^{3+}, Ce^{4+} is also accelerated by F^-; (c) [63] Tl^+, Tl^{3+} is accelerated by Cl^-.

The rates are commonly followed by taking low concentrations of the radioactive form of the higher valent ion in $HClO_4$ solutions and stopping the exchange by solvent extraction, precipitation, and other ways. It is assumed that no ion-pairs form with the ClO_4^- ion, and although this may not appear important in terms of dissociation constants, the ratio of cation to ClO_4^- is so low that a large fraction of the cations may be combined with this anion.

Ion-Molecule Reactions

According to (13.3), a reaction between a molecule and an ion should proceed at a rate that is independent of the ionic strength, provided that the reacting ion does not associate with any other ions which may be present. This has been confirmed by a number of studies [1, 7, 21] such as the alkaline hydrolysis of certain esters, and exchanges of the type RX + Y = RY + X in solvents of high dielectric constant, but if the molecule is polar, this is evidence (M — H)[1] that the ion attacks the dipole at right angles to the direction of its axis, so when the attack is not at right angles an electrolyte effect is possible; this is discussed later on.

As a first example of the effect of incomplete dissociation, a series of hydrolyses in hydroxide solutions undertaken and interpreted by Bell and his associates will be described. With stoichiometric concentrations of C_1 molecular reactant and C_2 of hydroxide, the main possible reactions lead to

$$k_{\text{obs.}}C_1C_2 = k_1C_1[\text{OH}^-] + k_2C_1[\text{MOH}^{(n-1)+}] + k_3C_1[\text{M}^{n+}] \tag{13.31}$$

or

$$k_{\text{obs.}}C_2 = k_1[\text{OH}^-] + [\text{M}^{n+}](k_2[\text{OH}^-]f_nf_1/Kf_{n-1} + k_3) \tag{13.32}$$

where K is the dissociation constant of the hydroxide.

With diacetone alcohol,[64] the rate of reaction (followed by dilatometer[64, 65] measuring the rate of expansion) at 25°C with RbOH, RbOH + RbCl, KOH and KOH + KCl is strictly proportional to C_2 (they chose conditions where $C_2 > C_1$ and evaluated $k'_{\text{obs.}} = k_{\text{obs.}}C_2$), i.e., $k'_{\text{obs.}}/C_2$ = constant. These results therefore establish k_1 and show that the right-hand part of (13.32) is unimportant, but with Na^+, Ca^{2+}, Ba^{2+} and Tl^+ hydroxides $k'_{\text{obs.}}/C_2$ decreases as C_2 increases, and by evaluating $[\text{OH}^-] = k'_{\text{obs.}}/k_1$, the dissociation constants of the hydroxides were derived. A similar procedure was adopted regarding the rate of hydrolysis of ethyl acetate [66] in NaOH + $CaCl_2$ or $BaCl_2$ but no retarding effects were observed; this is attributed to a more concentrated charge in the negative state of the transition state. On the other hand,[67] the rate constant when using a stirred-flow reactor [68] is lower with $Ba(OH)_2$ than with NaOH. In this method, the reactants are fed separately through capillaries into a 400 ml. vessel, stirred rapidly at 3000 r.p.m., and the overflow analysed at intervals until the steady-state concentrations are found. The expression for the second order rate constant is [68]

$$k = u(u_xx_0 - ux)/Vx(u_yy_0 - u_xx_0 + ux) \tag{13.33}$$

where x_0 and y_0 are the concentrations of OH^- and ester before entering the vessel of volume V, u_x and u_y are the flow rates (l./sec), $u = u_x + u_y =$ the rate of outflow, and x and y are the normalities in the reaction vessel at a particular moment, t. The expression assumes complete dissociation, and providing this is largely so, $[OH^-] = k_{obs.}x/k_{NaOH}$. The given five results (one for NaOH) thus lead to $K_{BaOH} = 0.3$, approx.

Two other series which have been used to examine hydroxide dissociation are the reaction between nitroethane and OH^- ions [69] in the presence of Na^+, Ca^{2+}, Ba^{2+} and Tl^+ ions (this fast reaction was measured by the "thermal maximum" method [70]) and diacetone alcohol hydrolysis in the presence of [71] $Co(NH_3)_6(OH)_3$. To summarize, the derive data from kinetics are (a) NaOH ~ 5, (b) $Ca(OH)^+ = 0.05 - 0.35$, (c) $Ba(OH)^+ = 0.3 - 0.21$, (d) $TlOH = 0.38 - 0.14$, (e) $Co(NH_3)_6(OH)^{2+} = 0.014$.

The reaction $S_2O_3^{2-} + RBr = RS_2O_3^- + Br^-$ where R is an alkyl or aryl group is [72] bimolecular, irreversible and relatively fast. To avoid too much hydrolysis, and to annul the effects of this, the reaction is studied [72, 73, 75] in mixed solvents containing acetate buffers. The chosen conditions can result in $S_2O_3^{2-}$, RSO_3^- and Ac^- forming ion-pairs, and since the ionic strength will change during the course of the reaction, allowance must be made for these for every calculated value of the rate constant.

Example 13.4. The reaction between 0.04431 M n-propyl bromide and 0.05250 M calcium thiosulphate in 44 % ethanol containing 0.005 M sodium acetate was followed [73] at 25°C. The required constants are $K_{CaS_2O_3} = 3.25 \times 10^{-4}$, $K_{NaS_2O_3^-} = 0.0144$ (from conductance [74]); $K_{CaAc^+} = 0.016$, $K_{CaPrS_2O_3} = 0.019$ (from solubilities [73]). After 65 min., $k_{obs.} = 3.83 \times 10^{-4}$/mole.sec., and total n-$PrS_2O_3 = 0.00325$. Applying successive approximations to obtain log [ion-pair] = log [cation][anion] — log K — $n\phi(I)$ where $\phi(I) = z_i^2 \times 0.932\{\sqrt{I}/(1 + \sqrt{I}) - 0.2I\}$, the concentrations of ion-pairs ($\times 10^3$) are $CaS_2O_3 = 30.83$, $NaS_2O_3^- = 0.85$, $CaAc^+ = 0.89$, $CaPrS_2O_3^+ = 0.50$; $I = 0.08395$, and assuming that the reaction does not involve the ion-pairs, $10^4k = 10.54$/mole. sec. Similar calculations [73] with other thiosulphates and over a range of ionic strengths average $10^4k = 10.0 \pm 0.2$ with no significant trends, which indicates the absence of ion-dipole effects.

Halide salts containing radioactive halide have been used extensively to derive information about the exchange reaction $RX + X_a^{*-} \rightleftharpoons RX_a^* + X^-$, where RX is an organic halide and X_a^{*-} is a radiohalogen ion not necessarily an isotope of X. The reaction is bimolecular, is not catalyzed by acids and bases,[76] but is reversible, although the equilibrium may lie well to the right. The majority of the reported work has been based upon acetone as solvent and there is the possibility that the salts present are not fully ionized. Evans and Sugden have analysed this situation in connection with RBr, LiBr systems. If the initial concentrations are $[RBr] = a$, $[X^{*-}] = c$, $[X^-] = b$, and if $x = [RX^*]$ at time t, then since $a > x$ and $b > c$,

$$kt(a + b) = 1/\ln\{1 - (x/c)(1 - b/a)\} \tag{13.34}$$

The reactants were equilibrated in the arms of a λ-shaped tube, mixed, cooled at time t, and a known fraction of the contents shaken with water and benzene (8 : 1). The layers were separated, the benzene layer washed then portions of this and the first aqueous extract counted in a suitable G.M. tube. After correcting for the effect of the organic solvent on the counting rate, the ratio of the activities gives x/c. With 90 % acetone, and with CH_3OH k scarcely changed with b, but with RBr and LiBr in acetone $k_{obs.}$ falls off rapidly as b increases. Now K_{LiBr} in acetone at 25°C, from conductance [77, 78] is 5.2×10^{-4} and from considerations of other data, is lower at higher temperatures. With *sec.* octyl bromide at 65.5°C, Evans and Sugden [77] found that by taking $K_{LiBr} = 2.4 \times 10^{-4}$, and $-\log f_i = 4.42\{\sqrt{I}/(1 + 2.116\sqrt{I}) - 5.2I\}$ (i.e., $a = 3.0$Å), $k_{calc.} = k_{obs.}/\alpha_c$ is constant:

$10^4 b$	2.61	8.61	15.7	58.0	78.0	250	390	760
$10^3 k_{obs.}$	7.21	4.94	4.15	2.34	2.34	1.37	1.14	0.83
$10^3 k_{obs.}/\alpha_c$	11.2	10.7	11.0	10.0	11.2	10.5	10.8	10.7

Le Roux and Swart [79] have used an extended version of the same idea which assumes that K_{LiBr} at higher temperatures can be derived from the values at 25°C by means of Bjerrum's theory (see Chapter 14); this gives $K_{LiBr} = 4.23 \times 10^{-4}$ at 40°C and 3.18×10^{-4} at 60°C (compared with 2.4×10^{-4} taken above). In a study of *t.* butyl bromide and LiBr at 40°C, they found it necessary to allow for both a first order as well as a second order reaction. The constant rate of exchange is given by

$$R = -\{ab/t(a + b)\} \ln\{1 - (x/c)(1 + b/a)\} \tag{13.35}$$

and

$$R/a = k_1 + k_2\alpha_c b \tag{13.36}$$

where k_1 and k_2 refer to the true first and second rate constants. From a plot of R/a against b, the intercept gave $k_1 = 32 \times 10^{-8}$ and $k_2 = 6.05 \times 10^{-5}$ (sec^{-1}) at 40°C. With *iso*-propyl bromide at 40°C, $k_1 = 22.6 \times 10^{-8}$, $k_2 = 44.3 \times 10^{-8}$ (data at 20°C and 60°C were also obtained). However, these answers are very dependent upon the actual values of K_{LiBr}. The essential difference between this and the Evans–Sugden treatment is that these latter authors assume that k_1 is zero and adjust K_{LiBr} accordingly, and in both treatments it is assumed that ion-dipole effects are negligible. Until accurate and reliable values of, e.g. K_{LiBr} at the reaction temperatures are known, these factors cannot be completely resolved.

Many other examples of halide exchange with RX compounds are referenced in the extensive and comprehensive series of investigations by de la Mare,[80,82,83] Hughes and Ingold [81–83] and their associates. With RBr, LiI exchanges, the equilibrium constants of the reaction have to be taken into consideration (see below). To obtain standard conditions so that allowance is made for the decrease in k with increasing salt concentration ("negative salt effect"), a fixed concentration of LiX (0.027 M) was taken. This provides an accurate comparative basis among data with the same X but different R at the same temperature, but is subject to some uncertainty when comparing reactions of the same RBr with different LiX salts [82] at the same temperature, or when the temperature changes in any reaction, owing to the differences in K_{LiX}.

Aziz and Moelwyn–Hughes,[84] in connection with the reversible exchange $CH_3Cl + I^- \rightleftharpoons CH_3I + Cl^-$ in acetone at different temperatures (using Li salts) have taken into account the ion-association of both LiI and LiCl as well as K, the equilibrium constant of the reaction. They also devised a reaction vessel which eliminates the vapour phase. It consists of an internally ground tube fitted with a ground-glass cylinder connected to a capillary tube and tap. Mercury is placed on top of the cylinder to act as a seal and to help express samples from the capillary at timed intervals. Their final expression for calculating the second order rate constant k_2^0 is

$$2\alpha_c\beta(1 - W)k_2^0t = \ln [\{1 - x/(\gamma + \beta)\}/\{1 - x/(\gamma - \beta)\}] \quad (13.37)$$

where

$$\gamma = (a + b)/(2 - 2W), \beta^2 = \{(a - b)^2 + 4\,Wab\}/4(1 - W)^2$$
$$W = K_{LiCl}/K\alpha_c^2(2b - x_\infty) \quad (13.38)$$

in which a = initial $[CH_3Cl]$, b = initial total [LiI], $x = [CH_3Cl]$ at time t, and α_c is the degree of dissociation of the lithium iodide (the average mean of the initial and final values—which are almost identical). The data for calculating K_{LiCl} and K_{LiI} were derived from published results [85] and expressed as formulae.

Another instance where the ionization of LiBr in acetone is of significance is in connection with the rate of racemization of the dimethyl ester of l-bromosuccinic acid. Olson [86] found the data could be fitted to $k_{obs.} = k'[Br^-] + k''[LiBr]$, using $K_{LiBr} = 3.6 \times 10^{-4}$ at 25°C; there is some latitude in this, but the uncertainty is limited to $\pm$ 10 %.

The addition of, e.g., $LiClO_4$ to acetic acid solutions of substituted benzene and toluene sulphonates (brosylates and tosylates) produces very marked increases in the first order rates of acetolysis. This has been surveyed by Winstein and his associates.[87] In some cases the plot of log k against the molarity of salt curves upwards followed by a linear rise which was extrapolated to obtain k for a "special" salt effect as compared with k for the ordinary salt effect. Now salts such as $LiClO_4$ are weak electrolytes in acetic acid (see Chapter 8) so that plots of the kind mentioned can only give a partial and arbitrary picture. Heldt [88] has found that $NaClO_4$ produces similar effects while LiCl, sodium brosylate and sodium tosylate have retard ing actions.

Among the many studies of the effects of salts on the rates of solvolyses of organic halides in partly aqueous or purely organic solvents,[89,90] those concerned with the tertiary butyl halides have attracted much attention. The mechanism is considered to be [91] the slow ionization process, $t\text{-BuX} = t\text{-Bu}^+ + X^-$ followed by the fast S_N1 reaction, $t\text{-Bu}^+ + H_2O = t\text{-BuOH} + H^+$ and the simultaneous fast $E1$ reaction, $t\text{-Bu}^+ = iso\text{butylene} + H^+$. The overall reaction can be followed by measuring the liberated acid. The unimolecular rate constant increases in the presence of salts, e.g., on adding bromides to a 90 % acetone solution of *t*-BuBr. A quantitative treatment of this is incorporated in the treatment of Hughes, Ingold and their associates,[91] the appropriate expression being

$$\log k = \log k^0 + 0.912 \times 10^{16}\, z^2 d/(\epsilon T)^2 = \log k^0 + b'\mathrm{I} \qquad (13.39)$$

It is assumed that in the transition state, half an electron is transferred from the R group to the Br atom, so $z^2 = 0.25$; d, the distance between the centres of R and Br, from the Morse potential energy function, is about 2.6 Å. For 90 % acetone at 25°C, $b' = 1.04$.

An alternative view [92] is to consider the reaction as the result of collision of the $t\text{-Bu}^+$ ion with a water molecule. An expression which relates constants between ions and polar molecules with I, which has been derived by Moelwyn–Hughes [1] can then be applied. This is

$$\log k = \log k^0 + (8\pi N e^3 \mu z \cos\theta/2303(\epsilon kT)^2 = \log k^0 + b''I \qquad (13.40)$$

and under the specified conditions, where μ, the dipole moment of the water molecule is 1.85 D, gives $b'' = 1.26$ if θ, the angle of approach of the ion to the polar axis of H_2O is taken as zero ($\cos\theta$ at a maximum of 1).

Assuming that the liberated HBr (in the absence of added electrolytes) is fully ionized, the average experimental value of b' or b'' is [92] 1.04, while by calculating the true ionic strength [93] ($K_{HBr} = 0.0063$), the experimental value of b' or b'' is raised to 1.65, and the average value from other series in the presence of LiBr, KBr, HBr and $MgBr_2$ is 1.5 ± 0.25. Under the selected conditions, these answers are in closer accord with (13.37) than with (13.36), but d of the latter need be only raised to 3.9 Å to get an exact agreement. Similar considerations to those just described apply to the effects of salts on the hydrolysis of substances such as *pp'*-dimethylbenzhydryl chloride in aqueous acetone [93] for which the rate is depressed by LiCl. A recently found complication [94] is that Cl exchange can occur, e.g., capture of $t\text{-Bu}^+$ by Cl^-.

Acid-Base Catalysis

The dependence of the rates of reaction of diacetone alcohol and of nitroethane upon $[OH^-]$ discussed in a previous section are examples of specific OH^- ion catalysis. There are likewise a number of reactions such as the inversion of sucrose and the hydrolysis of acetals that depend on $[H^+]_{hyd.}$ and have therefore been used to ascertain the dissociation constants of incompletely dissociated acids. However, these are particular examples of what is termed acid-base catalysis, i.e., the influence of the acid-base properties of substances upon rates of reactions. Bell [95] has given a detailed exposition of this in his monograph and a certain amount of information on the subject is to be found in more general books on reaction kinetics.[1]

Fast Reactions

Reactions with half-lives from a few seconds down to as low as 10^{-9} sec are studied by methods and techniques which make use of N.M.R., ultrasonic H.F. electric impulses, "stopped-flow" mixers, thermal effects and so on. Useful reviews on this material have been published by the Faraday Society[96] and by Ljunggren and Lamm.[97]

REFERENCES

1. Moelwyn-Hughes, E. A. "The Kinetics of Reactions in Solution." Clarendon Press, Oxford, 2nd edn. (1947). Glasstone, S., Laidler, K. J. and Eyring, H. "The Theory of Rate Processes." McGraw-Hill Book Co., Inc., New York and London (1941). Frost, A. A. and Pearson, R. G. "Kinetics and Mechanism." J. Wiley, and Sons Inc., New York (1953). Laidler, K. J. "Chemical Kinetics." McGraw-Hill Book Co., Inc., New York and London (1950).
2. Bronsted, J. N. *Z. phys. Chem.* **102,** 169 (1922); **115,** 337 (1925); *Chem. Rev.* **5,** 231 (1928).
3. Bjerrum, N. *Z. phys. Chem.* **108,** 82 (1924); **118,** 251 (1925).
4. Bronsted, J. N. and Livingston, R. *J. Amer. chem. Soc.* **49,** 435 (1927).
5. Livingston, R. *ibid.* **48,** 53 (1926).
6. La Mer, V. K. *ibid.* **51,** 3341, 3678 (1929). La Mer, V. K. and Fessenden, R. W. *ibid.* **54,** 2351 (1932). La Mer, V. K. and Kamner, M. L. *ibid.* **53,** 2832 (1931); **57,** 2662 (1955).
7. Guggenheim, E. A. and Prue, J. E. Ref. 51, Ch. 4.
8. Gimblett, F. G. and Monk, C. B. *Trans. Faraday Soc.* **51,** 793 (1955).
9. Wyatt, P. A. H. and Davies, C. W. *Trans. Faraday Soc.* **45,** 774 (1949).
10. Davies, C. W. and Williams, I. W. *ibid.* **54,** 1547 (1958).
11. von Kiss, A. and Vass, P. *Z. anorg. Chem.* **217,** 305 (1934). Kappana, A. N. *J. Indian chem. Soc.* **6,** 45 (1929); **9,** 379 (1932).
12. Denney, T. O. and Monk, C. B. *Trans. Faraday Soc.* **47,** 992 (1951).

13. Davies, C. W. and Wyatt, P. A. H. *ibid.* **45,** 770 (1949).
14. Olson, A. R. and Simonson, T. R. *J. chem. Phys.* **17,** 1167 (1949).
15. Davies, E. W. Thesis, University of Wales (1957).
16. Indelli, A. and Prue, J. E. *J. chem. Soc.* 107 (1959).
17. Davies, C. W. and Monk, C. B. *ibid.* 413 (1949).
18. Barrett, J. and Baxendale, J. H. *Trans. Faraday Soc.* **52,** 210 (1956).
19. Reynolds, W. L. *J. Amer. chem. Soc.* **80,** 1830 (1958).
20. Fudge, A. J. and Sykes, K. W. *J. chem. Soc.* 119, 124 (1952).
21. Basolo, F. and Pearson, R. G. "Mechanisms of Inorganic Reactions." J. Wiley and Sons, Inc., New York; Chapman and Hall, Ltd., London (1958). Basolo, F. *Chem. Rev.* **52,** 459 (1953).
22. Ingold, C. K. *Special Pub. Chem. Soc., London,* No. 1 (1954).
23. Taube, H. "Advances in Inorganic Chemistry and Radiochemistry." Academic Press Inc., New York and London **1,** (1959); *Chem. Rev.* **50,** 69 (1952). Basolo, F., Messing, A.F., Wilks, P. H., Wilkins, R. G. and Pearson, R. G. *Int. Symp. on Co-ord. Chem. Rome* (1957). Pearson, R. G. *J. phys. Chem.* **63,** 321 (1959).
24. Brown, D. D., Ingold, C. K. and Nyholm, R. S. *J. chem. Soc.* 2680, 2696 (1953).
25. Pearson, R. G., Henry, P. M. and Basolo, F. *J. Amer. chem. Soc.* **79,** 5382 (1957).
26. Ref. 1, Ch. 1.
27. Brasted, R. C. and Hirayama, C. *J. Amer. chem. Soc.* **80,** 788 (1958).
28. Pearson, R. G., Henry, P. M., Bergmann, J. G. and Basolo, F. *ibid.* **76,** 5920 (1954).
29. Nyholm, R. S. and Tobe, M. L. *J. chem. Soc.* 1707 (1956).
30. Caton, J. A. and Prue, J. E. *ibid.* 671 (1956).
31. Garrick, F. J. *Trans. Faraday Soc.* **34,** 1088 (1938).
32. Taube, H. and Posey, F. *J. Amer. chem. Soc.* **75,** 1463 (1953).
33. Pedersen, K. J. *ibid.* **53,** 18 (1931).
34. Evans, W. P. and Monk, C. B. *J. chem. Soc.* 550 (1954).
35. Prue, J. E. *ibid.* 2331 (1952).
36. Bell, R. P. and Trotman-Dickenson, A. F. *ibid.* 1288 (1949).
37. Harned, H. S. and Hickey, F. C. *J. Amer. chem. Soc.* **59,** 1284 (1937).
38. Pedersen, K. J. *Acta Chem. Scand.* **2,** 385 (1948).
39. Pedersen, K. J. *ibid.* **6,** 243, 285 (1952).
40. Pedersen, K. J. *ibid.* **12,** 919 (1958).
41. Steinberger, R. and Westheimer, F. H. *J. Amer. chem. Soc.* **71,** 4158 (1949); **73,** 429 (1951).
42. Gelles, E. and Clayton, J. P. *Trans. Faraday Soc.* **52,** 353 (1956).
43. Gelles, E. and Salama, A. *J. chem. Soc.* 3689 (1958).
44. Gelles, E. *ibid.* 4736 (1956).
45. Gelles, E. and Nancollas, G. H. *Trans. Faraday Soc.* **52,** 98 (1956).
46. Gelles, E. and Salama, A. *J. chem. Soc.* 3683 (1958).
47. Money, R. W. and Davies, C. W. *Trans. Faraday Soc.* **28,** 609 (1932).
48. Gelles, E. and Hay, R. W. *J. chem. Soc.* 3673 (1958).
49. Pedersen, K. J. *Trans. Faraday Soc.* **23,** 316 (1927); *J. phys. Chem.* **38,** 559 (1934); *J. Amer. chem. Soc.* **53,** 18 (1931); *Acta Chem. Scand* **1,** 437 (1947).
50. Pederson, K. J. *ibid.* **3,** 676 (1949).
51. Pedersen, K. J. *ibid.* **12,** 919 (1958).
52. Davies, C. W. *J. chem. Soc.* 349 (1939).
53. Pedersen, K. J. *J. Amer. chem. Soc.* **53,** 18 (1931).
54. Harned, H. S. and Hickey, F. C. *ibid.* **59,** 1284 (1937). Kilpatrick, M. and Eanes, R. D. *ibid.* **75,** 586 (1953).

55. Sutcliffe, L. H. and Weber, J. R. *Trans. Faraday Soc.* **52,** 1225 (1956); **55,** 1892 (1959).
56. Pedersen, K. J. *Acta Chem. Scand.* **9,** 1640 (1955).
57. Hoppé, J. I. and Prue, J. E. *J. chem. Soc.* 1775 (1957).
58. Swain, G. C. and Kaiser, L. E. *J. Amer. chem. Soc.* **80,** 4089 (1958).
59. von Halban, H. *Z. phys. Chem.* **67,** 129 (1909). Essex, H. and Gelormi, O. *J. Amer. chem. Soc.* **48,** 882 (1926). Corran, R. F. *Trans. Faraday Soc.* **23,** 605 (1927).
60. Gleave, J. L., Hughes, E. D. and Ingold, C. K. *J. chem. Soc.* 236 (1935).
61. Silvermann, J. and Dodson, R. W. *J. phys. Chem.* **56,** 846 (1952). Hudis, J. and Wahl, A. C. *J. Amer. chem. Soc.* **75,** 4153 (1953).
62. Hornig, H. C. and Libby, W. F. *J. phys. Chem.* **56,** 853 (1952). Duke, F. R. *J. Amer. chem. Soc.* **78,** 1540 (1957).
63. Prestwood, R. J. and Wahl, A. C. *ibid.* **71,** 3137 (1949). Harbottle, G. and Dodson, R. W. *ibid.* **73,** 2442 (1951).
64. Bell, R. P. and Prue, J. E. *J. chem. Soc.* 362 (1949).
65. Halberstadt, E. S., Hughes, E. D. and Ingold, C. K. *ibid.* 2441 (1950).
66. Bell, R. P. and Waind, G. M. *ibid.* 1979 (1950).
67. Ref. 51, Ch. 4.
68. Stead, B., Page, F. M. and Denbigh, K. G. *Disc. Faraday Soc.* **2,** 263 (1947).
69. Bell, R. P. and Panckhurst, M. H. *J. chem. Soc.* 2836 (1957).
70. Bell, R. P. and Clunie, J. C. *Proc. roy. Soc.* **212A,** 16 (1952). Bell, R. P., Gold, V., Hilton, J. and Read, M. H. *Disc. Faraday Soc.* **17,** 151 (1954).
71. Caton, J. A. and Prue, J. E. *J. chem. Soc.* 671 (1956).
72. Crowell, T. I. and Hammett, L. P. *J. Amer. chem. Soc.* **70,** 3444 (1948). Akagi, K., Oae, S. and MuraKami, M. *ibid.* **78,** 4034 (1956).
73. Bevan, J. R. and Monk, C. B. *J. chem. Soc.* 1396 (1956).
74. Bevan, J. R. and Monk, C. B. *loc. cit.* 1392.
75. Fuchs, R. *J. Amer. chem. Soc.* **79,** 6531 (1957). Fuchs, R. and Nisbet, A. *ibid.* **81,** 2371 (1959).
76. Le Roux, L. J. and Sugden, S. *J. chem. Soc.* 1279 (1939).
77. Evans, C. C. and Sugden, S. *ibid.* 270 (1949).
78. Dippy, F. J. F. *ibid.* 1386 (1939).
79. Le Roux, L. J. and Swart, E. R. *ibid.* 1475 (1955).
80. de la Mare, P. B. D. *ibid.* 3169, 3180, 3196 (1955).
81. Hughes, E. D., Ingold, C. K. and Mackie, J. D. H. *loc. cit.* 3173, 3177. Fowden, L. Hughes, E. D. and Ingold, C. K. *loc. cit.* 3187, 3193.
82. de la Mare, P. B. D., Fowden, L., Hughes, E. D., Ingold, C. K. and Mackie, J. D. H. *loc. cit.* 3200.
83. de la Mare, P. B. D. and Hughes, E. D. *ibid.* 845 (1956).
84. Aziz, F. and Moelwyn-Hughes, E. A. *ibid.* 2635 (1959).
85. Sokov, J. *J. Russ. Phys. Chem. Soc.* **40,** 413 (1908). Ross Kane, *Ann. Rep. Chem. Soc., Lond.* **27,** 351 (1930). Blokkery, P. C. *Rec. trav. chim.* **54,** 975 (1935).
86. Olson, A. R., Frashier, L. D. and Spieth, F. J. *J. phys. Chem.* **55,** 860 (1951).
87. Winstein, S. and others. *J. Amer. chem. Soc.* **76,** 2597 (1954). *Chem and Ind.* 664 (1954). *J. Amer. chem. Soc.* **78,** 328, 2767, 2777, 2780, 2784, (1956); **79,** 4146 (1957); **81,** 5511 (1959).
88. Heldt, W. Z. *ibid.* **80,** 5972 (1958).
89. Ingold, C. K. "Structure and Mechanism in Organic Chemistry." G. Bell and Son Ltd., London (1953). Streitwieser, A. *Chem. Rev.* **56,** 571 (1956).
90. Winstein, S. and Fainberg, A. H. *J. Amer. chem. Soc.* **78,** 2770 (1956); **79,** 5937 (1957).

91. Bateman, L. C., Church, H. G., Hughes, E. D., Ingold, C. K. and Taher, N. A. *J. chem. Soc.* 979 (1940).
92. Nash, G. R. and Monk, C. B. *ibid.* 1899 (1955).
93. Bateman, L. C., Hughes, E. D. and Ingold, C. K. *ibid.* 974 (1940).
94. Bunton, C. A. and Nayal, B. *ibid.* 3854 (1959).
95. Bell, R. P. "Acid-Base Catalysis" Clarendon Press, Oxford (1949).
96. *Disc. Faraday Soc.* **17** (1954).
97. Ljunggren, S. and Lamm, O. *Acta Chem. Scand.* **12,** 1834 (1958).

CHAPTER 14

INCOMPLETE DISSOCIATION, PART VII: THERMODYNAMICS OF IONS AND OF DISSOCIATION. SIZES OF IONS AND OF ION-PAIRS. DISSOCIATION CONSTANT RELATIONSHIPS

Thermodynamics of Ions and of Dissociation

FOR the general reaction $x_1X_1 + x_2X_2 = y_1Y_1 + y_2Y_2$, combination of (4.3), (4.5) and (4.8) gives

$$\Delta G^0 = G^0_{Y_1} + G^0_{Y_2} - G^0_{X_1} - G^0_{X_2} = -nE^0F = -RT \ln K \qquad (14.1)$$

Two other standard thermodynamic expressions are

$$(\mathrm{d} \ln K)/\mathrm{d}T = \Delta H^0/RT^2 \qquad (14.2)$$

$$\Delta G^0 = \Delta H^0 - T\Delta S^0 \qquad (14.3)$$

where ΔH^0 is the standard heat of reaction, or enthalpy change. It follows that

$$\Delta H^0 = -nE^0F + nFT(\mathrm{d}E^0/\mathrm{d}T) \qquad (14.4)$$

so that ΔG^0, ΔH^0 and ΔS^0 can be evaluated from a series of E^0 determinations over a range of temperatures, and if Y_1 is an electrolyte solution while X_1, X_2 and Y_2 are elements or pure compounds for each of which G^0, H^0 and S^0 are known, the corresponding data for Y_1 can be calculated. In addition, if arbitrary standards are fixed for a specific ion, the values for individual ions can be derived. One standard taken for entropies is $S^0_{H^+} = 0$ at 25°C in a hypothetical solution of unit molality with $\gamma_{H^+} = 1$.

Example 14.1. For [1] the reaction which takes place in cell (4.11), i.e., $\frac{1}{2}H_2 + AgCl = HCl + Ag$, at 25°C, $\Delta H° = -9600$ cal/mol, $\Delta G^0 = -5.25$ cal/mol, so $\Delta S^0 = -15.01$ cal/mol. deg. (e.u.). Also,[2] from the third law of thermodynamics, $S^0_{Ag} = 10.0$, $S^0_{AgCl} = 23.0$, $S^0_{H_2} = 31.2$, all at 25°C, so $S^0_{Cl^-} + 10.0 - (15.6 + 23.0) = -15.01$, i.e., $S^0_{Cl^-} = 13.6$ e.u.

Similar information can be extracted from solubilities. In a saturated solution, the solid form and its ions in solution are in equilibrium, so from (1.15)

$$-\Delta G^0 = RT \ln a_{\mathrm{sat.}} \qquad (14.5)$$

and from (14.3),

$$\Delta S^0 = \sum S^0_{\mathrm{ions}} - S^0_{\mathrm{cryst}} = \Delta H^0_{\mathrm{sat.}}/T + R \ln a_{\mathrm{sat.}} \qquad (14.6)$$

where $\Delta H^0_{\text{sat.}}$ is the increase in heat content or heat absorbed when 1 g.mol is dissolved. It can be measured directly if the solubility is high enough, or it can be obtained by calorimetric measurement of the heat of precipitation using two soluble electrolytes which provide the appropriate ions, allowing for any heat of dilution if one electrolyte is in excess. Another way is to measure the solubility at different temperatures and to apply

$$\mathrm{d} \ln a_{\text{sat.}}/\mathrm{d}T = \Delta H^0_{\text{sat.}}/RT^2 \tag{14.7}$$

(cf. (5.6)). If the solubilities are very low it is sufficient to take concentrations for activities.

Example 14.2. From calorimetric measurements of mixtures of KCl and $AgNO_3$ solutions [3] with the former in excess, and correcting for its heat of dilution,[4] $\Delta H^0_{\text{sat.}}$ for AgCl = 15.74 kcal/mol at 25°C. Also,[5] $a_{\text{sat.}} = a_{\text{Ag}}a_{\text{Cl}} = (1.33 \times 0.996 \times 10^{-5})^2$, so from (14.5), $\Delta G^0 = 12.98$ kcal/mole, and from (14.6), $\Delta S^0 = 9.39$ e.u. Also, $S^0_{\text{AgCl}} = 22.9$ and $S^0_{\text{Cl}^-} = 13.6$ from example (14.1), so $S^0_{\text{Ag}^+} = 23.0 + 9.4 - 13.6 = 18.8$ e.u. (a more accurate value is given in Table 14.1).

Latimer and his associates [1,6,7,8] have used these methods to obtain the $S^0_{\text{i.l.}}$ values of many simple, oxy- and complex ions; some of the figures are given in Table 14.1. The data for monoatomic ions can be expressed by the empirical expression [7]

$$S^0_{\text{i.l.}} = 1.5R \ln M + 37 - 270(z/r_e^2) + R \ln g \tag{14.8}$$

where M is the atomic weight, z is the positive value of the valence, $r_e = r_{\text{cryst}} + 2.0$ Å for cations, $r_e = r_{\text{cryst}} + 1.0$ Å for anions, and g is the multiplicity of the ground state (unity for most of the ions considered).

Using Gurney's estimate [9] of $S^0_{\text{H}^+} = -5.5$ e.u., another equation, by Laidler,[10] is

$$S^0_{\text{i.l.}} = 1.5 \text{ R} \ln M + 10.2 - 11.6z^2/r_{\text{cryst}} \tag{14.9}$$

Cobble [11] has given an equation for oxyanions, namely

$$S^0_{\text{i.l.}} = 1.5 \text{ R} \ln M + 66 - 81(zf/r_{12}) \tag{14.10}$$

where M is the molecular weight of the ion, and r_{12} is the X — O distance as calculated by a method of Pauling [12] using covalent radii [13] except for the first row elements (X), for which the values of Schomaker [14] were taken, and where covalent radii were not available, metallic radii were used. The factor f depends on the structural class; it is also valence-dependent for tetrahedral ions. Some figures are given in Table 14.1.

Laidler [10] has also formulated an expression for oxyanions, this is

$$S^0_{\text{i.l.}} = 1.5 \text{ R} \ln M + 40.2 - 27.2z^2/0.25n(r_{12} + 1.40) \tag{14.11}$$

where n is the number of charge bearing ligands. A similar form, devised by

Connick and Powell [15] for oxyanions of the type XO_n^{-z} and species containing — OH groups providing these are ignored in counting n, is

$$S^0_{\mathrm{i.l.}} = 43.5 - 46.5(z - 0.28n) \tag{14.12}$$

For complex inorganic ions, according to Cobble,[16]

$$S^0_{\mathrm{i.l.}} = 49 - 99(zf/r_{12}) + n_1 S^0_{H_2O} \tag{14.13}$$

is satisfactory. Here n_1 is the maximum co-ordination number, or number of water molecules displaced from the simple cation hydration shell (often six) in forming the complex ion, $S^0_{H_2O} = 16.7$ e.u., $f = 1$ for monoatomic ligands, and 0.65 for others such as NH_3 and CN^-; $r_{12} = r_M + r_L$ (data of Wyckoff,[17] M = central ion, L = ligand).

1. Thermodynamics of Ion-solvation

The energetics of ion-solvation are conventionally treated in terms of the process: Ion(gas)→Ion(solution). According to Born,[18]

$$\Delta G^{i0}_{.\mathrm{hyd.}} = G^0_{\mathrm{i.l.}} - G^0_{\mathrm{i.g.}} = -(Nz^2e^2/2r)(1 - 1/\epsilon) \tag{14.14}$$

but this is inadequate if [19] $r = r_{\mathrm{cryst}}$. However, Latimer *et. al.*[20] have shown that by taking $r' = r_{\mathrm{cryst}} + r''$, where $r'' = 0.85$ Å for cations and 0.1 Å for anions, plots of $\Delta G^0_{\mathrm{i.hyd.}}$ against z/r' are linear. They derived values of $\Delta G^0_{\mathrm{i.hyd.}}$ by the following steps and processes (others are described later on):

(1) $\Delta H^0_{\mathrm{e.hyd.}}$ for the electrolyte = L — U is calculated where, e.g.,

$$MX(s) = M^+(g) + X^-(g) \quad \Delta H = -U$$

$$MX(s) + \mathrm{aq.} = M^+(l) + X^-(l) \quad \Delta H = L$$

The crystal lattice energy—U is computed [2,13,21] and L, the heat of solution, is obtained by calorimetry as described for $\Delta H^0_{\mathrm{sat.}}$ earlier on; the data for many electrolytes are available.[22,23]

(2) $\Delta S^0_{\mathrm{i.hyd.}} = S^0_{\mathrm{i.l.}} - S^0_{\mathrm{i.g.}}$ is calculated for both cation and anion. The obtaining of $S^0_{\mathrm{i.l.}}$ has been described and exemplified earlier on, while the entropies of gaseous ions, $S^0_{\mathrm{i.g.}}$ are calculated by means of the Sackur-Tetrode equation.[2] This, and the ways of applying it are dealt with a little further on. Latimer [24] has plotted $\Delta S^0_{\mathrm{i.hyd.}}$ values against z/r'; the plot forms two straight lines, one for univalent ions and another for ions of higher valence (cf. his $G^0_{\mathrm{i.hyd.}}$ plot mentioned above).

(3) Following (2), $\Delta G^0_{\mathrm{e.hyd.}}$ is calculated via (14.3).

(4) For a first assessment, Latimer *et al.*[20] divided $\Delta G^0_{\mathrm{e.hyd.}}$ so that the values of $\Delta G^0_{\mathrm{i.hyd.}}$ for the ions of a number of alkali metal halides gave a linear plot against z/r'. This produced $\Delta G^0_{\mathrm{i.hyd.}} = -84.2$ kcal/g ion for Cl^-, but this was later modified [24] to — 85. At the same time, since it has been estimated from

TABLE 14.1

Ion Entropies,[7] $S^0_{i.l.}$ at 25°C in cal./mol. deg. (standard state = one molal solution, standard reference is $S^0_{H^+.l.} = 0$). Entropies of Hydration[20], $\Delta S^0_{i.hyd.}$ at 25°C in cal./mol. deg. (standard states = one molar gas ions, one molal aqueous ions, (a) $S^0_{H^+.l.} = 0$, (b) $S^0_{H^+.l.} = -2.1$ eu). Crystal radii[13] in Å.

Ion	Li^+	Na^+	K^+	Rb^+	Cs^+	Ag^+	NH_4^+	Tl^+	Mg^{2+}	Ca^{2+}	Sr^{2+}
$S^0_{i.l.}$	3.4	14.4	24.5	29.7	31.8	17.67	26.4	30.4	− 28.2	− 13.2	− 9.4
$-\Delta S^0_{i.hyd.}$(a)	22.2	14.5	6.0	3.1	2.2	15.8	(11.5)	7.0	59	46	44
$-\Delta S^0_{i.hyd.}$(b)	24.1	16.6	8.1	5.2	4.3	17.9	(13.6)	9.1	61	48	46
$r_{cryst.}$(Å)	0.60	0.95	1.33	1.48	1.69	1.26	1.48	1.44	0.65	0.99	1.13

Ion	Ba^{2+}	Zn^{2+}	Cd^{2+}	Mn^{2+}	Cu^{2+}	Fe^{2+}	Pb^{2+}	Sn^{2+}	Hg^{2+}	Al^{3+}	Fe^{3+}
$S^0_{i.l.}$	3.0	− 25.45	− 14.6	− 20	− 23.6	− 27.1	5.1	− 5.9	− 5.4	− 74.9	− 70.1
$-\Delta S^0_{i.hyd.}$(a)	33	60	50	54	61	—	35	—	—	108	108
$-\Delta S^0_{i.hyd.}$(b)	35	62	52	56	63	—	37	—	—	110	110
$r_{cryst.}$(Å)	1.35	0.74	0.97	0.80	0.70	0.75	1.27	—	1.10	0.50	0.60

Ion	Cr^{3+}	Ga^{3+}	U^{4+}	Pu^{4+}	F^-	Cl^-	Br^-	I^-	OH^-	IO_3^-	ClO_3^-
$S^0_{i.l.}$	− 73.5	− 83	− 78	− 87	− 2.3	13.17	19.25	26.14	− 2.45	28.0	39.4
$-\Delta S^0_{i.hyd.}$(a)	—	—	—	79	30.3	17.0	13.4	7.9	—	—	—
$-\Delta S^0_{i.hyd.}$(b)	—	—	—	81	28.2	14.9	11.2	5.8	—	—	—
$r_{cryst.}$(A)	0.64	0.62	0.95	1.00	1.33	1.81	1.95	2.16	1.40	—	—

Ion	BrO_3^-	CN^-	MnO_4^-	SO_4^{2-}	CrO_4^{2-}	Ox^{2-}	PO_4^{3-}	NO_3^-
$S^0_{i.l.}$	38.5	25	46.7	4.4	10.5	9.6	− 45	35
$-\Delta S^0_{i.hyd.}$(a)	—	—	—	(44.7)	—	—	—	(15)
r_{12}(Å)	1.67	—	1.81	—	1.86	—	1.56	1.21
f	0.74	1.34	0.74	0.74	0.74	0.77	0.83	0.68

TABLE 14.2

Values of Heats and Entropies of Ion Solvation[29] at 25°C. Standard States: (a) Gaseous Ion = 1 atm., (b) Solute = 1M(ideal).

Ion	H^+	Li^+	Na^+	K^+	Rb^+	Cs^+	Cu^+	Ag^+	
$\Delta H_{hyd.}$(kcal)	0	137.67	153.79	183.98	189.9	197.8	118.7	147.12	
$\Delta S_{hyd.}$(e.u.)	0	− 2.4	5.1	13.6	16.5	17.2	− 18.7	3.74	
Ion	Tl^+	NH_4^+	Be^{2+}	Mg^{2+}	Ca^{2+}	Sr^{2+}	Ba^{2+}	Ra^{2+}	Cr^{2+}
$\Delta H_{hyd.}$(kcal)	182.83	185	− 73	61.98	140.77	176.05	209.80	220	79.3
$\Delta S_{hyd.}$(e.u.)	14.6	—	—	− 11.7	1.8	—	—	—	—
Ion	Mn^{2+}	Fe^{2+}	Co^{2+}	Ni^{2+}	Cu^{2+}	Cd^{2+}	Zn^{2+}	Pb^{2+}	Sc^{3+}
$\Delta H_{hyd.}$(kcal)	80.4	62.5	30.4	18.1	19.45	89.76	32.84	167.68	− 164.4
$\Delta S_{hyd.}$(e.u.)	− 9.5	—	—	—	− 14.5	− 2.6	− 11.87	15.2	—
Ion	Y^{3+}	Ce^{3+}	La^{3+}	Ga^{3+}	In^{3+}	Tl^{3+}	Al^{3+}	Cr^{3+}	
$\Delta H_{hyd.}$(kcal)	− 83.1	− 67	− 2.5	− 337.6	− 199.9	− 217.9	− 331.6	− 680	
$\Delta S_{hyd.}$(e.u.)	—	− 10	—	− 44	− 24	—	− 32.7	—	
Ion	Fe^{3+}	Ce^{4+}	F^-	Cl^-	Br^-	I^-	OH^-	ClO_3^-	
$\Delta H_{hyd.}$ (kcal)	− 264	− 508	− 366.3	− 348.8	− 340.7	− 330.3	− 345.7	− 327.6	
ΔS_{hyd} (e.u.)	− 33.6	—	− 63.1	− 49.49	− 45.79	− 40.32	—	—	

thermocells [25] that $S^0_{\text{i.l.}} = -2.1$ e.u. for H^+, this was taken as reference when calculating $\Delta S^0_{\text{i.hyd.}}$. Some of Latimer's figures [24] are recorded in Table 14.1.

Frank and Evans [26] have also evaluated values of $\Delta S^0_{\text{i.hyd.}}$; their chosen standard states are $P = 1$ atm. for the gaseous ion, unit mole fraction for the solution and $S^0_{\text{i.l.}} = 18.1$ e.u. for Cl^-. Robinson and Stokes [27] have pointed out that to compare these figures with those of Laidler they must be increased by $R \ln (22.4 \times 298 \times 55.51/273) + y = 14.3 + y$. For Cl^-, Frank and Evans cite $\Delta S^0_{\text{i.hyd.}} = -26.6$, so from Table 14.1, $-26.6 + 14.3 + y =$ (a) -17.0, (b) -14.9, whence $y_a = -4.7$, $y_b = -2.6$ for anions; $y_a = 4.7$, $y_b = 2.6$ for cations. This gave the figures shown in brackets in the Table; the rest are about the same as in (a) and (b) except $Fe^{3+} =$ (a) 101, (b) 103.5: $Al^{3+} =$ (a) 114.5, (b) 116.6: $Mg^{2+} =$ (a) 65.2, (b) 67.3.

Eley and Evans [28] calculated values of $\Delta H^0_{\text{i.hyd.}}$ and $\Delta S^0_{\text{i.hyd.}}$ by a detailed method of analysis which considers the tetrahedral structure of water, ion-dipole interaction, the entropy of translation of aqueous ions in solution, the entropy change accompanying the orientation of water around an ion, and other factors. Their standard states are one atm. of gaseous ions and one molar aqueous solutions, while their figures for $\Delta S^0_{\text{i.hyd.}}$ of cations are about 15.5 e.u. lower than those in rows (a) of the Table, the anion values from one to 7 e.u. lower (e.g., $F^- = 35.2$, $Li^+ = 34.7$).

A Table of ΔH, ΔS and ΔG ion hydration values at 25°C has been compiled by Benjamin and Gold.[29] These are given in Table 14.2, and were derived by considering the hypothetical chemical changes: $El_{st} \rightarrow El^z_g + ze^-_g$, $El_{st} + zH^+_l \rightarrow \frac{1}{2}zH_{2,g} + El^z_l$, $El^z_g + zH^+_l \rightarrow El^z_l + zH^+$ where El and El^z represent an element and its ion, *st.* refers to the standard state, l and g to the aqueous and gaseous states. The respective changes in $X (X = H, S$ or $G)$ are represented by

$$\Delta X_g(El^z) = X_g(El^z) + zX_g(e^-) - X_{st}(El)$$

$$\Delta X_l(El^z) = X_l(El^z) - X_{st}(El) - z[X_l(H^+) - X_{st}(\tfrac{1}{2}H_2)]$$

$$\Delta X_{\text{hyd.}}(El^z) = X_l(El^z) - X_g(El^z) - z[X_l(H^+) - X_g(H^+)]$$

whence

$$\Delta X_{\text{hyd.}}(El^z) = \Delta X_l(El^z) - \Delta X_g(El^z) + z\Delta X_g(H^+) \qquad (14.15)$$

Values of $\Delta H_g(El^z)$, $\Delta H_l(El^z)$, $\Delta S_l(E^z)$ and $\Delta G_l(El^z)$ were taken from the compilation of Rossini and his co-workers,[23] and the values of $\Delta S_g(El^z)$ were calculated via the Sackur–Tetrode equation (14.16), taking values of $g = 2J + 1$ from Landolt–Bornstein.[30] $\Delta G_g(El^z)$ was derived from the appropriate form of (14.3). As an example of the use of the data, $\Delta G_{\text{hyd.}}(CaCl_2) = 140.77 + 2(-348.8)$. By subtracting $257z_i$ from the values of $\Delta H_{\text{hyd.}}$, values agreeing roughly with those of Latimer [20] and Verney [31] may be obtained, and likewise by substraction of $280z_i$ there is rough agreement with Gurney's figures.[9] There is no general agreement with the calculations of Eley and Evans.[28]

2. Sackur–Tetrode Equation

This equation is [2] used to calculate the entropy of the gaseous form of an ion:

$$S_{\text{i.g.}} = k\{\ln [g(2\pi mkT)^{\frac{3}{2}}kT/Ph^3] + 2.5\} \qquad (14.16)$$

where m is the mass of the ion, P is the pressure and g is the degeneracy of the ground state ($g = 1$ for the ground state of ions with inert gas structures). At 25°C, for one g. ion of mass M, and P in atms. (14.16) becomes

$$S^0_{\text{i.g.}} = 1.5\ \text{R} \ln M + 26.00 - \text{R} \ln P + \text{R} \ln g \quad \text{cal/mol. deg.} \qquad (14.17)$$

For P corresponding to 1 M, $R \ln P = R \ln (22.4 \times 298/273) = 6.35$ e.u.

A more general entropy expression which embraces polyatomic species is

$$S^0_{\text{g.}} = S^0_{\text{trans.}} + S^0_{\text{rot.}} + S^0_{\text{vib.}} \qquad (14.18)$$

The translational entropy term is expressed by (14.17) on omission of the last term, the rotational entropy term does not apply to monoatomic species, but for linear species it is expressed by [32]

$$S^0_{\text{rot.}} = \tfrac{1}{2}R(2 \ln T + 2 \ln I_B - 2 \ln \sigma) + 2(267.52)/3 \qquad (14.19)$$

if I_B, the moment of inertia, is expressed in g. cm. If I is in at. wts., Å, the last term is $-$ 2(5.384)/3. The σ term is often taken as unity.

For other rigid polyatomic species,

$$S^0_{\text{rot.}} = \tfrac{1}{2}R(3 \ln T + \ln I_A I_B I_C - 2 \ln \sigma) + 267.52 \qquad (14.20)$$

(I's in g. cm). The product $I_A I_B I_C$ can be evaluated from [33]

$$I_A I_B I_C = \begin{vmatrix} + I_{xx} & - I_{xy} & - I_{xx} \\ - I_{xy} & + I_{yy} & - I_{yz} \\ - I_{xz} & - I_{yz} & + I_{zz} \end{vmatrix} \qquad (14.21)$$

where I_{xx}, etc., are the products of the moments of inertia with respect to an x, y, z co-ordinate system with the centre of mass as the origin.

For polyatomic species containing one free internal rotation, $S^0_{\text{rot.}}$ contains an additional term represented by

$$S^0_{\text{f.i.r.}} = \tfrac{1}{2}R(\ln T + \ln I_m + \ln (8\pi^3 k/h^2) + 1 - 2 \ln n \qquad (14.22)$$

where n is the number of indistinguishable positions of any attached "top" of reduced moment of inertia I_m as defined by

$$I_m = I^0_m [1 - I^0_m(\lambda^2_{mA}/I_A + \lambda^2_{mB}/I_B + \lambda^2_{mC}/I_C)] \qquad (14.23)$$

in which I^0_m is the moment of inertia of the "top" and the λs are the cosines of the angle between the axis of the top and the axes of I_A, I_B, I_C. For a single *f.i.r.* of a symmetrical top,

$$S^0_{\text{f.i.r.}} = 2.287 \log T - 2.714 \qquad (14.24)$$

The vibrational entropy term is a complex one involving measurements of vibrational frequencies in the infra-red. It is usually quite small. When there is *f.i.r.*, the torsional oscillation terms are omitted.

Altshuller [34] has used the Sackur–Tetrode equation to evaluate $S^0_{\text{i.g.}}$ of a number of gaseous oxyanions and polyatomic ions, and hence from experimental values of $S^0_{\text{i.l.}}$, obtained the corresponding $\Delta S^0_{\text{i.hyd.}}$ data. He used known geometric configurations and interatomic distances derived from X-ray and infra-red measurements on salts;[35,36] a very comprehensive source has since been published by the Chemical Society, London.[37] The vibrational frequencies for $S^0_{\text{vib.}}$ were taken from available data,[32, 38] and so also were values of $S^0_{\text{i.l.}}$[8, 23] Since these refer to 1 M solutions, while the calculated values of $S^0_{\text{i.g.}}$ refer to 1 atm. at 25°C, 6.35 e.u. was subtracted from the latter [39] (see below (14.17)) before calculating $\Delta S^0_{\text{i.hyd.}}$ Within an uncertainty of 4 e.u., the results for oxyanions are expressed by

$$-\Delta S^0_{\text{i.hyd.}} = 5 + (29 - 4.5\,n)z^2 \quad (n = \text{no. of } 0) \qquad (14.25)$$

TABLE 14.3

Values of Gaseous Ion Entropies,[34] $S^0_{\text{i.g.}}$ at 25°C and 1 atm., and Entropies of Ion Hydration, $\Delta S^0_{\text{i.hyd.}}$ at 25°C, and 1M(ideal). H = height; I's × 10^{40}.

Ion	Configuration	X – O(Å)	I_A	I_B	I_C	$S^0_{\text{trans.}}$	$S^0_{\text{rot.}}$	$S^0_{\text{vib.}}$	$-\Delta S^0_{\text{i.hyd.}}$
NO_2^-	< ONO = 115°	1.24	7.2	58.1	65.3	37.40	18.86	0.25	20
ClO_2^-	< OClO = 114°	1.59	21	94	115	38.54	21.0	1.35	30
NO_3^-	Planar	1.24	122	61	61	38.30	19.47	1.02	17.4
CO_3^{2-}	Planar	1.26	126	63	63	38.20	19.57	1.02	65
ClO_3^-	Pyramidal, $H = 0.49$	1.48	194	105	105	39.18	22.4	2.2	18
BrO_3^-	Pyramidal, $H = 0.56$	1.68	250	141	141	40.45	23.2	3.8	23
SO_3^{2-}	Pyramidal, $H = 0.51$	1.39	175	96	96	29.06	22.1	2.2	64
ClO_4^-	Tetrahedral	1.52	$I_A = I_B = I_C = 164$			39.71	20.34	3.0	13.5
ReO_4^-	Tetrahedral	1.9	$I_A = I_B = I_C = 256$			42.46	21.67	6.8	16.2
SO_4^{2-}	Tetrahedral	1.50	$I_A = I_B = I_C = 159$			39.60	20.26	3.0	52.4
CrO_4^{2-}	Tetrahedral	1.60	$I_A = I_B = I_C = 181$			40.17	20.64	3.9	49
SeO_4^{2-}	Tetrahedral	1.65	$I_A = I_B = I_C = 193$			40.78	20.83	5.7	55
PO_4^{3-}	Tetrahedral	1.55	$I_A = I_B = I_C = 170$			39.57	20.26	411	109
AsO_4^{3-}	Tetrahedral	1.75	$I_A = I_B = I_C = 217$			40.70	21.18	5.6	96
OH^-		(0.97)			$I = 0.74$		$S^0_{\text{i.g.}} = 38.4$		– 34.5
CN^-		1.10			$I = 6.5$		$S^0_{\text{i.g.}} = 45.2$		– 11
N_3^-		2.30			$I = 61.5$		$S^0_{\text{i.g.}} = 50.7$		—
SCN^-	$r_{SC} = 1.61$, $r_{CN} =$	1.17			$I = 134$		$S^0_{\text{i.g.}} = 55.5$		—
NH_4^+		1.032			$I = 4.75$		$S^0_{\text{i.g.}} = 44.5$		– 11

3. Dissociation

ΔG^0, ΔH^0 and ΔS^0 for the dissociation process are derived from (14.1 – 3) by obtaining dissociation constants at several temperatures (usually 5° to 10° intervals). Since the K values of the stronger electrolytes generally decrease slowly with increasing T, ΔH^0 and ΔS^0 can be obtained by direct application of the equations. On the other hand, the relationship between K and T for weak electrolytes is often complex and the data are fitted to empirical equations which are then differentiated. Many of these equations have been classified by Feates and Ives [40] and tested with their results for cyanoacetic acid from 5° to 45°C. The equations can be arranged into the following groups; some or all of the terms in brackets are often omitted:

(a)[41,42] $$-\ln K = a + bT + cT^2 + (dT^3 + eT^4 + fT^5) \quad (14.26)$$

(b)[41,42] $$-\ln K = a + bT + cT\ln T + (dT^2 + eT^3) \quad (14.27)$$

(c)[42,43] $$-\ln K = a/T + b + cT + (dT^2) \quad (14.28)$$

(d)[44] $$-\ln K = a/T + b/T\ln T + c/T(\ln T)^2 \quad (14.29)$$

(e)[45] $$-\ln K = -\ln K_{max} + 5\times 10^{-5}(t-\theta)^2 \quad (14.30)$$

$(t = {}^\circ\mathrm{C},\ \theta = \text{value of } t \text{ at } -\ln K_{max})$

(f)[46,47,48,49] $$-\ln K = a/T + b + c\ln T + (dT + eT^2) \quad (14.31)$$

From (14.2) and, e.g. (14.26),

$$\Delta H^0/R = bT^2 + 2cT^3 + (3dT^4 + \ldots) \quad (14.32)$$

From (14.1), (14.3) and (14.32),

$$\Delta S^0/R = a + 2bT + 3cT^2 + (4dT^3 + \ldots) \quad (14.33)$$

and also from (14.32),

$$\Delta C_p^0 = d\Delta H^0/dT = 2bRT + 6cRT^2 + \ldots \quad (14.34)$$

The dissociation constants of many weak acids can be expressed satisfactorily by (14.31) since $-\ln K$ reaches a maximum (at which $\Delta H^0 = 0$); (14.28) is equally successful, and the values of the parameters for many acids have been computed by Robinson and Stokes.[27]

Denison and Ramsay [50] have derived the following expressions and found them to be satisfactory for a number of substituted phenyltrimethylammonium perchlorates in ethylene chloride and ethylidene chloride between 20° and 35°C; a is the ion-pair size parameter, and is discussed later on (see (14.59)).

$$\Delta G^0 = Ne^2/a\epsilon \quad (14.35)$$

$$\Delta S^0 = (\Delta G^0/T)(\mathrm{d}\ln\epsilon/\mathrm{d}\ln T) \quad (14.36)$$

$$\Delta H^0 = \Delta G^0(1 + \mathrm{d}\ln\epsilon/\mathrm{d}\ln T) \quad (14.37)$$

Evans and Uri [51], and Evans and Nancollas [52] have shown that plots of ΔS_{ass}, the entropy of association of some salts against $S^0_{i.1.}$ (anions) for a specific cation are linear. The terms entering the entropy of the association process $M^{3+} + X^- \rightleftharpoons MX^{2+}$ were investigated by means of the entropy cycle shown below (excluding the parts put into brackets):

$$\begin{array}{llll} M^{x+}(g) + X^{y-}(g) & \xrightarrow{\Delta S_3} MX^{(x-y)+}(g) & [+\, nH_2O(g)] & \\ \uparrow\Delta S_1 \qquad \uparrow\Delta S_2 & \quad \downarrow\Delta S_4 & [\downarrow\Delta S_5] & \qquad (14.38) \\ M^{x+}(l) + X^{y-}(l) & \xleftarrow{-\Delta S} MX^{(x-y)+}(l) & [+nH_2O(l)] & \end{array}$$

for which

$$\Delta S_1 + \Delta S_2 + \Delta S_3 + \Delta S_4 (+ \Delta S_5) = \Delta S \qquad (14.39)$$

The terms ΔS_1, ΔS_2 and ΔS_4 are $-\Delta S^0_{\text{i.hyd.}}$ for the cation, anion and ion-pair respectively, and are calculated by applying the Sackur-Tetrode equation as described earlier on (although the correction factor of 6.35 e.u. has often been omitted).

Nancollas and Nair [53] have made a number of detailed calculations relating to the above cycle; ion-pairs such as $PbCl^+$ are treated as linear species, while $PbNO_3^+$ is regarded as having an NO_2 "top" with free internal rotation. Examples of the kind of information that they have extracted are (a) For some Pb^{2+} systems, ΔS_{ass}(exptl.) is close to ΔS(calc.), e.g., 22,19; 16.4, 15; 3.5, 5. for $PbCl^+$, $PbBr^+$ and $PbNO_3^+$ respectively; (b) $\Delta S_{\text{hyd.}}$ (TlX) $= \Delta S_4$ can be obtained with the aid of ΔS_{ass}(exptl.), and for X = Cl^-, Br^-, OH^-, the answers are linearly related to $1/\{r_{\text{cryst}}(\text{anion}) + r_{\text{cryst}}(\text{cation})\}$; this is not so with MSO_4 ion-pairs.

By extending the entropy cycle (14.38) to include any water of ion-solvation which is displaced during ion-association (indicated by the portions in brackets), Austin, Mathieson, and Parton [54] derived an expression which predicts closely the found values of ΔS_{ass} for the ion-pairs formed by Fe^{3+}, Cd^{2+} and Pb^{2+} with Cl^-, Br^- and I^-. George [55], by using the form of (14.38) which includes $n\ H_2O(l)$ but excludes $n\ H_2O(g)$ so that $\Delta S_5 = 0$, has found that plots of ΔS_{ass}, i.e., $-\Delta S$, against (ΔS_1 or ΔS_2 + constant) are linear for certain series of ion-pairs having one ion in common. The constant varies with the series. The alkaline earth hydroxides do not conform to this scheme, and the evidence for MSO_4 ion-pairs is that they do not lose water of hydration during association for although ΔS_1 varies from cation to cation, ΔS is practically constant.

The Sizes of Ions and Ion-Pairs

If $\bar{u}_j^0$ is the velocity of an ion j at zero concentration in an applied field of 1 V/cm, and ω_j^0 is its velocity under unit applied force in c.g.s. units, then from (3.73),

$$\omega_j^0 = \lambda_j^0 10^{-8} c^* / F z_j e \qquad (14.40)$$

Also, from Stokes's law,

$$\omega_j^0 = 1/6\pi\eta r_s \tag{14.41}$$

where r_s is the ion radius in solution. Thus

$$r_s(\text{cms}) = Fz_je/6\pi\eta\lambda_j^0 10^{-8}c^* = 8.20 \times 10^{-9}z_j/\lambda_j^0\eta \tag{14.42}$$

Stokes's law is meant to apply to spherical particles which are relatively large compared to the solvent particles so it is not surprising that (14.42) has not proved to be a reliable way of obtaining the radii of ions in solution. At 25°C in water, $r_s(\text{Å}) = 91.6z_j/\lambda_j^0$, and as Table 14.4 shows, at least a number of the values must be too small since in some cases $r_s < r_{\text{cryst}}$. Robinson and Stokes [27] have endeavoured to find a suitable modification to (14.42) by comparing the r_s values of a series of NR_4^+ ions with their radii as derived from bond lengths and molal volumes; their treatment has been explored by Nightingale.[56] It is concluded that (14.42) is valid only when $r_s \geqslant 5$ Å, which is larger than the radii of most ordinary ions. The necessary correction is found from a plot of r_s against the "true" NR_4^+ radii. Other evidence given later on indicates that these corrections are too large.

Stern and Amis,[57] who have reviewed the methods that have been developed for ascertaining the sizes of ions in gases, crystals and in solution have, at one stage, used what is in effect (14.42). There is some confusion for two of their sets of figures should be identical with those labelled r_s in Table 14.4.

A modified form which involves V_m, the molar volume of the solvent, has been suggested by Kortüm,[104a] namely

$$r_j(1 - 0.21V_m/r_j^3) = 0.820z_j/\lambda_j^0\eta \quad (r_j \text{ in Å}) \tag{14.43}$$

At 25°C, $V_m = 18$ cm³ for water, and r_j values are 10 to 50 % larger than r_s.

There are numerous empirical ways of estimating the radii of hydrated ions by treating them as adjustable parameters. One such example from the preceding section is Latimer's assignment of $r_{\text{cation}} = r_{\text{cryst}} + 2.0$ Å, $r_{\text{anion}} = r_{\text{cryst}} + 1.0$ Å in connection with $S_{\text{i.l.}}^0$, and $r_{\text{cation}} = r_{\text{cryst}} + 0.85$ Å, $r_{\text{anion}} = r_{\text{cryst}} + 0.1$ Å in connection with $\Delta G_{\text{i.hyd.}}$ estimations. Other examples are given later on.

1. Hydration Numbers

The hydration number of an ion attempts to give numerical expression to the number of solvent molecules which, through ion-dipole forces, form one or more layers around and in association with an ion. This layer, it is believed, may contain as many as three "thicknesses" of solvent molecules although the outer molecules are less firmly bound than those adjacent to the ion. There are many ways whereby it should be possible to ascertain hydration numbers and most of them have been explored at various times. The answers for a particular electrolyte often depend on the method employed and this reflects, to a certain extent, that the exact nature and forces of ion-solvation still present a number of unsolved features. A useful review on many of the current ideas will be found in the monograph of Robinson and Stokes.[27]

One of the earliest attempts to evaluate hydration numbers consisted of obtaining Hittorf transference numbers in the presence of a reference non-electrolyte such as sugar. In this way, the net transfer of water could be determined.[58] Washburn made considerable efforts to exploit this approach,[59] but it has since been found [60] that, to some extent, the reference materials also migrate with the ions. Nevertheless, by ignoring this source of error, Haase has [61] evaluated a number of hydration numbers by this method taking as reference answers $h = 7$ for Li^+ and 5 for Na^+ (h = hydration number), these being values deduced by Robinson and Stokes [27] from their modified form of (14.42).

Some other experimental procedures have been reviewed by Conway and Bockris.[62] These include Ulich's expression [63] which makes use of $\Delta S^0_{\text{i.hyd.}}$ data, ultrasonic measurements of the compressibilities of solvent and solution,[64] Darmois's analysis[65] of the densities of electrolyte solutions as modified by Conway and Bockris,[62] dielectric constant and loss measurements,[66] and Debye's theory [67] that the ultrasonic irradiation of electrolyte solutions produces periodic differences of potential which are related to the difference in the weights of the solvated cation and anion.

By means of dielectric studies of electrolyte solutions [69] at microwave frequencies, the number of H bonds broken by each ion, n_{bb}, and the number of water molecules prevented by the ion from rotating in an electric field, n_{rr}, can be ascertained. Hasted and Roderick [66,68] identify these quantities with hydration numbers. Some figures are: (n_{bb}) $NaCl = 5$, $RbCl = 9.5$, $BaCl_2 = 12$, $MgCl_2 = 12$, $HCl = 9$; (n_{rr}) $NaCl = 6$, $RbCl = 6$, $BaCl_2 = 9$, $MgCl_2 = 14.5$, $LaCl_3 = 16$, $HCl = 11$. These are of the same order as those obtained in other ways.[27]

Debye's theory [67] is difficult to test, but experimental evidence in its favour has been obtained [70] although the interpretations are impeded by the fact that water and other solvents also produce ultrasonic vibration potentials.[71]

A few hydration numbers have been deduced from the X-ray diffraction patterns of electrolyte solutions. According to Brady,[72] $h = 4$ for K^+ and Li^+, and 6 for OH^-. Other interpretations [73] put $h=6$ for K^+ and Ca^{2+}, and 4 for OH^-.

The hydration number of the H^+ ion has been assessed by Bascombe and Bell [74] from considerations of the equilibrium $B + H_h^+ \rightleftharpoons BH^+ + hH_2O$, where B is an uncharged base and H_h^+ is a proton solvated by h water molecules. The equilibrium was used to redefine Hammett's acidity function:

$$- H_0 = \log [H_h^+] - h \log a_{H_2O} + \log (f_B f_{H_h} / f_{BH}) \qquad (14.44)$$

whence, with c_H for $[H_h^+]$, since $f_{H_h} \simeq f_{BH}$,

$$-H_0 - \log c_H + h \log a_{H_2O} = \log f_B = A(c_H) \quad (14.45)$$

where A is a coefficient. Using known data [75] for a_{H_2O} in HCl, $HClO_4$ and H_2SO_4 solutions, plots of the left side against c_H are linear up to $c_H = 8$ if h is taken as 4. An alternative, equally valid expression is obtained by putting $a_{H_2O} = [H_2O] = (1 - 0.018\, hC)c_H/C$, where C is the acid molarity. This value of h agrees with that derived by Glueckauf [76,77] from activity coefficients (see later) and the uptake of water by acidic ion-exchange resins, and that deduced by Wicke, Eigen and Ackermann [78,79] from the specific heats of acid solutions. Bell [74] has suggested that in spite of the approximations involved, his method can be applied to other ions, e.g., $M(H_2O)_3Cl_3 + Cl^- \rightleftharpoons M(H_2O)_2Cl_4^- + H_2O$. From other considerations of H_0 data, Wyatt [80] has also suggested that h is 4 to 5 for the H^+ ion.

The mean distance of closest approach, or mean diameter of anion and cation, a, is treated as a parameter in certain conductance, transference number and activity coefficient expression, and it is selected (for completely dissociated electrolytes) so that the experimental results fit the expressions. These topics form part of the subject matter of Chapters 3 and 5. In only one treatment, namely, the extended Fuoss–Onsager conductance equation (3.103, 3.112) does the value of a emerge from the calculations free of arbitrary selection. As a rule, the D.H. activity coefficient expression in the form (3.40), (3.41) rarely applies up to unit ionic strength even with 1 : 1 electrolytes, and it is necessary to take into account ion-solvation in order to obtain activity coefficient expressions which are satisfactory up to high concentrations. The equations of Robinson and Stokes [81] and of Glueckauf,[77] which are founded on this concept have been given in Chapter 5. Glueckauf [77] has tabulated his derived values of h for a large number of halides and by assuming that the hydration of the halide ions correspond to $h = 0.9$, has evaluated the h values of some cations and anions. These are shown in Table 14.4. The plots of h for monoatomic cations against r_{cryst} are linear although M^+, M^{2+} and M^{3+} ions give rise to separate lines. This is also true of plots of $S^0_{i.1.}$ against h; the gradients give the entropy loss per molecule of hydration water as 5.6, 12.0 and 6.0 e.u. for M^+, M^{2+} and M^{3+} ions respectively.

If h for both cation and anion are known, the D.H. mean ion size parameter a can be derived. Stokes and Robinson [82] have analyzed the relationship between h and a in terms of the random close-packed volumes of ions in solution, but before outlining their treatment, some reference will be made to ionic volumes. Using the definitions of Chapter 6, $\phi_v^0 = \bar{V}_2^0$ can be divided into its ionic contributions, $\bar{V}_i^0$, if a standard reference is selected. Owen and Brinkley [83] took $\bar{V}^0_{H^+} = 0$, while according to Stern and Amis,[57] Eucken [84] derived values of $\bar{V}_i^0$ by assuming that Cs^+ and I^- are only slightly hydrated and their volumes in solution are proportional to r^3_{cryst}; Zan [55] has calculated many $\bar{V}^0_i$ values on this basis. Fajans and Johnson [86] have taken yet another standard; they assume that $\bar{V}^0_{NH_4^+} = \bar{V}^0_{Cl^-} = 18.0$ ml. at 25°C. Some results by the above authors are listed in Table 14.4.

Formulae for evaluating $\bar{V}^0_i$ have been devised by Couture and Laidler:[87,89]

$$(\bar{V}^0_{H^+} = 0) \qquad \bar{V}^0_+ = 16 + 4.9\, r^3_G - 20\, z_+; \quad \bar{V}^0_- = 4 + 4.9\, r^3_G - 20\, |\, z_-\, | \tag{14.46}$$

$$(\bar{V}^0_{H^+} = -\,6) \qquad \bar{V}^0_\pm = 16 + 4.9\, r^3_G - 26\, |\, z_\pm\, | \tag{14.47}$$

where r_G are Goldshmidt's crystal ionic radii.[88] For oxyanions,[89]

$$\bar{V}^0_- = 58.8 + 0.89\,(0.25\, n r_L)^3 - 26\, |\, z_-\, | \tag{14.48}$$

where $r_L = r_{12}$ [see below (14.10)] + 1.40 Å, the latter figure being taken as the radius of the O^{2-} ion.

Stokes and Robinson [82] have calculated ionic volumes by making use of Alder's observations [90] that when spheres of equal size, or a 50 : 50 mixture with diameters in the ratio of 1 : 1.7 are packed randomly in a liquid of nearly the same density they occupy 0.58 of the total volume so [82]

$$\bar{V}^0_- = 4\pi b^3 N/(3 \times 0.58) = 4.35 \times 10^{24}\, b^3_- \tag{14.49}$$

where b_- is the anion crystalline radius. Values of $\bar{V}^0_+$ were derived from

$$\bar{V}^0_+ = \nu_+ \bar{V}^0(\text{exptl.}) - \nu_- \bar{V}^0_-\ (\text{calc.}) \tag{14.50}$$

It was found that chlorides, but not bromides or iodides, show a small electro-striction effect. The values of $\bar{V}^0_+$ so obtained are about 4 ml. larger than those of Bernal and Fowler.[91]

From this information, a can be calculated. The molar volume of a hydrated ion is expressed by

$$\bar{V}^0_{h,i} = \bar{V}^0_i + 18\, h_i \qquad \text{cm}^3/\text{mole} \tag{14.51}$$

so from (14.49) and (14.50), and $a = r_{h^+} + r_{h^-}$, the radii of the hydrated ions are given by

$$r^3_{h^+} = (\bar{V}^0_+ + 18\, h_+)/4.35 \times 10^{24}; \quad r^3_{h^-} = (\bar{V}^0_- + 18\, h_-)/4.35 \times 10^{24} \tag{14.52}$$

One way of finding $h = \nu_+ h_+ + \nu_- h_-$ and the associated value of a is to find that value which when put into the activity coefficient expression (5.73) best fits the data.

Example 14.3. For NaCl, $\gamma_\pm$ for a 1.0 M solution at 25°C is [27] 0.657. Also,[92] 1.0 M = 0.9787 C. From ${}^a\bar{V}^0_i$ of Table 14.4 and $\bar{V}^0_{H_2O} = 17.97$, $r = \phi_v/v^0_w = \bar{V}^0_{NaCl}/V^0_{H_2O} = 0.92$. By approximations between (5.73) for h and (14.52) for a, taking $h = 0$, the answers are $h = 3.0$, $a = 3.98$ Å.

If h_i is known for one ion, the procedure of Example 14.3. may be followed for deriving values of h_i of other ions. This has been done, taking data for 1 M solutions, by considering h_- to be zero for the Br^- ion, whence from NaBr, $h_+ = 2.9$ for Na^+, whence, from NaCl data, h_- for $Cl^- = 0.1$. These results

TABLE 14.4

Ion Mobilities in Water at 25°C. Crystalline Ion Radii (Å). Partial Molal Volumes of Ions, $\bar{V}_i^0$ (ml.). Solvated Ion Radii: r_s from (14.42), r_h from (14.52) via h_i of present work using $\bar{V}_i^0$ of Ref. (82). Hydration Numbers of Glueckauf,[77] h_G, and h_i of present work. Densities from Refs. (93), (94) and (95). Refs: $a = 27, 92$, $b = 96$, $c = 97$, $d = 98$, $e = 99$, $f = 100$, $g = 41, 40$, $h = 101$, $i = 102$, $j = 103$, $k = 104$, $l = 139$, $m = 13$, $n = 83$, $p = 86$, $q = 82$, $r = 77$.

Ion	λ_j^0	${}^{m}r_{\text{cryst.}}$	${}^{n}\bar{V}_i^0$	${}^{p}\bar{V}_i^0$	${}^{q}\bar{V}_i^0$	${}^{r}h_G$	h_i	r_h(Å)	r_s(Å)
H^+	349.8[a]	—	0	0.2	− 7.24	3.9	3.9	2.44	0.26
Li^+	38.66[a]	0.60	− 1.0	− 0.9	− 8.25	3.4	4.25	2.50	2.37
Na^+	50.10[a]	0.95	− 1.5	− 1.7	− 8.75	2.0	2.9	2.17	1.83
K^+	73.52[a]	1.33	8.7	8.4	1.50	0.6	1.2	1.75	1.25
Rb^+	77.8[a]	1.48	13.7	13.7	6.47	0	0.5	1.53	1.18
Cs^+	77.3[a]	1.69	21.1	21.7	13.93	0	0	1.47	1.19
Ag^+	62.1[b]	1.26	− 1.0	− 1.2	− 8.61	—	—	—	1.48
NH_4^+	73.5[a]	1.48	17.9	18.1	10.67	0.2	1.05	1.88	1.25
Tl^+	74.7[a]	1.44	—	—	—	—	—	—	1.23
Mg^{2+}	53.06[a]	0.65	− 20.9	− 21.4	(− 36.8)	5.1	8.25	2.96	3.45
Ca^{2+}	59.50[a]	0.99	− 17.7	− 19.0	− 32.36	4.3	7.2	2.72	3.08
Sr^{2+}	59.46[a]	1.13	− 18.2	− 18.4	− 32.67	3.7	6.7	2.74	3.08
Ba^{2+}	63.64[a]	1.35	− 12.3	− 13.2	− 27.01	3.0	5.2	2.48	2.88
Cu^{2+}	54[a]	0.70	—	—	− 26.6	—	—	—	3.4
Ni^{2+}	53.7[c]	0.70	—	—	—	—	—	—	3.41
Co^{2+}	53.5[c]	0.72	—	− 18	(− 32.6)	—	—	—	3.42
Mn^{2+}	53.1[c]	0.80	—	− 14	(− 28.6)	—	—	—	3.45
Zn^{2+}	52.8[c]	0.74	—	− 26	(− 40.6)	5.3	9.3	3.07	3.47
Cd^{2+}	52.7[c]	0.97	− 13	− 13.2	—	—	—	—	3.48
Pb^{2+}	69.45[d]	1.27	—	—	− 28	—	—	—	2.64
La^{3+}	69.6[e]	1.15[l]	− 38.3	—	− 59.9	7.5	11.0	3.19	3.95
Nd^{3+}	69.35[f]	1.00[l]	—	—	− 65.45[k]	—	11.2	3.15	3.96
$Co(NH_3)_6^{3+}$	99.2[h]	—	—	—	—	—	—	—	2.77
$Co(en)_3^{3+}$	74.7[h]	—	—	—	—	—	—	—	3.67
$Co(pn)_3^{3+}$	65.1[h]	—	—	—	—	—	—	—	4.22
OH^-	197.8[a]	1.4	− 5.3	− 4.8	1.8	4.0	3.5	2.46	0.46
Cl^-	76.35[a]	1.81	18.1	18.0	25.31	0.9	0.1	1.84	1.20
Br^-	78.20[a]	1.95	25.0	25.1	32.23	0.9	0	1.92	1.17
I^-	76.9[a]	2.16	36.6	36.7	43.85	0.9	0	2.16	1.19
NO_3^-	71.34[b]	(1.21)	29.3	29.4	36.62	0	0	2.03	1.28
BrO_3^-	55.84[h]	(1.67)	44	36.4	—	—	—	—	1.64
IO_3^-	40.75[h]	—	—	25.1	—	—	—	—	2.25
ClO_4^-	67.3[h]	—	—	44.5	51.1	0.3	0.15	2.31	1.36
Ac^-	40.9[a]	—	41.5	40.5	—	2.6	—	—	2.24
$ClAc^-$	42.2[g]	—	—	—	—	—	—	—	2.17
$CNAc^-$	43.4[g]	—	—	—	—	—	—	—	2.11
CNS^-	—	—	—	—	40.6	0.3	—	—	—
SO_4^{2-}	80.0[h]	—	14.5	15.4	29.15	—	—	—	2.29
Ox^{2-}	74.15[a]	—	—	—	16	—	—	—	2.47
$Fe(CN)_6^{3-}$	99.1[e]	—	—	—	—	—	—	—	2.77
$P_3O_9^{3-}$	83.6[i]	—	—	—	—	—	—	—	3.29
$Fe(CN)_6^{4-}$	113.1[j]	—	—	—	—	—	—	—	3.24
$P_4O_{12}^{4-}$	93.7[i]	—	—	—	—	—	—	—	3.91

and those for other ions, together with the correponding r_h values are shown in Table 14.4. The figures for ClO_4^-, NO_3^- and OH^- are based on $LiClO_4$, $LiNO_3$ and KOH, it being assumed that these are completely dissociated. The radii calculated by Nightingale [56] are about 1.4 Å larger than r_h.

For the alkali metals, r_h is 0.13 to 0.5 Å larger than r_s, for the alkaline earths r_h is about 0.4 Å smaller than r_s, while for the two lanthanides examined r_h is 0.8 Å smaller than r_s. Though it is hazardous to use such differences to discuss "unknowns," by taking $r_h = r_s - 0.4 = 3.07$ Å for Zn^{2+}, from (14.52). $h_+ = 9.3$ and the calculated value of $\gamma_\pm$ for $Zn(ClO_4)_2$ at 1.0 M is only 1 % higher than the experimental value of [27] 0.929. On the same tentative basis, the halides and nitrates of the divalent transition metals are not quite completely dissociated and neither are the sulphates and nitrates of the alkali metals (except $LiNO_3$) for the calculated activity coefficients of 1 M solutions are higher than those obtained in practice. Although the evidence is incomplete, there is enough [93] to give general support to these conclusions, and it would be easy to use the results for quantitative information (see e.g., 9.29).

2. Ion-Pairs

Shortly after the publication of the Debye–Hückel treatment of interionic effects, Bjerrum,[106] from considerations of the probability of the distance of approach, r, between ions of opposite charge z_ie and z_je, derived a relationship correlating the dissociation constant, K, of an ion-pair, with a certain minimum value of r and the average diameter or distance of closest approach, a, of the two ions concerned. The details of Bjerrum's treatment are given in numerous standard works,[2,107] and only the final expressions are given here. These are

$$1/K = (4\pi N/1{,}000)(\,|z_iz_j|\,e^2/\epsilon kT)^3Q(b) \tag{14.53}$$

$$Q(b) = \int_2^b \exp(x)\,.\,x^{-4}dx \tag{14.54}$$

$$x = |z_iz_j|\,e^2/r\epsilon kT, \qquad b = |z_iz_j|\,e^2/a\epsilon kT \tag{14.55}$$

The treatment is limited to very dilute solutions and to ions which associate through purely electrostatic or electrovalent forces. In addition, all ions are regarded as being associated in the rather special sense of being no longer "free" when they approach within the arbitrary distance of $r_{min} = |z_iz_j|e^2/2\epsilon kT$. There are defects arising from this choice but it has been shown [108] that the inter-relationship of K and a is changed very little on replacing the 2 by a slightly different numerical factor. Values of $Q(b)$ for various values of b have been calculated.[106,107] Those of Guggenheim [109] and of Fuoss and Kraus [110] ($b = 15$ to 80) are shown in Table 14.4a.

There are two main practical applications of (14.53). Firstly, if K is known, a may be evaluated (or vice versa), and secondly, if K in one solvent is known and a is assumed to remain constant, it should be possible to calculate K in solvents of other dielectric constants.

TABLE 14.4a

Values of b Corresponding to $Q(b)$ of (14.53)

b	$\log Q(b)$	b	$\log Q(b)$	b	$\log Q(b)$	b	$\log Q(b)$
2.0	—	5.5	− 0.047	11	0.884	30	7.19
2.5	− 0.726	5.6	− 0.034	12	1.127	40	11.01
2.8	− 0.561	6.0	+ 0.017	14	1.668	50	14.96
3.0	− 0.488	7.0	0.151	15	1.964	60	18.98
3.5	− 0.355	8.0	0.300	18	2.92	70	23.05
4.0	− 0.260	9.0	0.470	20	3.59	80	27.15
5.0	0.113	10.0	0.665	25	5.35		

Example 14.4. In numerical terms, (14.53) becomes

$$-\log K = 13.5474 + 3\log|z_i z_j|/\epsilon T + \log Q(b) \tag{14.56}$$

$$= 0.4386 + 3\log|z_i z_j| + \log Q(b), \text{ at } 25°\text{C in water} \tag{14.57}$$

Also, from (14.55),

$$\log a = \log|z_i z_j| - 7.1468 - \log b, \text{ at } 25°\text{C in water} \tag{14.58}$$

Taking [111] $K = 1.82 \times 10^{-4}$ for La[Fe(CN)$_6$] (in water at 25°C), $\log Q(b) = 0.493$, and by interpolation on a plot of $\log Q(b)$ against b, $b = 9.1$, whence $a = 7.1$ Å, {cf. $r_s = 6.72$ Å, $r_{h^+} + r_{s^-} = 5.96$ Å from Table 14.4}.

By taking $a = 6.40$ Å, Fuoss and Kraus [110] found that (14.53) can reproduce the trends in K of tetraisoamylammonium nitrate in dioxan-water solvents ($\epsilon = 38$ to 2.4), and James [111] also found this treatment to be satisfactory for La[Fe(CN)$_6$] in a variety of mixed solvents with $a = 7.2$ Å (see Example 14.4) except [112] in aqueous glycine in which a steadily increases with ϵ. On the other hand, the Bjerrum equation cannot predict the found trends in the dissociation constants of [112] $MgSO_4$, of [111] $AgNO_3$ or of [111] $PbCl^+$ with solvents of changing ϵ. For the first two of this trio a plausible explanation is that the radii of the solvated ions are somewhat dependent on the nature and composition of the solvent, while for the last one, since a for $PbCl^+$ in water from (14.56) is 1.85 Å, which is much smaller than the sum of the r_{cryst} values, it is concluded that ion-association in this case involves covalence forces.

If similar comparisons are made with other electrolytes, a is found to be sometimes larger and sometimes smaller than mean ion-pair diameters calculated by other methods. One reason for these fluctuations can be attributed to the basic assumption that any ions of opposite charge lying in the region r_{min} to a are associated, and while r_{min} in water at 25°C is quite small for 1 : 1 electrolytes, namely 3.6 Å, it increases rapidly as the product $z_i z_j$ is raised, being 14.3 Å for 2 : 2 electrolytes and 32.1 Å for 3 : 3 valent types. Such large values imply that the Debye-Hückel formula for activity coefficients is not valid for higher valent electrolytes unless its ion-size parameter is greater than r_{min}. Guggenheim,[109] in a detailed analysis of the position, has concluded that the Debye-Hückel treatment is useless for 2 : 2 and higher valent electrolytes since most ions have a diameter of about 4 Å. By combining Müller's more exact solution [113] (see Chapter 3) of the potential with Bjerrum's concepts, he has, from a comparison of calculated osmotic coefficients with those derived from freezing-points and e.m.f. measurements, reached the conclusion that some sulphates with divalent cations are completely dissociated in aqueous solution.

Assuming that two oppositely charged ions must be in contact before they have associated to form an ion-pair, Denison and Ramsay [50] have reasoned that (14.56) reduces to

$$-\ln K = e^2/a\epsilon kT \tag{14.59}$$

for 1 : 1 electrolytes in solvents of low dielectric constant. By taking $a =$ 6.67 Å, they found that the expression will reproduce the dissociation constants of [110] tetraisoamylammonium nitrate in dioxan-water solutions (c.f. $a = 6.40$ Å mentioned earlier on when applying (14.56)).

By applying a method of Boltzmann [114] and the Denison-Ramsey concept of an ion-pair, Fuoss [115,116] has, with the aid of a modification suggested by Gilkerson,[117] given the following expression for 1 : 1 electrolytes in solvents of low ϵ:

$$1/K = (4\pi \text{N}a^3/3{,}000) \exp\{(e^2/a\epsilon - \text{E}_s)/kT\}$$
$$= 2.524 \times 10^{-27} a^3 \,.\, \exp\{(e^2/a\epsilon - \text{E}_s)/kT\} \tag{14.60}$$

where a is in cm. and E_s/kT allows for ion-dipole interaction. Putting $K_A = 1/K$,

$$K_A = K_A^0 \exp\,(e^2/a\epsilon kT) \tag{14.61}$$

whence a is obtained by plotting $\log K_A$ against $1/\epsilon$. In general, it is claimed that the values of a so obtained are in good accord with those derived during the course of applying the Fuoss-Onsager conductance equations [118,119] (see Chapter 8). For instance, with $Bu_4N^+ \,.\, BPh_4^-$ in CH_3CN/CCl_4 mixtures where $\epsilon = 36$ to 4.8, $K_A = 14.3$ to 2.96×10^6, $a = 8.5$ Å from (14.61) and [119] $a = 9.0$ Å from conductances.

A combination of (14.5) and (14.14) produces the expression

$$\ln\,(S_1^0/S_2^0) = (\,|\,z_i z_j\,|\,e^2/2rkT)(1/\epsilon_2 - 1/\epsilon_1) \tag{14.62}$$

where S^0 represents the activity solubility product of a sparingly soluble salt and r is the mean radius of the ions of the salt with charges z_ie and z_je. This expression was obtained by both Born [18] and Scatchard.[120] In general, the values of r, as deduced from plots of log S^0 against $1/\epsilon$ are too small and are characteristic of the solvent; some studies with [121,122] $AgBrO_3$, AgAc, and the iodates of Ca^{2+}, Ba^{2+} and La^{3+} indicate r tends to increase with increase in the size of the solvent molecules but that these effects depend on the class of substance, all of which points to the importance of ion-dipole interactions. A relationship that can be obtained from (14.1) and (14.14) is

$$\ln(K_1/K_2) = (\,|\,z_iz_j\,|\,e^2/rkT)(1/\epsilon_2 - 1/\epsilon_1) \tag{14.63}$$

Under test with weak acids, plots of log K against $1/\epsilon$ are curved [123] for a specific acid in mixtures of water with a specific organic solvent, or on a more general basis, show considerable scatter [27] around a straight line corresponding to a value of r which is far too small. A study of silver acetate in mixed solvents [121] suggests that like solubilities, the variation of K with $1/\epsilon$ is very dependent on the nature and type of solvent used.

Dissociation Constant Relationships

In the opinion of Fuoss and Kraus [124] "association can in principle and does in fact occur in any solvent." While this is not meant to imply that no electrolyte undergoes complete dissociation, it is in line with the general conclusion that the number which do so in water and other solvents of high dielectric constant appears to be relatively small. The chief members of the group which are fully ionized, or almost so, comprise the acids $HClO_4$, HCl, HBr, HI, HNO_3, and their alkali metal, alkaline earth metal, lanthanide and some of their transition metal salts. Even for some of these large dissociation constants have been reported, e.g., 25 for HNO_3 (Chapter 12), 3.9 for [125] $NaNO_3$. Also belonging to this group are the alkali metal hydroxides (except [126] LiOH), and many of the alkali metal salts of organic acids.

There are a few solvents with dielectric constants higher than that of water. Apart from H_2SO_4, the chief ones are liquid HCN, aqueous solutions of aminoacids and peptides and certain substituted amides. From the limited available evidence,[127] electrolytes which are strong in water are also strong in these other media. This is also true of the alkali metal halide salts and their acids in aqueous and pure methanol [128] and ethanol [128] although the trends towards association are more marked.

In a particular series of inorganic electrolytes, the dissociation constants diminish as the product of the charges on the associating ions increases, e.g.,[129] $NaP_3O_9^{2-} = 0.07$, $CaP_3O_9^- = 0.00033$, $LaP_3O_9 = 0.000002$, providing the non-common ions are about the same size. As expected, the general effect of increasing size for a fixed product of ion charges is that K increases, e.g.,[129] $Co(NH_3)_6P_3O_9 = 3.65 \times 10^{-5}$, $Co(en)_3P_3O_9 = 3.94 \times 10^{-5}$, $Co(pn)_3P_3O_9 = 22.6 \times 10^{-5}$. However, since the radii of hydrated ions, r_h, may decrease for a particular series as $r_{cryst.}$ increases (e.g., the alkali and alkaline earth metals), the trends in K for a given series of cations may be in opposite directions with different ligands, or the series may be irregular, depending on the extent of hydration of the ion-pair or complex.[130] There are other possible interpretations, which are described further on, particularly with respect to the divalent transition metals, but considering only the M^+ and M^{2+} ions of the alkali metals and alkaline earths, a discussion in terms of ion sizes can be made by reference to Table 14.5. This shows that for each of the M^+ and M^{2+} series the $K_{ass.}$ values of (a) the weakly associating iodates, nitrates, sulphates and thiosulphates increase as r_h increases, (b) the strongly associating ligands oxalate, malonate, glycinate and aspartate increase as $r_{cryst.}$ increases. The hydroxides also belong to the latter group, and it has been shown by Davies [131] that a plot of their log $K_{ass.}$ values together with that of $La(OH)^{2+}$ (4.3) against $z^2_{cat.}/r_{cryst.}$ is fairly linear (although the hydroxides of Ag^+, Tl^+ and Zn^{2+} do not fit into this scheme). The conclusion drawn is that the coulombic forces of association are not strong enough between the ions of group (a) for them to break through the solvation shell of the cation, while they are strong enough for the ions of group (b) and the hydroxides. In the remainder of Table 14.5 where there are no specific trends, association involves partly solvated cations.

Williams,[132] who has tabulated many other series of alkaline earth complexes where $K_{ass.}$ does not follow a uniform pattern with either r_h or $r_{cryst.}$, has given a detailed account of the factors which can affect their stabilities. Mg^{2+}, which has the smallest $r_{cryst.}$ is usually the member responsible for anomalies, and by analogy with the lattice energies of its salts, when Mg^{2+} associates with anions the latter tend to approach one another so closely that short-range repulsion force between them become operative (cf. the EDTA and nitriloacetate series). With chelating ligands, the stability diminishes as the ring size grows (cf. oxalates and malonates) and if certain groups such as — OH and — NH_2 are present, although owing to the small size of Mg^{2+} "crowding" effects may impose limitations (this is apparent in the lactates but not in the glycinates). Williams [132] has also made reference to the consequences of partial ion hydration along the lines already mentioned.

The divalent cations of the first transition metal series tend to associate or complex with anion ligands to a greater extent than do the alkali and alkaline earth metal cations. Examinations of their stability orders shows that they follow general patterns and there have been numerous attempts to establish the full range of these orders and to find the physical factors which govern them. Irving and Williams [135] have reviewed earlier observations in their own

TABLE 14.5

Values of log $K_{ass.}$ for some Alkali Metal and Alkaline Earth Metal Electrolytes in Water ($I = 0$, 25°C unless specified otherwise). Data from Ref. 133 unless indicated by (a) = Ref. 134. A = 18°C, B = 20°C, C = 30°C, $D = I$ of 0.1, $E = I$ of 1.0 (See Refs, 160 and 176 for further and more recent data)

	Li	Na	K	Mg	Ca	Sr	Ba
Iodate	—	− 0.47	− 0.24	0.72	0.89	0.98	1.1
Nitrate	− 1.0	− 0.4	− 0.22	0.0^{a}	0.28^{A}	0.82^{A}	0.92^{A}
Sulphate	0.64^{A}	0.72	0.96	2.3	2.31	—	—
Thiosulphate	—	0.68	0.92	1.8	2.0	2.0	2.2
Trimetaphosphate	—	1.17	—	3.31	3.46	3.35	3.35
Hydroxide	0.17	− 0.7	—	2.6	1.37	0.82	0.64
Oxalate	—	—	—	3.4^{A}	3.0^{A}	2.55^{A}	2.3^{A}
Malonate	—	—	—	2.85	2.50	2.35	2.15
Lactate	− 0.20	—	—	1.37	1.42	0.98	0.64
Glycinate	—	—	—	3.44	1.40	0.9	0.77
EDTA	$2.9^{B,D}$	$1.7^{B,D}$	—	9.12	11.0	8.80	7.78
En	—	—	—	$0.37^{C,E}$	—	—	—
Acetylacetone	—	—	—	$3.63^{C}(K_1)2.54(K_2)$		—	—
Aspartate	—	—	—	2.43^{D}	1.60^{D}	1.48^{D}	1.14^{D}
Methyliminodiacetate	—	—	—	$3.44^{B,D}$	$3.75^{B,D}$	$2.85^{B,D}$	$2.59^{B,D}$
Nitriloacetate	3.28^{B}	2.1^{B}	—	7.00^{B}	8.17^{B}	6.73^{B}	6.41^{B}
1st. I.P. (eV)[129]	5.36	5.12	4.32	7.61	6.09	5.67	5.19
2nd. I.P. (eV)[129]	—	—	—	14.96	11.82	10.98	9.95
$-\Delta E$(kcal)[129]	124	118	100	521	413	384	349
$-\Delta H^{0}_{i,hyd.}$[24,129]	120	96.5	76.5	456	377	348	308

discussion of a large series of Mn^{2+}, Fe^{2+}, Co^{2+}, Ni^{2+}, Cu^{2+} and Zn^{2+} complexes, starting with the noting by Mellor and Maley [136] that the order of decreasing stability of some M^{2+} complexes of salicylaldehyde, glycine, oxine and ethylenediamine (*en*) is (with occasional reversal of neighbours) Pd > Cu, Ni > Pb > Co, Zn > Cd > Fe > Mn > Mg, in agreement with a previous but smaller sequence [137] for some pyridine complexes. Shortly after the first note of Mellor and Maley, Irving and Williams [138] reported that if the successive stability constants of some complexes with NH_3, *en*, *pn* and salicylaldehyde are plotted against the atomic number of M^{2+}, there is a monotonic increase to a maximum at Cu^{2+}, and these increases can be correlated with the electronic structures and fundamental properties of the cations, especially their ionization potentials. The Irving-Williams series and order is Mn < Fe < Co < Ni < Cu > Zn. Some figures illustrating this order are shown in Table 14.6. In Tables 14.5 and 14.6,

1st I.P. refers to $M^+(g) + e = M(g)$, 2nd I.P. refers to $M^{2+}(g) + e = M^+(g)$

$$\Delta E = \text{1st I.P.} + \text{2nd. I.P.},\ 1\ \text{eV} = 23.06\ \text{kcal.}$$

The relationship between the stabilities of transition metal complexes and ionisation potentials has also been discussed by others.[139,140,141] In some discussions [92,140] only the 2nd I.P.s have been used, but Irving and Williams[135] regard ΔE as being a better measure of complexing capacity; ΔE indicates the loss of potential energy when electrons are donated from the high energy levels of the ligand to the low energy levels of the cation. It is a simple way of taking account of the levels of electron availability of the ligand and of the available levels in the cation, and also the effect of bond polarity.[142]

Attempts to extend the Irving-Williams order have not been successful. The order [143] Ba < Ca < Sr < Mg < Mn < Co, Pb < Zn < Ni < Cu holds for glycine, alanine, oxalic and malonic acids, but it fails [135] with, e.g., salicylaldehyde, nitriloacetic acid, EDTA, etc. Even the more limited form Mn < Fe < Cd < Co < Zn < Ni < Cu, which is valid for most aminoacids and peptides fails for, e.g. oxine.

Irving and Williams [135] have also discussed stability orders in terms of cation sizes. A general interpretation cannot be made for although the $r_{\text{cryst.}}$ order is Mg < Mn, and Ca < Cu, yet the Mg^{2+} and Ca^{2+} complexes are invariably less stable than those of Mn^{2+} and Cu^{2+}. This difference is attributable to the larger $-\Delta E$ values of Mn^{2+} and Mg^{2+} in these respective pairs, but this is not the only cause for although the ΔE values of Cd^{2+} and Ni^{2+} are identical they have different $r_{\text{cryst.}}$ values (0.97 Å and 0.70 Å respectively) and the stability orders are Ni < Cd with weak electron donors such as Cl^-, Br^- and I^-, and Ni > Cd with strong electron donor ligands. The significance of cation size also appears to be shown by certain hydroxides, as mentioned earlier on. However, Davies [131] observation that a plot of the

TABLE 14.6.

Values of log $K_{ass.}$ for some Divalent Metal Electrolytes in Water. $I = 0$, 25°C unless otherwise specified. Data from Ref. 133 except (*a*) Ref. 53, (*b*) Ref. 147, (*c*) Ref. 97. Other conditions: A = 6°C, B = 18°C, C = 20°C, D = 30°C, E = I of 0.1, F = I of 0.4, G = I of 0.5, H = I of 1.0, J = log β_2. (See Refs. 160 and 176 for further and more recent data).

	Pb	Mn	Fe	Cd	Co	Zn	Ni	Cu
Sulphate	—	2.28	—	2.29	2.36[a]	2.38[a]	2.32[a]	2.35
Thiosulphate	—	1.95	2.17[A]	3.93	2.05	2.35	2.06	12.3[J]
Trimetaphosphate	—	3.57	—	—	—	—	3.22	—
Oxalate (log K_1)	—	3.82	—	4.00	4.7	4.68	5.3[B]	6.2[b]
Oxalate (log K_2)	—	1.43	—	—	—	2.36	—	3.6
Oxalate (log β_2)	6.54	5.25	—	5.77	6.7	7.04	6.51[B]	8.5
Malonate (log K_1)	—	3.29	—	3.29	3.72	3.68	4.00	5.55
Lactate	2.78[c]	1.43[c]	—	1.70[c]	1.90[c]	2.24[c]	2.22[c]	3.02
Glycinate (log K_1)	5.47	3.44	4.3	4.74	5.02	5.52	6.12	8.62
Glycinate (log K_2)	3.39	2.05	—	3.86	3.97	4.44	5.03	6.97
Glycinate (log β_2)	8.86	5.5	7.8	8.60	8.99	9.96	11.15	15.59
EDTA (log K_1)[C, E]	18.3	13.58	14.33	16.59	16.21	16.26	18.56	18.79
En (log K_1)[D, H]	—	2.73	4.28	5.63	5.89	5.71	7.66	10.72
En (log K_2)[D, H]	—	2.06	3.25	4.59	4.83	4.66	6.40	9.31
En (log β_2)[D, H]	—	4.79	7.53	10.22	10.72	10.37	14.06	20.03
Acetylacetone (log K_1)[D]	—	4.18	5.07	3.83	5.40	4.98	5.92	8.22
Acetylacetone (log K_2)[D]	—	3.07	3.60	2.76	4.11	3.83	4.46	6.73
Acetylacetone (log β_2)[D]	—	7.25	8.67	6.59	9.51	8.81	10.38	14.95
Aspartate (log K_1)[D, E]	—	—	—	4.37	5.90	5.84	7.12	8.57
Aspartate (log K_2)[D, E]	—	—	—	3.11	4.28	4.31	5.27	6.78
Aspartate (log β_2)[D, E]	—	—	—	7.48	10.18	10.15	12.39	15.35
Methyliminodiacetate[C, E] (log K_1)	8.02	5.40	6.65	6.77	7.62	7.66	8.73	11.09
Methyliminodiacetate[C, E] (log K_2)	4.10	4.16	5.37	5.75	6.29	6.43	7.22	6.83
Methyliminodiacetate[C, E] (log β_2)	12.12	9.56	12.02	12.52	13.91	14.09	15.95	17.92
Nitriloacetate[C]	11.8	7.44	5.84	9.54	10.6	10.45	11.26	12.68
1st I.P.(eV)	7.38	7.41	7.83	8.96	7.81	9.36	7.61	7.68
2nd I.P.(eV)	14.96	15.70	16.16	16.84	17.3	17.89	18.2	20.34
$-\Delta E$(kcal.)	515	533	553	595	579	628	595	646
$-\Delta H^{\circ}_{i.hyd.}$(kcal.) [24, 151]	350	438	460	428	489	485	499	499

$pK_{ass.}$ of the hydroxides of the metals in Table 14.5 against $z^2_{cation.}/r_{cryst.}$ is linear is rendered less convincing by the fact that a plot of $pK_{ass.}$ against ionization potentials has the same characteristics.[141]

A comparison of the stabilities of some stable complexes of the lanthanides which have (so far as is known) very similar ionization potentials should give a relatively unambiguous picture of the influence of cation size on stabilities. The values of the association constants of their EDTA complexes and the cation crystalline radii are listed in Table 14.7, and a plot of $pK_{ass.}$ against $1/r_{cryst.}$ does tend towards linearity. Close inspection shows that the plot either has an inflection in the middle or that two straight lines can be drawn, one from La to Eu and the other from Gd to Lu so that the trend towards increasing stability is not exactly proportional to $1/r_{cryst.}$.

Several estimates of the lanthanide EDTA complexes have been made [133]; those in the Table refer to 0.1 M KCl + 0.01 M NH_4Ac at 25°C and a pH of 5.3. They were obtained by a radio-tracer, ion-exchange method,[146] taking Spedding's value for Nd as standard.[145] The relative ionic radii [146] are based primarily on the cubic oxides taking O = 1.380 Å and a co-ordination number of six. The eight nitroacetates [133] and the twelve 1 : 2 diaminocyclohexane tetra-acetates [148] ($\log K_1 = 16.26$ for La and 21.12 for Yb at 20°C in 0.1M KNO_3) which have been reported give similar plots to those of the EDTA complexes, but the glycollate plots [149] are, owing to the low stabilities, too scattered to ascertain if the association is too weak to disturb the cations hydrations shells.

TABLE 14.7

Values of $\log K_{ass.}$ for the Trivalent Lanthanide EDTA Complexes [144] at 25°C and $I = 0.11$, and Relative Cation Crystalline Radii.[146]

	La	Ce	Pr	Nd	Pm	Sm	Eu	Gd
$\log K_1$	15.19	15.45	15.76	16.04	—	16.53	16.66	16.82
r(Å)	1.061	1.034	1.013	0.995	0.979	0.964	0.950	0.938
	Tb	Dy	Ho	Er	Tm	Yb	Lu	
$\log K_1$	17.32	17.78	18.04	18.37	18.65	19.00	19.14	
r(Å)	0.923	0.908	0.894	0.881	0.869	0.858	0.848	

The limited available evidence [150] suggests that the stability constants of the M^{3+} and M^{4+} ions of the actinides with strongly complexing ligands are also related to the crystalline cation radii which, like those of the lanthanides, decrease with increasing atomic number.

EDTA is a strong electron donor ligand so the increase in $K_{ass.}$ with decreasing $r_{\text{cryst.}}$ found in the lanthanide series might suggest a generalization on what has already been stated in connection with Ni and Cd complexes earlier on, namely that with this class of ligands this is the order of stability in a series at constant ΔE, while with weak electron donors such as Cl^-, Br^- and I^-, the stability order at constant ΔE is one of decrease as $r_{\text{cryst.}}$ increases. However, this is not always so. The suggestion of Williams [132] (stated earlier on) that Mg^{2+} has $K_{\text{ass.}}$ values that are often smaller than expected because owing to its small $r_{\text{cryst.}}$, anions repel one another through a tendency to crowd around the cation has been extended by Williams [152] to interpret unexpected differences in other series. It serves to explain why the halide and NH_3 complexes of Pb^{2+} are more stable than those of $Mg^{2}{}_{+}$ ($r_{\text{cryst.}} = 1.27$ and 0.65 Å respectively, ΔEs about the same), why these complexes of Hg^{2+} are more stable than those of Cu^{2+} ($r = 1.10$ and 0.70 Å respectively, ΔEs about the same), and why the Cd^{2+} halides are more stable than those of Zn^{2+} ($r = 0.97$ and 0.74 Å respectively, ΔEs in favour of stronger Zn^{2+} complexes). On the other hand, this view predicts a converse order of stability to that which is found with strong electron donor ligands, namely that $K_{\text{ass.}}$ increases as $r_{\text{cryst.}}$ decreases at constant ΔE (cf., Ni^{2+} and Cd^{2+}, and lanthanide complexes). This might be because the stronger interactions of these ligands swamp any repulsion tendencies, but this is not evident on consideration of the Mg^{2+} complexes either among themselves or by comparison with those of Pb^{2+}.

The electronegativity values of metals are a measure of their electron-attracting powers, and it has been shown [153] that there is an approximate relationship between this function and the stabilities of some dibenzoylmethane complexes in 75 % dioxan, and [184] of some amine complexes. Critical appraisals of this theme are restricted because electronegativity values have not been fixed with sufficient precision. This can be seen from the article of Gordy and Thomas [154] who have reviewed the answers that have been obtained by four different methods. No one series is complete and the differences between answers for the same element limit their selected values to two significant figures. As these authors have observed, the quantity depends to some extent upon the type of bonding and upon the electrical environment.

An important practical aspect of stability constant determinations is illustrated by a study of the effects of salt anions upon some results in 75 % dioxan.[155] The K_{ass} values of ten M^{2+} complexes with acetylacetone were determined at constant I in the presence of chlorides nitrates and perchlorates. Both $K_{\text{ass},1}$ and $K_{\text{ass},2}$, in general, were in the apparent order $ClO_4^- > NO_3^- > Cl^-$ for a particular M^{2+} cation although the ClO_4^-, NO_3^- differences were often very small (e.g., for Mg^{2+} and Zn^{2+}). These differences can be ascribed to the incomplete dissociation of the M^{2+} inorganic salts and show the importance of taking into account all subsidiary ion-pairs which form. This is also illustrated by a number of studies of Hg^{2+} complexes [133] by glass electrode methods where I was kept constant with, e.g., 0.1 M KCl. As a result, the K_{ass} values are markedly lower than those of the corresponding Cu^{2+} complexes, while if, e.g., 0.1 M KNO_3 is used, the order is reversed (the "normal" order). The cause is the extreme weakness of mercuric chloride complexes.[133] Even nitrates and perchlorates are not necessarily fully dissociated even in aqueous media [133] and although the extent of ion-association may be small it can be sufficient to influence the accuracy of comparative series especially when the ratio inorganic anion : cation is very large.

A qualitative survey of the relative affinities of ligand atoms for acceptor molecules or ions which embraces the whole of the Periodic Table has been undertaken by Ahrland, Chatt and Davies.[156] Their chief conclusion is that acceptors are of two well defined types:

Class (a). These comprise H, Li, Be, B, C, N in the 1st Period; Na, Mg, Al to Cl in the 2nd Period; K to Br in the 3rd Period; Rb to I in the 4th Period; Cs to At in the 5th Period; and Fr to U in the 6th Period. These form their most stable complexes with the first ligand atom of each Group.

Class (b). These comprise B, C; Mn to Cu in the 3rd Period; Mo, Tc, Ru and Cd in the 4th Period; W, Re, Os, Tl to Po in the 5th Period. These form their most stable complexes with the second or a subsequent ligand atom of each Group.

Most metals in their common valency state belong to class (a), which also contains H^+ so the affinities for class (a) often tend to run parallel to basicities. Class (b) occupies a triangular section of the Periodic Table and many of its members belong to both classes, which one depending on the valency state, e.g., Cu^+ to (a), Cu^{2+} to (b). These metals form stable olefin complexes. Class (b) character appears to depend on the availability of electrons from the lower d orbitals of the metal for π-bonding.[157]

The conclusions regarding the relative affinities of ligand atoms of Groups are:

(i) Group V. With class (a) complexes, the order is $N \gg P > As > Sb > (Bi)$. With class (b) acceptors the sequence is $N \ll P > As > Sb > (Bi)$.

(ii) Group VI. With metals of strong class (a) character, the order is $O \gg S > Se > Te$, e.g., $AlMe_3$. With class (b), almost any order occurs with S, Se and Te.

(iii) Group VII. With acceptor ions of class (a) the order is $F \gg Cl > Br > I$, with class (b), $I > Br > Cl \gg F$.

The general trends in the dissociation constants of organic acids have been interpreted by Ingold,[158] Dippy [159] and others in terms of the electron-displacement capacities of the anion components and substituents. To take a simple example, some pK_{diss} values in water at 25°C are [133]: $CCl_3COOH =$

0.70, $CH_2ClCOOH = 1.30$, $CH_2OH.COOH = 3.88$, $H.COOH = 3.77$, $CH_3COOH = 4.76$. These show that the electron-attracting capacity order is halogens > OH > H > Alkyl; the greater the separation of the charge on the anion from the proton, the easier it is for dissociation to occur. Conjugative and steric effects, and H-bonding are other examples of induction effects which govern the extent of dissociation. The strengths of organic bases can also be interpreted along the same lines.

Many efforts have been made to find quantitative relationships between the dissociation constants of acids, bases and their metal complexes. Much of this work, particularly those parts dealing with the ratios between log k_{H^+} of weak bases and the complexity constants of metal complexes have been adequately reviewed by Smith [160] and by Bjerrum.[161] Since, on statistical grounds,[162] the first ligand is more strongly bound to a cation than is the second one, and so on, log β_n/n should be used for these comparisons. For a given cation of the series Ag^+, Hg^{2+}, Cu^{2+}, Zn^{2+} and Ni^{2+} associating with different ligands, the log ratios show only approximate constancy, being subject to specific ligand effects.

A comparison of the dissociation constants of a series of monocarboxylic acids with those of their Cu^{2+}, Zn^{2+}, Ca^{2+} and Ba^{2+} complexes [163] shows there is an approximate relationship between the pKs of the acids and of these salts except when chelation occurs such as with hydroxyacids. Good linear relationships between the first and second log $K_{ass.}$ values of the cupric complexes of glycyl and sarcosyl aminoacids and peptides and the corresponding pK_A values (see 9.10) have been found by Datta and Rabin.[164]

1. Ligand-Field Theories and Electrolytic Dissociation

A fuller knowledge of the forces that cause cations and ligands to associate or complex to such widely varying extents will undoubtedly emerge from close considerations founded on theories of valence. The particular aspects that are concerned with the interaction of ligands with transition metal cations are termed ligand-field theories and two main approaches have been pursued. The first, termed the crystal-field or electrostatic theory (Bethe, Van Vleck and others [165]) considers the interaction of cation charges and the negative charges of ligand anions or the negative ends of the dipoles of ligand molecules; the emphasis is on the electrostatic nature of this interaction, that is to say, upon the ionic character of the bonding. The second approach has two main branches both of which regard complexing as being at least partly covalent, and comprise the valence-bond treatment of Pauling [166] which emphasizes the available empty orbitals of the metal cation which can accept ligand electrons, and Van Vleck's development [167] of the molecular orbital method in which any d orbital of a cation together with a σ orbital of a ligand occupied by two electrons when not bonding gives a σd bonding orbital and a σd^* antibonding orbital.

Reviews and discussions on the applications of these ligand-field theories, including the application of them to interpretations of the stabilities of the transition metal complexes have been prepared by Griffith and Orgel,[168] Williams [169] and by Moffitt and Ballhausen.[170] Only an outline of a few of their observations can be made here, and these are mainly in connection with absorption spectra and heats of formation.

The crystal-field theory regards the energies of the d orbitals of a transition metal cation as being affected in different degrees by the field created by its ligands. Thus for an octahedral complex MX_6, where M^{2+} belongs to the first transition series, the d_{z^2} and $d_{x^2-y^2}$ orbitals have substantial amplitudes in the direction of the ligands while the other three d orbitals tend to avoid them. From symmetry considerations, the degeneracy (i.e., same energy values) of these three orbitals is maintained in the complex, and this, from group theory,[171] is also true of the d_{z^2} and $d_{x^2-y^2}$ orbitals. Since these lie in the direction of the ligand field, their electrons are raised to a higher energy level, so there is an upper energy level group (e_g, γ_3 or d_γ) and a lower energy group (t_{2g}, γ_5 or d_ε). To take an example, Ti^{3+} has one d electron which occupies one of the t_{2g} orbitals in its ground state. In solution, $Ti(H_2O)_6^{3+}$ forms and the H_2O ligands can cause a transition which transfers this electron to an e_g orbital at an energy corresponding to 20,400 cm^{-1} as shown by a weak absorption band in this region.

Complications occur if there are more than three electrons; if there are four or seven, either as many as possible are placed in t_{2g} orbitals or they are distributed so as to maintain a maximum number of parallel spins. The former is favoured if Δ, the gain in orbital energy, is large, and the latter is favoured if Δ is small. The value of Δ depends mainly on the nature of the ligand and the cation charge. Generalizations can be made [168,172,174] including the one that ligands can be arranged in a sequence so that Δ for their complexes with any given M^{n+} increases along the sequence, e.g., I^-, Br^-, Cl^-, F^-, H_2O, Ox, *py*, NH_3, *en*, NO_2^-, CN^-.

Williams [169] has criticized the crystal-field theory from this aspect for the usual calculation [172,174] of the energy states of the d electrons from absorption spectra leaves one parameter, dQ (a function of Δ), undetermined (see, e.g., Heidt *et al.*,[175] Orgel,[172] Bjerrum and Jørgensen [173] for detailed examples and further references on such calculations).

A major source of discussion concerning crystal-field theory and the relative stability constants of transition metal complexes centres around considerations of heats of formation. Writing

$$M^{2+}(g) + nH_2O(l) = M(H_2O)_n^{2+}(l) + \Delta H_s(= -\Delta H^0_{i.hyd.}) \qquad (14.64)$$

if the d^x-ions were spherically symmetrical in complexes, ΔH_s would be expected to increase uniformly from $x = 0$ (Ca^{2+}) to $x = 10$ (Zn^{2+}) through the gradual contraction of the $3d$ shell with increasing nuclear charge. As the figures for $\Delta H^0_{\text{i.hyd.}}$ in Tables 14.2 and 14.6 show, this is not so. This is because the t_{2g} levels in octahedral complexes lie below the e_g levels so that greater stabilization is achieved the greater the number of t_{2g} electrons and the smaller the number of e_g electrons.[168] Considering only "high spin" type complexes (i.e., Δ small so with four to seven electrons there is a maximum tendency to orient spins, as shown by magnetic susceptibility), the first three d electrons confer extra stability, but the next two, by entering e_g orbitals, cancel out this stabilization. The same pattern is followed as the next five electrons are added, thus in Fe^{2+} complexes, five electrons have parallel spin and each occupy one orbital but the sixth one enters a t_{2g} orbital and confers extra stability on the complex. In terms of Δ, the theoretical values of the stabilizations are:[168]

No. of d electrons	0, 5, 10	1, 6	2, 7	3, 8	4, 9
Stabilization	0	0.4	0.8	1.2	0.6

Orgel [177] has estimated experimental values of the stabilization energies of M^{2+} ions in their hydrated salts from absorption spectra.[179] To the nearest 5 kcal they are: V = 35, Cr = 20, Mn = 0, Fe = 10, Co = 20, Ni = 25, Cu = 20, Zn = 0. If these values are subtracted from the experimental values of ΔH_s (Table 14.6), the "corrected" values lie on a uniform curve which is nearly a straight line (it can be extended to Ca^{2+}).

Equation (14.64) refers to the special case of complex formation between M^{2+} and H_2O ligands. George [180] has expanded the treatment and established a correlation between the variation in bonding energies of complexes and the extra stabilization arising from crystal-field splitting.

From the following steps:

$$M^{2+}(g) + nL(g) = ML_n(g) + \Delta H_g \tag{14.65}$$

$$M(H_2O)_n^{2+}(l) + nL(l) = ML_n^{2+}(l) + nH_2O(l) + \Delta H_c \tag{14.66}$$

$$M^{2+}(g) + nL(l) = ML_n^{2+}(l) + \Delta H_L \tag{14.67}$$

it follows from (14.64) that

$$\Delta H_L = \Delta H_c + \Delta H_s \tag{14.68}$$

Now although ΔH_s data are available (Tables 14.2, 14.5, 14.6), ΔH_c has only been established for a few ligands, so few experimental values of ΔH_L are available. However, values of ΔH_c for M^{2+} complexing relative to Mn^{2+} (zero stabilization energy) can be calculated if the association constants have been determined, if $\bar{S}^0_{i.l.}$ data are available (Table 14.1), and if use is made of the observation that, to a first approximation, $\Delta S^0_{c.l.}$ (corresponding to (14.66)) is independent of M^{2+} and determined only by the charge, number and nature of the ligand group.[181] In this way the relative ΔH_L values for complexes of 18 ligands were obtained, e.g., aminoacids, salicylaldehyde, malonate, oxalate, OH^-. The results show that plots of ΔH_L for each ligand against the atomic number of M have the same characteristics as a similar plot using ΔH_s. The variation of ΔH_L from one ligand to another is characterized by the maximum stabilization energy E_m, which is the maximum of the curve through the ΔH_L values for the d^6 to d^9 ions above the straight line joining the values for the d^5 and d^{10} ions, and E_r, the ΔH_L difference for these last two. For M^{2+} complexes, E_m ranges from 35 to 46 kcal/mole, and E_r from 46 to 53. Within experimental error, both E_m and E_r are the same for all complexes with O-atom bonding to M^{2+}, including H_2O. They both increase progressively as O-atom bonding is replaced by N-atom bonding (although EDTA does not quite fit into this scheme).

Leussing has [182] applied George's method to the first transition metal thioglycollates (excluding Cu^{2+}). He has calculated that the relative crystal-field stabilizations for these and O-atom and N-atom bonding complexes are about the same, the values being 0.18, 0.65, 0.72 and 0.56 for Fe, Co, Ni and Cu respectively, based on E_r from Mn to Zn = 1.00 (these figures should be multiplied by 1.2 to be comparative with Orgel's theoretical values given earlier on). Thus, from Table 14.2, E_r ($Mn^{2+} \rightarrow Zn^{2+}$) = 47.6, hence for Fe^{2+}, ΔH_s corrected is 80.4 — 0.2 × 47.6 = 70.9. Accordingly the stabilization energy is 8.4 kcal, and the stabilization factor = 8.4/47.6 = 0.18.

To summarize the conclusions of the crystal-field theory, since, from the above discussions, together with (14.1) and (14.3), ΔH_L (corrected) = ($RT \ln K_{diss.} + T\Delta S^0_{c.l.}$—stabilization energy), although the middle right-hand term may be practically constant, since the last term varies with M^{2+} even for a fixed ligand, it upsets the relative order of the dissociation constants of the transition metal complexes, this being particularly marked with Zn^{2+}.

Williams,[152] who has shown that there is an approximate linear relationship between ΔH_s and $1/r_{cryst.}$ for the alkaline earth metals, has proposed an empirical relationship between ΔH_s for a number of divalent cations, $r_{cryst.}$ ionization potentials and constants which represent factors due to covalent and repulsion forces. He has also [169] pointed out that the molecular-orbital description of cation-ligand interaction overcomes some of the disadvantages of the crystal-field theory since it takes into account covalence bonding. Detailed discussions are given in support of this. There are theoretical and empirical grounds for believing that such bonding does occur in the transition metal complexes.[183] According to the M.O. treatment,[168] electrons are placed in molecular orbitals formed from the 3*d*, 4*s* and 4*p* cation orbitals, and the ligand orbitals. Although the M.O. theory leads to the same kind of splitting of the five levels available to d electrons as does the crystal-field theory, the difference is that the e_g electrons have a certain probability of being on the ligands. Infra-red studies of metal complexes[185] indicate that as cation-ligand bonding becomes more covalent, there are certain increases in frequency shifts of particular bonds such as that of C=O in EDTA.

REFERENCES

1. Latimer, W. M., Schutz, P. W. and Hicks, J. F. *J. chem. Phys.* **2,** 82 (1934). Latimer, W. M. *Chem. Rev.* **18,** 349 (1936).
2. Moelwyn-Hughes, E. A. "Physical Chemistry." Pergamon Press, London, New York, Paris, (1957).
3. Lange, E. and Fuoss, R. M. *Z. phys. Chem.* **125,** 431 (1927).
4. Pitzer, K. S. and Smith, W. V. *J. Amer. chem. Soc.* **59,** 2633 (1937).
5. Ref. 51, Ch. 4.
6. Latimer, W. M., Pitzer, K. S. and Smith, W. V. *J. Amer. chem. Soc.* **60,** 1829 (1938).
7. Powell R. E. and Latimer W. M. *J. chem. Phys.* **19,** 1139 (1951).
8. Latimer, W. M. "The Oxidation States of the Elements and their Potentials in Aqueous Solution." Prentice-Hall, New York, 2nd edn. (1952).
9. Gurney, R. "Ionic Processes in Solution." McGraw-Hill Book Co., Ltd., London New York (1953).
10. Laidler, K. J. *Canad. J. Chem.* **34,** 1107 (1956). Couture, A. M. and Laidler, K. J. *ibid.* **35,** 202 (1957).
11. Cobble, J. W. *J. chem. Phys.* **21,** 1443 (1953).
12. Pauling, L. *J. phys. Chem.* **56,** 361 (1952).
13. Pauling, L. "Nature of the Chemical Bond." Cornell Univ. Press, Ithaca, New York (1939) (1950).
14. Schomaker, V. and Stevenson, D. P. *J. Amer. chem. Soc.* **63,** 37 (1941).
15. Connick, R. E. and Powell, R. E. *J. chem. Phys.* **21,** 2206 (1953).
16. Cobble, J. W. *ibid.* **21,** 1446 (1953).
17. Wyckoff, R. W. G. "Crystal Structure." Interscience Inc., New York (1951).
18. Born, M. *Z. Phys.* **1,** 45 (1920).
19. Webb, T. J. *J. Amer. chem. Soc.* **48,** 2589 (1926). Bernal, J. D. and Fowler, R. H. *J. chem. Phys.* **1,** 515 (1933). Voet, A. *Trans. Faraday Soc.* **32,** 1301 (1936).
20. Latimer, W. M., Pitzer, K. S. and Slansky, C. M. *J. chem. Phys.* **7,** 108 (1939).
21. Huggins, M. *ibid.* **5,** 143 (1937).
22. Bichowsky, F. R. and Rossini, F. D. "Thermochemistry of Chemical Substances." Reinhold Publishing Corp., New York (1936).
23. Rossini, F. R., Wagman, D. D., Evans, W. H., Levine, S. and Jaffe, I. "Selected Values of Chemical Thermodynamic Constants." Nat. Bur. Stds., Washington, Circ. No. 500 (1952).
24. Latimer, W. M. *J. chem. Phys.* **23,** 90 (1955).
25. Eastman, E. D. reported by Goodrich, J. C., Goyan, F. M., Morse, E. E., Preston, R. G. and Young, M. B. *J. Amer. chem. Soc.* **72,** 4411 (1950).
26. Frank, H. S. and Evans, M. W. *J. chem. Phys.* **13,** 507 (1945).
27. Ref. 11, Ch. 1.
28. Eley, D. D. and Evans, M. G. *Trans. Faraday Soc.* **34,** 1093 (1938).
29. Benjamin, L. and Gold, V. *ibid.* **50,** 797 (1954).

30. Landolt-Bornstein, "Zahlenwerte und Funktionen." J. Springer, Berlin. Vol. I, Pt. 1 (1950).
31. Verney, E. J. W. *Rec. trav. chim.* **61,** 127 (1942).
32. Herzberg, G. "Infra-red and Raman Spectra." Van Nostrand Co., Inc., New York (1951).
33. Wilson, E. B. *Chem. Rev.* **27,** 17 (1940). Hirschfelder, J. O. *J. chem. Phys.* **8,** 431 (1940).
34. Altshuller, A. P. *ibid.* **24,** 642 (1956); **26,** 404 (1957); **28,** 1254 (1958).
35. Claassen, H. H. and Zielen, A. J. *ibid.* **22,** 707 (1954).
36. Hückel, W. "Structural Chemistry of Inorganic Compounds." trans. by Long, L. H. Elsevier Pub. Co., Inc., Amsterdam, Vols. I and II (1951). Wells, A. E. "Structural Inorganic Chemistry." Clarendon Press, Oxford, 2nd edn. (1950).
37. Sutton, L. E. "Tables of Interatomic Distances and Configurations in Molecules and Ions." Chemical Society, London, Spec. Pub. No. 11 (1958).
38. Ta-You Wu, "Vibrational Spectra and Structural and Polyatomic Molecules." Edward Bros., Inc., Ann Arbor (1946). Mathieu, J. P. *C.R. Acad. Sci., Paris,* **234,** 2272 (1952).
39. Latimer, W. M. and Slansky, C. M. *J. Amer. chem. Soc.* **62,** 2019 (1940).
40. Feates, F. S. and Ives, D. J. G. *J. chem. Soc.* 2798 (1956).
41. Ives, D. J. G. and Pryor, J. H. *ibid.* 2104 (1955).
42. Harned, H. W. and Robinson, R. A. *Trans. Faraday Soc.* **35,** 1380 (1939).
43. Harned, H. S. and Done, R. S. *J. Amer. chem. Soc.* **63,** 2579 (1941).
44. Jenkins, H. O. *Trans. Faraday Soc.* **41,** 138 (1945).
45. Harned, H. S. and Embree, N. D. *J. Amer. chem. Soc.* **56,** 1050 (1934).
46. Harned, H. S. and Ehlers, R. W. *ibid.* **55,** 652 (1933). Harned, H. S. and Hamer, W. J. *loc. cit.* 2194.
47. Everett, D. H. and Wynne-Jones, W. F. K. *Trans. Faraday Soc.* **35,** 1380 (1939).
48. Ref. 10, Ch. 1.
49. Pitzer, K. S. *J. Amer. chem. Soc.* **59,** 2365 (1937). Briegleb, G. and Bieber, A. *Z. Elektrochem.* **55,** 250 (1951).
50. Denison, J. T. and Ramsay, J. B. *J. Amer. chem. Soc.* **77,** 2615 (1955).
51. Evans, M. G. and Uri, N. *Symp. Soc. Exp. Biol.* **5,** 130 (1951).
52. Evans, M. G. and Nancollas, G. H. *Trans. Faraday Soc.* **49,** 363 (1953).
53. Nancollas, G. H. *J. chem. Soc.* 1458 (1955); 745 (1956); *Disc. Faraday Soc.* **24,** 108 (1957). Nair, V. S. K. and Nancollas, G. H. *J. chem. Soc.* 318 (1957); 3706, (1958); 3934 (1050).
54. Hamer, W. J. "The Structure of Electrolytic Solutions." J. Wiley and Sons, Inc., New York (1959). Article by Austin, J. M., Matheson, R. A. and Parton, H. N.
55. George, J. H. B. *J. Amer. chem. Soc.* **81,** 5530 (1959).
56. Nightingale, E. R. *J. phys. Chem.* **63,** 1381 (1959).
57. Stern, K. H. and Amis, E. S. *Chem. Rev.* **59,** 1 (1959).
58. Glasstone, S. "The Electrochemistry of Solutions." Methuen and Co., Ltd., London (1937).
59. Washburn, E. W. and Millard, E. B. *J. Amer. chem. Soc.* **31,** 322 (1909); **37,** 694 (1915).
60. Fischer, P. Z. and Koval, T. E. *Bull. Sci. Univ. Kiev.* No. 4, 137 (1939). Longsworth, L. G. *J. Amer. chem. Soc.* **69,** 1288 (1947).
61. Haase, R. *Z. Elektrochem.* **62,** 279 (1958).
62. Conway, B. E. and Bockris, J. O'M. "Modern Aspects of Electrochemistry" Butterworths Scientific Pubs., London (1954).

63. Ulich, H. *Z. Elektrochem.* **36,** 497 (1930); *Z. phys. Chem.* **168,** 141 (1934).
64. Passynsky, A. H. *Acta Physicochem.* **8,** 835 (1903).
65. Darmois, E. *J. Phys. Radium* **2,** 2 (1942).
66. Hasted, J. B., Ritson, D. M. and Collie, C. H. *J. chem. Phys.* **16,** 1 (1948); Haggis, G. H., Hasted, J. B. and Buchanan, T. J. *ibid.* **20,** 1452 (1952).
67. Debye, P. *ibid.* **1,** 13 (1933).
68. Hasted, J. B. and Roderick, G. W. *ibid.* **29,** 17 (1958).
69. Lane, J. A. and Saxton, J. A. *Proc. roy. Soc.* **214A,** 531 (1952).
70. Yeager, E., Bugosh, J., Hovorka, F. and McCarthy, J. *J. chem. Phys.* **17,** 411 (1949). Yeager, E. and Hovorka, F. *J. acoust. Soc. Amer.* **25,** 443 (1953).
71. Rutgers, A. J. and Rigole, W. *Trans. Faraday Soc.* **54,** 139 (1958).
72. Brady, G. W. and Krause, J. T. *J. chem. Phys.* **27,** 304 (1957). Brady, G. W. *ibid.* **28,** 464 (1958).
73. Van Panthalon van Eck, C. V., Mendel, H. and Boog, W. *Disc. Faraday Soc.* **24,** 158 (1957).
74. Bascombe, K. N. and Bell, R. P. *loc. cit.* 158.
75. Paul, M. A. and Long, F. A. *Chem. Rev.* **57,** 1 (1957).
76. Glueckauf, E. and Kitt, G. P. *Proc. roy. Soc.* **228A,** 322 (1955).
77. Glueckauf, E. *Trans. Faraday Soc.* **51,** 1235 (1955).
78. Wicke, E., Eigen, M. and Ackermann, Th. *Z. phys. Chem.* (*N.F.*) **1,** 340 (1954).
79. Ackermann, Th. *Disc. Faraday Soc.* **24,** 180 (1957).
80. Wyatt, P. A. H. *loc. cit.* 162, 235.
81. Stokes, R. H. and Robinson, R. A. *J. Amer. chem. Soc.* **70,** 1870 (1948).
82. Stokes, R. H. and Robinson, R. A. *Trans. Faraday Soc.* **53,** 301 (1957).
83. Owen, B. B. and Brinkley, S. R. *Chem. Rev.* **29,** 461 (1941).
84. Eucken, A. *Z. Elektrochem.* **51,** 6 (1948). "Lehrbuch der Chem. Physik." Akad. Verlag, Geest and Partig, K-G, Leipzig, II (1949).
85. Zan, E-An, *Geochim. et Cosmoch. Acta* **12,** 103 (1957).
86. Fajans, K. and Johnson, O. *J. Amer. chem. Soc.* **64,** 668 (1942).
87. Couture, A. M. and Laidler, K. J. *Canad. J. Chem.* **34,** 1209 (1956).
88. Goldshmidt, V. M. "Geochemische Verteilungeteze." Skrift d. Norsk Vidensk, Akad. Oslo, Math-Nat., **8,** Kl. I (1926); *Chem. Ber.* **60,** 1263 (1927). Gurney, E. "Ionic Processes in Solution." McGraw-Hill Pub. Co., Ltd., London, New York (1953).
89. Laidler, K. J. *Canad. J. Chem.* **34,** 1107 (1956). Couture, A. M. and Laidler, K. J. *ibid.* **35,** 202 (1957).
90. Alder, B. J. *J. chem. Phys.* **23,** 263 (1955).
91. Bernal, J. D. and Fowler, R. H. *ibid.* **1,** 515 (1933).
92. Ref. 2, Ch. 1.
93. Ref. 14, Ch. 8.
94. "International Critical Tables." McGraw-Hill Book Co., Inc., New York (1928).
95. Shedlovsky, T. and Brown, A. S. *J. Amer. chem. Soc.* **56,** 1066 (1934).
96. See Ch. 8.
97. Evans, W. P. and Monk, C. B. *J. Chem. Soc.* 550 (1954).
98. Nancollas, G. H. *ibid.* 1458 (1955).
99. James, J. C. and Monk, C. B. *Trans. Faraday Soc.* **46,** 141 (1950).
100. Spedding, F. H. and Jaffe, S. *J. Amer. chem. Soc.* **76,** 884 (1954). Also other lathanides data.
101. Jenkins, I. L. and Monk, C. B. *J. chem. Soc.* 68 (1951).
102. Davies, C. W. and Monk, C. B. *ibid.* 413 (1949).

103. James, J. C. *Trans. Faraday Soc.* **45,** 855 (1949).
104. Spedding, F. H. and Atkinson, G. Article in Ref. 54.
104a. Kortüm, G. and Weller, A. *Z. Naturf.* **5a,** 590 (1950).
105. Templeton, D. H. and Dauben, C. H. *J. Amer. chem. Soc.* **76,** 5237 (1954).
106. Bjerrum, N. *K. danske vidensk Selsk.* **7,** No. 9 (1926).
107. Refs. 2 and 11, Ch. 1., Ref. 8, Ch. 3.
108. Fuoss, R. M. *Trans. Faraday Soc.* **30,** 967 (1934).
109. Guggenheim, E. A. *Disc. Faraday Soc.* **24,** 53 (1957).
110. Fuoss, R. M. and Kraus, C. A. *J. Amer. chem. Soc.* **55,** 1019 (1933).
111. James, J. C. *J. chem. Soc.* 1094 (1950).
112. Dunsmore, H. S. and James, J. C. *ibid.* 2925 (1951).
113. Müller, H. *Phys. Z.* **28,** 324 (1927). See Ch. 3.
114. Boltzmann, L. "Gastheorie." Barth, Leipzig, II, 177 (1898).
115. Fuoss, R. M. *J. Amer. chem. Soc.* **79,** 3301 (1957); **80,** 5059 (1958).
116. Sadek, H. and Fuoss, R. M. *ibid.* **81,** 4507 (1959).
117. Gilkerson, W. R. *J. chem. Phys.* **25,** 1199 (1956).
118. Fuoss, R. M. and Accascina, F. "Electrolytic Conductance." Interscience Pub., Inc., New York, London (1959).
119. Accascina, F., Petrucci, S. and Fuoss, R. M. *J. Amer. chem. Soc.* **81,** 1301 (1959).
120. Scatchard, G. *Chem. Rev.* **3,** 383 (1927).
121. Davies, P. B. and Monk, C. B. *J. chem. Soc.* 2718 (1951).
122. Monk, C. B. *loc. cit.* 2723.
123. Harned, H. S. *J. phys. Chem.* **43,** 275 (1939).
124. Fuoss, R. M. and Kraus, C. A. *J. Amer. chem. Soc.* **79,** 3304 (1957).
125. Banks, W. H., Righellato, E. C. and Davies, C. W. *Trans. Faraday Soc.* **27,** 621 (1931).
126. Gimblett, F. G. R. and Monk, C. B. *ibid.* **50,** 965 (1954).
127. Ref. 42, Ch. 3.
128. Ref. 43, Ch. 3, Ref. 32, Ch. 8.
129. Davies, C. W. and Monk, C. B. *J. chem. Soc.* 413, (1949). Monk, C. B. *ibid.* 1317 (1952).
130. Jones, H. W., Monk, C. B. and Davies, C. W. *ibid.* 2693 (1949). Monk, C. B. *ibid.* 1431 (1952).
131. Davies, C. W. *ibid.* 1256 (1951).
132. Williams, R. J. P. *ibid.* 3770 (1952).
133. Bjerrum, J., Schwarzenbach, G. and Sillen, L. G. "Stability Constants." Parts I and II, Special Pubs. Nos. 6 and 7, Chemical Society, London (1957, 1958).
134. Righellato, E. C. and Davies, C. W. *Trans. Faraday Soc.* **26,** 592 (1930).
135. Irving, H. and Williams, R. J. P. *J. chem. Soc.* 3192 (1953).
136. Mellor, D. P. and Maley, L. *Nature, Lond.* **159,** 370 (1947); **161,** 436 (1948).
137. Pfieffer, P., Thielert, N. and Glaser, H. *J. prakt. Chim.* **152,** 145 (1939).
138. Irving, H. and Williams, R. J. P. *Nature, Lond.* **162,** 746 (1948).
139. Schwarzenbach, G., Ackermann, H. and Prue, J. E. *Nature, Lond.*, **163,** 723 (1949).
140. Calvin, M. and Melchior, N. C. *J. Amer. chem. Soc.* **70,** 3270 (1948).
141. Gimblett, F. G. R. and Monk, C. B. *Trans. Faraday Soc.* **50,** 965 (1954).
142. Walsh, A. D. *Proc. roy. Soc.* **207A,** 13 (1951).
143. Monk, C. B. *Trans. Faraday Soc.* **47,** 297 (1951).
144. Betts, R. H. and Dahlinger, O. F. *Canad. J. Chem.* **37,** 91 (1959).
145. Wheelwright, E. J., Spedding, F. H. and Schwarzenbach, G. *J. Amer. chem. Soc.* **75,** 4196 (1953).

146. Templeton, D. H. and Dauben, C. H. *ibid.* **76**, 5237 (1954).
147. Peacock, J. M. and James, J. C. *J. chem. Soc.* 2232 (1951).
148. Schwarzenbach, G., Gut, R. and Anderegg, G. *Helv. chim. acta* **37**, 936 (1954).
149. Sonesson, A. *Acta Chem. Scand.* **13**, 998 (1959).
150. Katz, J. J. and Seaborg, G. T. "The Chemistry of the Actinide Elements." Methuen and Co., Ltd., London; T. Wiley and Sons, Inc., New York (1957). Seaborg, G. T. "The Trans-Uranium Elements." Methuen and Co., Ltd., London; Yale Univ. Press, Inc. (1958).
151. Data for Fe^{2+}, Co^{2+} and Ni^{2+} from Ref. 135, subtracting 8 kcal. to conform to data of Ref. 24.
152. Williams, R. J. P. *J. phys. Chem.* **58**, 121 (1954).
153. Van Uitert, L. G., Fernelius, W. C. and Douglas, B. D. *J. Amer. chem. Soc.* **75,** 2736 (1953).
154. Gordy, W. and Orville Thomas, W. J. *J. chem. Phys.* **24,** 439 (1956).
155. Van Uitert, L. G., Fernelius, W. C. and Douglas, B. D. *J. Amer. chem. Soc.* **75,** 2739 (1953).
156. Ahrland, S., Chatt, J. and Davies, N. R. *Quart. Rev.* **12,** 265 (1958).
157. Chatt, J. *Nature, Lond.* **165**, 637 (1950). Craig, D. P., Maccoll, A., Nyholm, R. S., Orgel, L. E. and Sutton, L. E. *J. chem. Soc.* 332 (1954). Jaffe, H. H. *J. phys. Chem.* **58,** 185 (1954).
158. Ingold, C. K. *Chem. Rev.* **15,** 225 (1934); "Structure and Mechanisms of Organic Chemistry." G. Bell and Sons, Ltd., London (1953).
159. Dippy, J. F. J. *Chem. Rev.* **25,** 151 (1939), and many articles indicated by Ref. 38, Ch. 8.
160. Smith, R. L. "The Sequestration of Metals," Chapman and Hall, Ltd., London (1959).
161. Bjerrum, J. *Chem. Rev.* **46,** 386 (1950).
162. Bjerrum, N. *Ergeb. exact. Naturw.* **5,** 125 (1926).
163. Lloyd, M., Wycherley, V. and Monk, C. B. *J. chem. Soc.* 1786 (1951).
164. Rabin, B. R. *Trans. Faraday Soc.* **52,** 1130 (1956). Datta, S. P., Leberman, R. and Rabin, B. R. *ibid.* **55,** 2141 (1959).
165. Bethe, H. *Ann. Phys.* **3,** 133 (1929). Van Vleck, J. H. *Phys. Rev.* **41,** 208 (1935). Schlapp, R. and Penney, W. G. *ibid.* **42,** 666 (1932).
166. Pauling, L. *J. Amer. chem. Soc.* **53,** 1367 (1931). *J. chem. Soc.* 1461 (1948). "The Nature of the Chemical Bond." Cornell Univ. Press (1950).
167. Van Vleck, J. H. *J. chem. Phys.* **3,** 807 (1935).
168. Griffith, J. S. and Orgel, L. E. *Quart. Rev.* **11,** 381 (1957).
169. Williams, R. J. P. *J. chem. Soc.* 8 (1956).
170. Moffitt, W. and Ballhausen, C. J. *Ann. Rev. phys. Chem.* **7,** 107 (1956).
171. Eyring, H., Walter, J. and Kimbale, G. E. "Quantum Chemistry." J. Wiley and Sons, Inc., New York (1950).
172. Orgel, L. E. *J. chem. Phys.* **23,** 1004, 1819, 1824, 1958 (1955). Tsuchida, R. *Bull. chem. Soc., Japan* **13,** 388, 436, 471 (1938).
173. Jørgensen, C. K. *Proc. 10th Solvay Council, Brussels.* (1956).
174. Ilse, F. E. and Hartmann, H. *Z. Naturf.* **6A,** 751 (1951).
175. Heidt, L. J., Koster, G. F. and Johnson, A. M. *J. Amer. chem. Soc.* **80,** 6471 (1958).
176. Schwarzenbach, G. and others, *Helv. Chim. Acta* **40,** 1886 (1957); **42,** 827 (1959). Holloway, J. H. and Reilly, C. N. *Anal. Chem.* **32,** 249 (1960).
177. Orgel, L. *J. chem. Soc.* 4756 (1952).

178. Bjerrum, J. and Jørgensen, C. K. *Rec. trav. chim.* **75,** 658 (1956). Jørgensen, C. K. *Acta Chem. Scand.* **12,** 1537, 1539 (1958). Ballhausen, C. J. *Rec. trav. chim.* **75,** 665 (1956).
179. Bose, D. M. and Mukherjee, P. C. *Phil. Mag.* **26,** 757 (1938).
180. George, P. *Rec. trav. chim.* **75,** 671 (1956).
181. George, P., Hanania, G. I. H. and Irvine, D. H. *J. chem. Phys.* **22,** 1616 (1954).
182. Leussing, D. L. *J. Amer. chem. Soc.* **80,** 4180 (1958).
183. Bowers, K. D. and Owen, J. *Rep. Progr. Phys.* **18,** 304 (1955).
184. Venkatachalan, K. A. and Kabadi, M. B. *J. Ind. chem. Soc.* **35,** 503 (1958).
185. Busch, D. H. and Bailar, J. C. *J. Amer. chem. Soc.* **75,** 4574 (1953). Sawyer, D. J. and Paulson, P. J. *ibid.* **80,** 1597 (1958); **81,** 816 (1959). Nakamoto, N., Fujita, J. Murata, H. *ibid.* **80,** 4817 (1958). Beattie, I. R. and Tyrrell, H. J. V. *J. chem. Soc.* 2849 (1956).

APPENDIX

Fundamental Constants for Physical Chemistry

(Rossini, F. D., Gucker, F. T. Jr., Johnston, H. L., Pauling, L. and Vinal, G. W. *J. Amer. chem. Soc.* **74**, 2699 (1952).)

In absolute units unless otherwise stated.

Faraday F	96.493.1 $\pm$ 1.0 coulombs/g. equiv.
	96.509.0 international units
Electron charge e	1.601864×10^{-19} coulombs
	1.601864×10^{-20} e.m.u.
	4.80223×10^{-10} e.s.u.
Gas Constant R	8.31439 $\pm$ 0.00034 joules/deg. mole
	1.98719 cal./deg. mole
	82.0567 cm.^{3} atm./deg. mole
	0.0820544 l. atm./deg. mole
Calorie	4.1840 joules
	4.18331 international joules
Velocity of light c*	$2.997902 \pm 0.000013 \times 10^{10}$ cm/sec.
Avogadro Number N	$6.02380 \pm 0.00016 \times 10^{23}$ molecules/mole
Ice Point 0°C	273.16 $\pm$ 0.010 °K
Planck's Constant h	$6.62377 \pm 0.00027 \times 10^{-27}$ erg. sec/molecule
Boltzmann Constant k	$1.380257 \pm 0.000067 \times 10^{-16}$ erg/deg. molecule
Litre	1,000.028 $\pm$ 0.004 cm^3

Absolute Ampere or Coulomb = 1.000165 $\pm$ 0.000025 international Amp. or Coulomb
Absolute Volt = 0.999670 $\pm$ 0.000029 international Volt
Absolute Ohm = 0.999505 $\pm$ 0.000015 international Ohm
π = 3.141593

Numerical Data

Dielectric Constant of Water, ϵ:

Ref. (*a*) $\epsilon = 87.74 - 0.4008t + 9.398 \times 10^{-4}t^2 - 1.410t^3 \quad (t = °C)$
Ref. (*b*) $\epsilon = 78.54 - 4.579 \times 10^{-3}(t-25) + 11.9 \times 10^{-6}(t-25)^2 + 28 \times 10^{-9}(t-25)^3$

Viscosity of Water, η (poise):

°C	0	5	10	15	20	25	30	35	40
Ref. (*c*) ($10^3\eta$)	17.92	15.20	13.069	11.383	10.020	8.903	7.975	7.193	6.531
Ref. (*d*) ,,	—	—	—	—	10.02	8.885	7.960	—	6.518
Ref. (*e*) ,,	17.94	15.19	13.10	11.45	10.09	8.949	8.004	7.208	6.536

REFERENCES

(*a*) Malmberg, C. G., and Maryott, A. A., *J. Res. nat. Bur. Stand.* **56,** 1 (1956); (*b*) Wyman Jr., J. and Ingalls, E. N. *J. Amer. chem. Soc.,* **60,** 1182 (1938); (*c*) Weber, W. *Z. angew. Phys.* **7**, 96 (1955); (*d*) Swindells, J. F., Coe, J. R. and Godfry, T. B. *J. Res. nat. Bur. Stand.* **48,** 1 (1952); (*e*) "International Critical Tables", McGraw-Hill Book Co., Inc., New York (1929).

AUTHOR INDEX

Numbers in parentheses are reference numbers and are included to assist in locating references when the authors' names are not mentioned in the text.
Numbers in italics refer to the page on which the reference is listed.

C

D

L

N

O

S

T

SUBJECT INDEX